Good Statistical Practice for Natural Resources Research

Good Statistical Practice for Natural Resources Research

Edited by

Roger Stern,[1] Richard Coe,[2] Eleanor Allan[1] and Ian Dale[1]

[1]Statistical Services Centre, University of Reading, UK
[2]World Agroforestry Centre (ICRAF), Nairobi, Kenya

CABI Publishing

CABI Publishing is a division of CAB International

CABI Publishing
CAB International
Wallingford
Oxfordshire OX10 8DE
UK

Tel: +44 (0)1491 832111
Fax: +44 (0)1491 833508
E-mail: cabi@cabi.org
Website: www.cabi-publishing.org

CABI Publishing
875 Massachusetts Avenue
7th Floor
Cambridge, MA 02139
USA

Tel: +1 617 395 4056
Fax: +1 617 354 6875
E-mail: cabi-nao@cabi.org

A catalogue record for this book is available from the British Library, London, UK.

Library of Congress Cataloging-in-Publication Data
Good statistical practice for natural resources research / edited by
Roger Stern … [et al.].

"Statistical Services Centre, University of Reading, UK; World
Agroforestry Centre (ICRAF), Nairobi, Kenya."
Includes bibliographical references and index.
ISBN 0-85199-722-8 (alk. paper)
1. Natural resources--Management--Research--Statistical methods. 2.
Natural resources--Manangement--Research. I. Stern, Roger.
HC85 .G66 2004
333.7'07'27--dc22
2003025318

ISBN 0 85199 722 8

Printed and bound in the UK by Biddles Ltd, King's Lynn, from copy supplied by the author.

Contents

Preface

Three lines of activity led to the writing of this book.

Between 1998 and 2001 the Statistical Services Centre (SSC)[1] wrote a series of 'statistical good practice' guides for scientists conducting natural resources research. The first was a general overview, written by Professor Roger Mead, who at the time was head of our team contracted by the UK government's Department for International Development (DFID) to provide statistical support to Natural Resources programmes and projects. An updated version of this overview is within Chapter 2. Subsequent guides were on study design, data management and analysis. Our aim was to write short booklets – none of more than about 20 pages – that explained key ideas without mathematics. We received positive feedback from many readers, and found that the guides were useful in support of other activities, such as training courses for research scientists.

During the same period the SSC and the Natural Resources Institute (NRI)[2] worked on a joint project, funded by DFID, to help bridge the 'qualitative-quantitative divide' separating statisticians and practitioners of participatory studies. Some outputs from this project added to the stock of guides.

The third area of activity is a long-standing, informal collaboration between the SSC and the research support unit at the World Agroforestry Centre (ICRAF)[3] in Nairobi. Recently this has included joint work on the preparation of training materials dealing with the design and analysis of agroforestry experiments and with the design of effective systems for research data management. Key ideas from these materials have been incorporated into the book, specifically into Chapter 13, on strategies for data management, and Chapter 19, on statistical models.

These ideas and guides seemed useful enough that we succumbed to our own feelings and the views of others that it would be quick, simple and worthwhile to make these guides into a book. We now hope, even more strongly, that the results are worthwhile, given that the work has been neither quick, nor simple!

---

[1] Statistical Services Centre, a consultancy group within the University of Reading.
[2] Natural Resources Institute, based in Chatham and part of the University of Greenwich.
[3] World Agroforestry Centre (the International Centre for Research in Agroforestry).

The book is divided into five parts. Part 1 (Chapters 1 to 3) sets the scene. Managing natural resources is complex, so research usually proceeds by way of projects that involve multiple activities. The guides gave suggestions mostly for individual activities, principally surveys and experiments. Since a successful project is more than just a set of individual activities, in Part 1 we have added suggestions that concern the project as a whole.

In Part 2 (Chapters 4 to 9) we look at issues concerned with sampling and other aspects of the planning of the research activities. Part 3 (Chapters 10 to 14) is concerned with organising and archiving the data. Part 4 (Chapters 15 to 21) – the largest part – is devoted to ideas concerned with analysis. Part 5 (Chapters 22 and 23), called 'Where next', includes other resources that readers might find useful.

While it is logical first to consider planning, then data management, and then analysis, this does not have to be the order followed by readers. For example, those who currently have data to process may wish to start with the analysis chapters. We have tried to keep the self-contained nature of each chapter, so that readers can dip into the book at points that are relevant for their current interest. This has led to some points being repeated, but this is arguably an advantage where ideas are complex or difficult.

We have added considerably to the material in the original guides, so some guides are just a part of a chapter, while others are split between chapters. By the end of our editing, only five of the original guides escaped almost unscathed and these are Chapters 11 and 12 on data management, 17 and 18 on analysis, and 21 on the presentation of the results.

About a third of the chapters are new, and arose from our realization that gaps in the material, which might be acceptable in separate guides, gave a distorted view of NRM research when collected into a book. These affected all parts of the book, so most of Parts 1 and 5 are new, as are Chapters 4, 8, 13, 14, 15 and 19, and much of Chapters 16 and 20. Much of the new material followed the addition of Richard Coe as an editor. Our ideas on the integration of participatory methods in NRM research have also evolved recently, partly through work by Carlos Barahona and Sarah Levy in Malawi, and these ideas have been incorporated.

The large number of authors and contributors to the materials that led to this book means that there are differences in tone and style between (and occasionally within) the chapters. Some sections are didactic and prescriptive, others more discursive. Where eliminating these differences would detract from the sense of the chapter, we have largely retained the original wording. We hope that these discontinuities do not make the book as a whole less acceptable or less useful than we intended.

Roger Stern
Richard Coe
Eleanor Allan
Ian Dale
October 2003

Acknowledgements

We are very grateful to DFID for supporting various activities that led to this book. DFID funded the production of the 'statistical good practice' guides, and supported the 'QQA' project that was run jointly with NRI.

ICRAF's training activities – and thus the development of materials used in this book – were assisted by a contribution of the Ministry of Foreign Affairs of the Netherlands' Government through its Programme for Cooperation with International Institutes (SII). We gratefully acknowledge the SII's support.

We acknowledge the key contributions in the area of data management made by Peter Muraya of ICRAF. We are also grateful for the help given by other ICRAF staff, in particular Jan Beniest.

We thank all our colleagues in Reading, Nairobi and elsewhere for the many fruitful discussions that have widened the scope of the book. We are particularly indebted to Savitri Abeyasekera, Carlos Barahona, Robert Curnow, Sandro Leidi, Pat Norrish and Ian Wilson.

The datasets used in Chapter 12 were a component of socio-anthropological studies conducted within the DFID-funded Farming Systems Integrated Pest Management (FSIPM) project in Malawi. We thank Julie Lawson-McDowall (Social Anthropologist) and Mark Ritchie (Team Leader, FSIPM) for permission to use the data.

Many of the other datasets used in examples come from ICRAF projects: we are grateful for permission to use them. Details of some of the studies have been altered in order to make particular points.

We are very fortunate to have patient, imaginative and cheerful support staff in Reading, who at least pretended to enjoy the new challenges involved in producing this book. Alex Owen deserves our special thanks.

Contributors

Many people contributed, as authors, editors and critics, to the 'statistical good-practice' guides that formed most of the source material for this book. Most of the guides had at least three authors and several additional contributors. Their production was a team effort involving Roger Stern, Savitri Abeyasekera, Ian Wilson, Eleanor Allan, Cathy Garlick, Sandro Leidi, Carlos Barahona, and Ian Dale, with contributions from Richard Coe, Bob Burn, Sian Floyd, James Gallagher, Joan Knock, Roger Mead, Clifford Pearce, John Rowlands and John Sherington. Final editing followed by web and CD production was by Ian Dale. The full list of the guides, together with their principal author(s), is given in Section 23.4.

Several chapters are edited versions of material arising from collaborative work between the SSC and other organizations. Chapter 13 is based on ICRAF's training course notes on research data management, prepared largely by Peter Muraya and Richard Coe, with contributions from Jan Beniest, Janet Awimbo, Cathy Garlick and Roger Stern. Chapter 19 originated as material prepared jointly by the SSC and ICRAF for a course on the analysis of experimental data. Chapter 21 was extensively adapted, with permission, from notes prepared by John Sherington, when he was head of statistics at NRI.

For reference and citation purposes, the table overleaf names the principal author(s) of the material used in each chapter of this book.

Citation

To refer to this book, you are welcome to cite it as a whole, as follows:

Stern, R.D., Coe, R., Allan, E.F. and Dale, I.C. (eds) (2004) *Statistical Good Practice for Natural Resources Research.* CAB International, Wallingford, UK.

Reference to individual chapters should name the relevant authors from the table below. For example, to refer to Chapter 12, we suggest you use:

Garlick, C.A. (2004) The Role of a Database Package. In: Stern, R.D., Coe, R., Allan, E.F. and Dale, I.C. (eds) *Statistical Good Practice for Natural Resources Research.* CAB International, Wallingford, UK, pp. 167–186.

Table showing principal author(s) for each chapter in this book.

Chapter	Author(s)
1	R. Coe
2	R. Mead, R.D. Stern
3	R.D. Stern, R. Coe
4	R. Coe, R.D. Stern
5	R.D. Stern, S. Abeyasekera, R. Coe, E.F. Allan
6	I.M. Wilson, E.F. Allan
7	I.M. Wilson, R.D. Stern, E.F. Allan
8	R. Coe, R.D. Stern
9	R.D. Stern, E.F. Allan, R. Coe
10	C.A. Garlick, R.D. Stern, P. Muraya
11	C.A. Garlick, R.D. Stern
12	C.A. Garlick
13	P. Muraya, C.A. Garlick, R. Coe
14	R.D. Stern, R. Coe, E.F. Allan
15	R. Coe
16	R. Coe, R.D. Stern, I.M. Wilson, S. Abeyasekera, E.F. Allan
17	I.M. Wilson, R.D. Stern
18	R.D. Stern, E.F. Allan, R. Coe
19	E.F. Allan, R. Coe, R.D. Stern
20	R. Coe, S. Abeyasekera, R.D. Stern
21	J. Sherington, R.D. Stern, E.F. Allan, I.M. Wilson, R. Coe
22	R.D. Stern, R. Coe
23	E.F. Allan, R.D. Stern, I.C. Dale

Abbreviations

AHI	African Highlands Initiative (a project to promote good research).
ANOVA	Analysis of variance.
BACI	Before-after control-impact (a type of study).
CBO	Community-based organization.
CDC	US Government Centers for Disease Control and Prevention.
CIDA	Canadian International Development Agency.
CPP	Crop Protection Programme (a component of NRI).
CV	Coefficient of variation.
DBH	Diameter at breast height (a standard measurement of tree size).
DBMS	Database management system.
DF or d.f.	Degrees of freedom.
DFID	UK Government's Department for International Development.
DIY	Do-it-yourself (with the meaning of 'being self-sufficient').
EPA	Extension planning area.
FRP	Forestry Research Programme (a component of DFID).
FTR	Final technical report.
GIS	Geographic information system.
GLM	Generalized linear model.
GPS	Global positioning by satellite.
ICRAF	World Agroforestry Centre (the International Centre for Research in Agroforestry), based in Nairobi, Kenya.
IDS	Institute of Development Studies, based in Brighton, UK.
IFT	Indigenous fodder trees.
IPM	Integrated pest management.
LSD	Least significant difference.
MS or m.s.	Mean square.
MSTAT	Statistical software package from Michigan State University, USA.
NGO	Non-governmental organization.
NLM	Non-linear models/modelling.
NR	Natural resources.

NRI	Natural Resources Institute, based in Chatham and part of the University of Greenwich, UK.
NRM	Natural resources management.
NRSP	Natural Resources Systems Programme (a component of DFID).
ODBC	Open Data Base Connectivity (a standard for accessing databases).
PLA	Participatory learning and action.
PRA	Participatory rural appraisal.
QQA	Qualitative and quantitative approaches.
RBM	Results-based management.
REML	Residual maximum likelihood *or* Random Effects Mixed Models.
RNRRS	Renewable natural resources research sector.
RSS	Royal Statistical Society.
SAS	Statistical Analysis System (a software package).
SE or s.e.	Standard error.
SED	Standard error of differences.
SPSS	Statistical Package for the Social Sciences (a software package).
SD or s.d.	Standard deviation.
SS or s.s.	Sums of squares.
SSC	Statistical Services Centre, a consultancy group within the School of Applied Statistics, University of Reading, UK.
UKDA	United Kingdom Data Archive.
UNDP	United Nations Development Programme.

Part 1: Introduction

Chapter 1

What is Natural Resources Research?

1.1 Key Features

Natural resources include land, water, forests, germplasm and wildlife. The management of natural resources therefore embraces areas such as agriculture, forestry, fisheries and wildlife management. For many years, research into the management and use of natural resources has been done separately within these applied sciences. Over the last decade however, researchers have found that certain problems require an integrated approach if sustainable solutions are to be found. These integrated approaches are often called 'natural resource management research' (NRM research). The good-practice guidelines in this book apply to research in the component disciplines of NRM as well as to NRM studies themselves. We first explain some of the common features of NRM research to set the context for these guidelines.

A key element is to start the research with a well-defined problem (Izac and Sanchez, 2001). Potential solutions or interventions to address the problem are then evaluated. The productivity implications are assessed as in conventional studies in agriculture, livestock and forestry. Two more dimensions are added. These are the implications for human livelihoods (income, health, security of tenure, etc.) and environmental services (water supply, soil fertility, pests and diseases, etc.).The possible trade-offs are then evaluated since there are unlikely to be options that are attractive across all these dimensions.

Options have to be considered from the viewpoints of various stakeholders. These range from farmers and forest users to the global public concerned about issues such as climate change and biodiversity conservation (Van Noordwijk *et al.*, 2001). The trade-offs are likely to be very different for these stakeholders, so negotiation between them is inevitable. The role of science and research includes helping stakeholders understand the current status of the system, making better predictions of how it will respond to changes in management, and carrying out these changes in a way that will enhance understanding.

From this and other descriptions of NRM research we can identify the following common features of the approach which take it beyond traditional agricultural or land use research:

- **Problem focus**. The research aims to solve problems that are well-defined with a clear geographical boundary. The first step of the research may well be that of improving problem definition. However, the research must retain specific objectives to enable effective design of the study and analysis of the resulting data.
- **Management, action and learning**. The research usually aims to understand the impacts of changing resource management, and also to bring about improvements. The research is therefore linked to action. The impacts of these actions can be assessed and fed back into the problem for further examination. A common aim of many projects is to improve the effectiveness of the problem-action-impact cycle of all stakeholders, often described as improving adaptive management capacity.
- **Systems based**. The research looks at systems of land use, rather than at individual components. In any systems analysis there is an issue of 'bounding the problem' (deciding what can be considered as external) which is part of the problem definition. In NRM research, the system will contain elements relating to the human, economic and environmental implications of resource management decisions.
- **Integrative**. Implicit in the systems approach to problem solving is the idea of integrated solutions. There are unlikely to be single technological fixes for the problems which meet the multiple requirements of multiple stakeholders. Instead, we usually have to integrate a number of different approaches: for example, an agricultural pest problem may require integrated use of chemicals, biological control and changed crop management. A water quality problem may require the integrated use of advice, rewards and local legislation.
- **Multiple scales**. Also implicit in the systems approach is the use of multiple space and time scales. For example, changing tree cover in farm land may have crop growth implications measured at the plot scale, farm income implications measured at the household level, and water supply implications measured at the catchment scale.

All these features will have implications for the research design, data processing and analysis, the subjects of this book.

It is important to distinguish between projects that aim to solve problems for the relatively small numbers of people or limited land areas directly involved in the work (perhaps a few hundred households, a couple of catchments or a designated forest area) and projects that aim to produce information with wider potential application. If the project has only local goals then any approach which reaches them is acceptable; many participatory and action-oriented projects have been successful in this. However, many projects promise information which can be extrapolated with some degree of confidence beyond the immediate participants; indeed, this is often the only justification for funding a research project, and is usually what distinguishes 'research' from other projects. If the project is to deliver on this promise then careful attention must be paid to the design and analysis of the activities.

1.2 Examples

We assume readers are familiar with examples of agronomic experiments and standard surveys. Here we give three examples of larger studies. We have chosen examples that illustrate different elements of good practice. They are described only briefly, but each has a website, for readers who would like further information.

1.2.1 Malawi

To increase food security in Malawi, a starter-pack of maize seed, legume seed and fertilizer was provided to 2.8 million rural households in 1998-1999 and 1999-2000. In the following two years the programme was scaled down first to half and then one-third of the households; and was targeted only at the poorer households in the villages.

A survey was undertaken partly to assess whether the targeting was effective. This used standard survey methods to select villages, so the results could be generalized to the country. Participatory methods were used for the within-village data collection. This was because the questions were too complicated and delicate to be undertaken with standard questionnaires.

The archive, with a description of all the studies, plus the data and all reports are available from the website www.ssc.rdg.ac.uk

1.2.2 Hillside farming in South West Uganda

South West Uganda is an area of highlands with steeply sloping hillsides. The attractive climate has led to high population densities, with rural people trying to make a living from very small farms. Soils are degraded, resulting in low productivity. Much of the landscape is treeless and people are short of fuel wood and other tree products.

Projects therefore looked for technologies to address these problems. Several years of research involving on-station and on-farm trials resulted in the development of some promising options. However attempts to facilitate their use by farmers showed that their problems were not solved simply by technology. Issues such as protection of terraces, soil erosion and management of water required an understanding of the off-site effects of change within a farm. Hence the research moved to a landscape-scale of study. Bringing about change in the systems also needed an understanding of community action and the role of local and national policies. What started as a technology project using classical experimental methods grew to a natural resource management project with a series of connected activities using a range of formal and informal survey research methods.

More information can be found at www.worldagroforestrycentre.org/Sites/ Mountain AgroEcoSystems/mtagroeco.htm

1.2.3 Alternatives to slash and burn in the Peruvian Amazon

Deforestation in the Amazon and other tropical forest areas has repeatedly made headlines. However there is still much to understand about the driving forces of deforestation, its effects and alternatives. A project in Peru was part of a pan-tropical network aiming to understand this.

The project started with a detailed characterization of the problems using remote sensing together with formal and informal surveys. Some alternative land use systems were devised and tested in experiments. Part of the problem was found to be a reduction in diversity of some key forest species, so methods of conserving them had to be devised. The most effective were found to be community based conservation approaches. Those studies had to link detailed genetic investigation with studies of the ability and willingness of communities to be involved. It also became clear that there are important trade-offs between the interests of many stakeholders, from local farmers trying to make a living, to global society concerned about carbon balance or biodiversity. The studies therefore ranged from detailed measurements of gas emissions within some long-term experiments, to studies of biodiversity in a range of different land uses. All had to be conducted in a way that would allow the results from each component to be linked.

More details are available from the website www.asb.cgiar.org/home.htm

1.3 Where Does Statistics Fit In?

Why statistical guidelines for NRM research? What do the subject, the principles and practices of statistics, have to contribute to the research process?

From the brief description and examples of NRM research above, it should be clear that the data used to describe the current status of systems are likely to be complex. We might for example, have some measures of livelihood status from many households, measures of soil fertility and crop production from many plots, measures of biodiversity in surrounding forest and data on the viability of local institutions from each community. Each element may be measured at several time points with the whole study being carried out at several contrasting locations. Statistical ideas and tools are then needed for the following:

- **Providing summaries**. There are far too many numbers to interpret individually, so the information has to be reduced or compressed. Even if this simply requires calculation of some averages, there will be questions of how this should be done. Usually the objectives of the study will require further summaries, which can show, for example, the variation as well as the averages.
- **Understanding patterns and relationships**. Many study objectives require an understanding of the relationships between the multiple variables that have been measured. The question 'Do poorer farmers tend to have less fertile plots?' will require examination of the relationship between these variables, and will usually have to allow for the confusing or obscuring effects of other variables such as the soil type, time since forest conversion, or effectiveness of extension services. Much statistical modelling has been developed for use in these situations.
- **Understanding cause and effect**. Predicting the effect of changes in resource management requires understanding cause and effect. Will reductions in poverty lead to increased soil fertility (as farmers are able to invest more in the land), or is the relationship the other way around (soil fertility drives poverty)? Statistical ideas can help us understand the limits to causal interpretation of relationships, and can show the way to getting clearer understanding of causal relationships through experimentation.
- **Quantifying uncertainty**. All the data in the study will be 'noisy': subject to variation due to many unknown factors. Soil fertility varies due to local variation in soil type, unknown variation in land use history, the complex patterns of erosion and deposition, as well as being subject to measurement error. In addition, we can only ever measure a sample of the soil. The result is that every summary, relationship and pattern described is uncertain. The interpretation of the information will depend on the level of uncertainty, so this has to be determined.
- **Designing data collection**. The effectiveness of each of the above depends critically on how the data are collected, and statistical ideas can help optimize this. In some situations it may be possible to experiment. In others, we have to rely on survey methods, observing and measuring what is happening. In either case, careful planning of the study is needed to ensure, for example, that we will be able to estimate all the required quantities together with their uncertainty.
- **Improving research quality**. Two important dimensions of the quality of applied research are validity and efficiency. We have to be sure that results are valid, meaning that the conclusions are fully justified. We have to do this efficiently, meaning we want to maximize the information obtained with a given expenditure of research resources, or meet the objectives of the study at minimum cost. The statistical ideas presented in this book are key to achieving both of these.

1.4 Types of Study

Natural resources research uses a wide variety of methods, from focus group discussions in villages to interpretation of satellite images, from small plot experiments to long term resource monitoring. Despite this variety, there are a few basic research design types and it is important to recognize these.

The important distinction in research design is between surveys and experiments. Surveys measure the status of a system without deliberately intervening. They cover many different types and contexts, including conventional household, ecological and soil surveys. Typically a sample of 'units' (households, forest quadrats, sampling points) is selected and each one is measured (with a questionnaire, a species inventory or soil chemical analysis). Investigations such as environmental monitoring schemes and participatory rural appraisal (PRA) studies are also surveys, albeit with some unusual characteristics. For example, in a PRA study the 'unit' may be a village, the measurement tool the protocol for conducting a focus group discussion, and the observation a complex narrative. Nevertheless, the study still has the basic characteristics of a survey with the corresponding requirements for good design and limitations on what can be interpreted from the results.

A survey measures what is there. Thus, a survey to study soil macro fauna diversity can be used to show the current status. If we took measurements at several times, we could see the changes over time. We could see variation in fauna diversity between different locations, and show this is related to (correlated with) changes in soil organic matter content. However, we cannot conclude that a change in soil organic matter content causes a change in soil macro fauna diversity, in the sense that changing the soil organic matter at a particular place would lead to corresponding changes in fauna. This is because there are alternative explanations – such as both organic matter and fauna diversity depending on the length of time since forest conversion. The effect of such a variable could perhaps be controlled for, by sampling only plots of a constant age since conversion. However, as soon as one variable is controlled we can think of another that might be responsible for the observed correlation. By eliminating factors we become more and more convinced that it is soil organic matter that causes the variation in fauna diversity. Our reasoning becomes even more convincing if accompanied by a credible explanation of how soil organic matter interacts with soil macro fauna diversity. These studies do not amount to a 'proof' of cause and effect.

Ultimately the only way to show that changing soil organic matter changes macro fauna diversity is to change the organic matter and look for the response: that is, to do an experiment. The experiment may be tricky to design (for example, it may be difficult to change organic matter content without adding nutrients, and the new fauna communities may take a long time to establish). However, if we start with a number of units (e.g. soil plots) and randomly select some to which organic matter will be added and leave others unmodified, we can, to a known degree of uncertainty, determine the causal relationship between the variables. This is the fundamental characteristic of an experiment. Many of the statistical ideas and practical applications of experimental design were traditionally developed for agricultural and forestry research. These methods are still important in the broader NRM context. The research takes a systems approach, but classical experiments are needed to understand the properties and functioning of components of the system. NRM research is also likely to require experiments of other types: for example with catchments, farms or communities as the experimental unit. An important message in this book is that the same principles of sound design and analysis apply in these new contexts, and the same care needs taking if research is to be useful.

Within NRM research, we often come across studies that might be called pseudo-experiments. For example, we are interested in the effect of deforestation on soil organic matter and find plots which have been deforested, with neighbouring plots left undisturbed. Comparison of the deforested and control plots looks much the same as in an experiment. However, the situation differs from an experiment in that the decision as to which plots to deforest was not made at random, and not even made by the researcher. The interpretation of the data depends on whether there were any systematic differences between deforested and undisturbed plots other than the deforestation. A researcher may be convinced that there were none, but the validity of the conclusions depends on the extent to which this is the case rather than on the logic of the design. The pseudo-experiment occurs in many guises and must be recognized.

One type of design in which statistical ideas are less important is the case study. A single instance of a 'unit' (perhaps a household, a village or a watershed) is studied in considerable detail to gain insights into processes that were previously obscure. These insights might be valuable, but this style of investigation does not produce results that can be generalized. If only a single instance is investigated we have no way of knowing whether it represents the general situation or a unique case. However, case studies are important in various stages of research such as problem definition and formulation of hypotheses and models, which will then be used to design further steps using other methods.

Much of the data used in NRM research comes from routinely measured sources or monitoring systems. Included here would be satellite images and climate data from long-term weather stations. The important characteristic of this type of data is that the researcher does not design the data collection plan, but takes what is available. Appropriate methods of analysis will, as in all cases, depend on both the objectives and the assumptions it is reasonable to make about the collection process.

1.5 Is It Research?

The ideas discussed in this book are relevant to research (systematic generation of new understanding) about social, economic or biophysical aspects of natural resource management. However, the boundaries between research and other objectives are blurred. Resource management activities may have aims of introducing updated management policies and practices, of demonstrating how alternatives may be implemented or of empowering people to make changes. These are all important objectives in a development process. However, almost all activities will also have a learning aspect, hence the blurring of 'research' and 'development'. There are two important consequences of this.

Often an activity or a project is justified (and funded) because of its 'development' aspects; with anticipated positive changes in environmental or farmer well-being. However, there is a desire to learn from the experience, whether positive or negative, so that projects in other locations can benefit. The problem is that this often leads to a generalization, which cannot be justified. Observations made in one 'case study' (of a community, a watershed, a national park and so on) cannot be reliably generalized without some evidence that they apply in other contexts. Identifying the situations about which general conclusions are to be drawn, and making sure you collect evidence to support that generalization should be part of the project planning. And this planned, systematic learning is exactly what research is!

Secondly, an activity or project may have multiple objectives. For example, an on-farm trial might have objectives of evaluating performance of a technology over a range of conditions while at the same time acting as demonstrations. These two objectives do not always work well together. For example, demonstrations may have higher than typical levels of inputs (to ensure they 'work'), or may be modified during the season in a way that is not appropriate for a trial. The mixing of research and other objectives can lead to compromises that limit the ability to meet any of them.

The message is this: the mixing of 'research' and 'development' activities is often inevitable, but make sure they will really meet the joint objectives.

1.6 Who Is This Book For?

This book is designed for the members of the research team in an NRM project. The team may be of size two, if the work is for an MSc or research degree, where the student and the supervisor form the team. Or it may be a larger team, including an economist, social scientist, ecologist and agronomist such as in the Ugandan and Peruvian examples in Section 1.2.

By describing the common principles in the methods of data collection and processing in the activities that make up an NRM study, this book can support the team members who would like to understand and contribute to the project. If we remain within the limitations of research in a single discipline, we will make some progress, but not as much as if we recognize the arbitrary nature of some of the disciplinary boundaries. These boundaries are sometimes just as artificial and restricting as the geographical boundaries that divide countries.

Statisticians may find some of the ideas will support their role in project teams. A team's statistician should have skills that span the project activities, so he or she can act as the 'glue' that binds the project as a whole, helping to make it more than just the sum of its individual parts. Some statisticians will be horrified at the lack of formulae, even when we are trying to introduce ideas that are traditionally called 'advanced'. We are unashamed. If, through avoiding theory, we err on the side of over-simplification, it will perhaps help to redress the balance for all those occasions when statisticians have inflicted unnecessary theory on researchers, while omitting to put across the key ideas.

References

Izac, A.-M. and Sanchez, P.A. (2001) Towards a natural resource management paradigm for international agriculture: the example of agroforestry research. *Agricultural Systems* 69, 5-25.

Van Noordwijk, M., Tomich, T.P. and Verbist, B. (2001) Negotiation support models for integrated natural resource management in tropical forest margins. *Conservation Ecology* 5(2), 21 www.consecol.org/journal/vol5/iss2/art21/ (accessed 15 April 2003).

Chapter 2

At Least Read This…

2.1 Statistical Considerations

Research activities can often be enhanced by improvements in the statistical components of the work.

Researchers who lack confidence in statistics tend to plan conservatively, partly to ensure a simple analysis. While this is sometimes appropriate, simple improvements can often, for the same cost, result in a more informative survey or experiment. Measurements may be made that are inappropriate for the objectives of the study, and are subsequently not analysed. Data management is often not planned at all and becomes an unexpected nightmare as the study progresses. The analysis may be rushed and superficial because of the pressure to produce results quickly. The study is then concluded, with the realization that there is much more that can still be learned from the data. However, funding is at an end, and problems in the data management have made it difficult to allow easy access to the data, even for future researchers who have local knowledge of the country where the study was made.

In the following sections of this chapter we consider all the stages from planning through to reporting. They correspond to the parts of this book and hence provide a summary of some of the key points. If this chapter provides the answers you need then you can progress with your research. If you find it uses terms and concepts that you are not familiar with, or raises more questions than it answers, then we believe the rest of the book will help.

2.2 The Planning Phase

2.2.1 Specifying the objectives

The exact objectives of a research study should determine all aspects of its design, execution and analysis. Hence, the first step is to specify those objectives. If the research aims to solve real problems, as is usually the case in NRM research, start by assessing the nature of the problem and intervention points. Knowledge gaps will then determine research objectives.

Clarifying the objectives will also determine the type of research. For instance, if we need to know more about the constraints to adoption of a new technology, then a survey or participatory trial may be indicated. More knowledge, for example on critical processes affecting water use by proposed crops, might necessitate on-station, or laboratory work or the use of a process-based simulation model.

2.2.2 Units of observation

The units of observation are the individual items on which measurements are made. Examples include:

- Community
- Household
- Farmer
- Watershed
- Group of plants on an area of land (plot) or in a controlled environment
- Individual plant, single leaf or section of leaf
- Tissue culture dish
- Individual animal or group of animals (grazing flock, pen, hive)
- Fish pond
- Individual tree, sample plot or area of forest

Some studies involve more than one type of unit. For example, a participatory study may be partly at the village and also at the informant level. A survey may collect data on households and individual farmers. An agroforestry experiment may apply treatments to whole plots and make some of the measurements on individual trees.

2.2.3 Scope of the study

A key concept is the 'problem domain', namely the extent of the problem (geographically, or in terms of biophysical, social and economic factors) for which the conclusions of the study are to be relevant. A problem domain might be 'Farms in Western Kenya of less than 1 ha with a serious striga problem and no livestock on the farm'.

It is also important to have an idea of the precision of the answers before embarking on an investigation, since the degree of precision will determine the size of the investigation. The precision could be measured by the standard error of a difference between mean values, or between two proportions. It could also include precision of population parameters, such as the proportion of arable land used effectively or the percentage increase in uptake of new technologies by farmers.

The investigation should not be started unless it is expected to provide answers to an acceptable degree of precision.

2.2.4 Planning a survey or a participatory study

The key elements of both a survey and a participatory study are a well-designed sample, plus a questionnaire or some other data collection procedure that satisfies the study objectives. Important requirements of the sample design are that the sample is representative and that there is some element of objectivity in the selection procedure. This usually implies a need for some structuring of the survey, often involving stratification, to ensure representation of the major divisions of the population. Deliberate (systematic) selection of samples can give, in general, the greatest potential accuracy of overall answers but has the disadvantage of giving no information on precision.

Some random element of selection is needed if we are to know how precise the answers are. In most practical situations, clustering or multistage sampling will be a cost-effective method of sampling. A balance of systematic and random elements in sampling strategy is usually necessary. In multistage sampling the largest or primary units, such as the districts, are often selected purposively, but the ultimate sampling units are then selected at random within primary units.

For baseline studies the definition of sensible sampling areas, stratification scheme and the capacity for integrating data from varied sources are important.

In environmental sampling, where the spatial properties of variation are an important consideration, the spatial distribution of samples should provide information about variation at very small distances, at large distances and at one or two intermediate distances.

2.2.5 Planning an experiment

Key characteristics of an experiment are the choice of experimental treatments to satisfy the study objectives and the choice of the units to which the treatments are to be applied. A good experimental design ensures that the effect of treatments can be separated from the 'nuisance' effects with maximum efficiency.

There is usually some control of variation between the units and this is usually achieved by blocking. A randomization scheme is used to allocate the treatments to the experimental units.

For the treatments, the questions include what treatment structure to use, whether a factorial treatment structure is appropriate, and how levels of a quantitative factor should be chosen.

On controlling variation, the questions include the extent to which researchers should dictate management of experimental areas, what form of blocking should be used and what additional information about the experimental units (plots) should be recorded?

In certain types of trial there are particular aspects of design which are important. For example crop variety trials often include a large number of treatments and incomplete block designs are often used. Animal pasture trials are likely to involve multiple levels of variation: for example herds, individual animals within the herds, and the timings of observations on these animals.

2.2.6 Other types of study

Some studies, such as on-farm trials, combine elements of surveys and experiments. For example, farmers may respond through a semi-structured questionnaire, while observations are also made on plots where treatments have been applied. In pilot studies and more empirical appraisal systems the general concepts of experiments and surveys are relevant although the particular detailed methods may not be applicable. Thus representativeness and some element of random selection are desirable. It is also important to have some control, or at least recognition of potential sources of major variation.

Some studies also require access to routinely collected data, such as climatic records or aerial photographs. It is important to verify that these data are appropriate for the research task.

2.3 The Data

2.3.1 Types of measurement

Measurements can be made in many different forms ranging from open-ended responses to continuous measurements of physical characteristics such as weight, length, or cropped area. The number of insects or surviving trees may be counted, or scores of disease or quality of crops may be made. Options may be ranked, or a yes/no assessment can be given. The importance of including a particular measurement has to be assessed in the context of the objectives of the research.

There are some general points that apply to all measurements. First of all, the particular form of measurement does not change the relative precision of the results from alternative designs for an investigation. Secondly, for a given design, assessment in the form of continuous measurement gives greater precision than an ordered score. Thus, for a given precision, recording weights needs a smaller sample size than low/medium/high weight classes. Scores are generally to be preferred to ranks, though the latter may be easier for the respondent to give.

2.3.2 Collecting the data

Data collection forms will usually have to be prepared for recording the observations. In social surveys, the design and field-testing of the questionnaire are critical components of the study plan. Simple data collection forms often suffice in experiments and observational studies. In participatory research studies, the training and debriefing of the field staff are often critical components to ensure a consistency of approach by the researcher, as well as freedom of expression by the respondent.

2.3.3 Data entry and management

Part of the consideration of measurements is how the data are to be managed. At some stage, the data will normally be held in one or more computer files. The form of the data files and the software to be used for data entry, management and analysis are normally determined before data collection begins.

Data are sometimes collected directly into a portable or hand-held computer. Where data collection sheets are used, the data entry should be done directly from these sheets. Copying into treatment order or the hand calculation of plot values or values in kg/ha should not be permitted prior to the data entry. All data should be entered: if measurements are important enough to be made, they are important enough to be stored electronically. Studies where 'just the most important variables are entered first' inevitably result in a more difficult data entry process and the remaining measurements are then rarely computerized.

Data entry should normally use a system that has facilities for validation. Double entry should be considered, because it is often less time consuming and less error-prone than other systems for data checking.

It is desirable to follow the basic principles of good database management. In some studies the management is limited to simple transformations of the data into the units for analysis. It can, also present real challenges, particularly in multistage surveys and other studies where the data have a multilevel structure, as we show in Chapter 12.

2.4 The Analysis

In this section, more than previous ones, the reader may encounter unfamiliar statistical terminology. This is perhaps unavoidable, since we are trying to distil the elements of data analysis from Chapters 15 to 21 into a few short paragraphs. Our summary here is only a brief overview of some key steps, tools and concepts. We hope that the section will encourage NRM researchers with data to analyse to look further into the book.

2.4.1 Initial analysis: Simple graphs and tables

It is important to distinguish between exploratory graphics, which are undertaken at this initial stage, and presentation graphics. Exploratory graphics, such as scatterplots or boxplots, are to help the analyst understand the data, while presentation graphics are for showing the important results to others.

For surveys and participatory studies, simple tables are produced, often showing the results of each question in turn. For experiments, initial analyses usually include simple tabulations of the data in treatment order, with summary statistics, such as mean values. These initial results are only partly for analysis; they are also a continuation of the data checking process.

With some surveys, most of the analysis may consist of the preparation of appropriate multiway tables, giving counts or percentages, or both. Caution must be exercised when presenting percentages, making clear both the overall number of survey respondents and the number who responded to the specific question. Percentages should be avoided when the overall total count is low, because they can give a misleading impression of accuracy.

2.4.2 Analysing sources of variation

Particularly for experimental data, an important component of the analysis of measurements is often an analysis of variance (ANOVA), the purpose of which is to sort out the relative importance of different causes of variation. For simple design structures, the ANOVA simply calculates the sums of squares for blocks, treatments, etc. and then provides tables of means and standard errors, the pattern of interpretation being signposted by the relative sizes of mean squares in the ANOVA. For more complex design structures (incomplete block designs, multiple level information) the concept of ANOVA remains the same and provides the relative variation attributable to different sources and treatment means adjusted for differences between blocks.

In studies with less well-controlled structure it may be appropriate to split up the variation between the various causes by regression or, when there are multiple levels of variation, by the REML method that is discussed in 2.4.4.

The particular form of measurement will not tend to alter this basic structure of the analysis of variation, although where non-continuous forms of measurement are used the use of generalized linear models (a family of models which includes logistic regression and log-linear models) will usually be appropriate. Such methods are particularly appropriate for binary (yes/no) data and for many data in the form of counts.

All these methods are useful for all types of study – experiments, surveys and monitoring studies.

2.4.3 Modelling mean response

There are two major forms of modelling which may occur separately or together. The first form, which has been used for a long time, is the modelling of the mean response: for example, the response of a crop to different amounts of fertilizer, or the response of an animal's blood characteristic through time. The objective of such modelling is to summarize the pattern of results for different input levels. Frequently, the objectives also indicate the need to estimate particular comparisons or contrasts between treatments or groups.

Modelling of mean response may also include regression modelling of the dependence of the principal variable on other measured variables, though care should be taken to ensure that the structure of the study is properly reflected in the model.

2.4.4 Modelling variance

More recently, modelling of the pattern of variation and correlation of sets of observations has led to improved estimation of the treatment comparisons or the modelling of mean response. Particular situations where modelling variation has been found to be beneficial are when variation occurs at different levels of units, for spatial interdependence of crop plots or arrays of units in laboratories, or for temporal correlations of time sequences of observations on the same individual animals or plants.

The REML method, mentioned earlier, is relevant in all such cases (see Chapter 20). Multilevel modelling is also important for the correct analysis of most survey data whenever a modelling approach is required.

2.5 The Presentation of Results

The appropriate presentation of the results of the analysis of an experiment is usually in the form of tables of mean values, or as a response equation. A graph of change of mean values with time or with different levels of quantitative input can be informative. Survey results presentation takes the form of tables of numbers and percentages responding to questions. Graphical presentation should be clear and relatively simple, avoiding some of the 'fancy' effects (e.g. a third dimension to a two-dimensional plot) that some software produces. Standard errors (and degrees of freedom) should, as a general rule, accompany tables of means or percentages, or response equations.

2.6 Is This Enough?

This chapter has summarized the elements of good statistical practice in study design, data management and analysis. All are elaborated in further chapters. They are necessary but not sufficient conditions for best practice in research. Researchers have to bear in mind other aspects such as keeping focus on the real objectives rather than doing what comes easily, the need for integrating different activities, and the need to analyse and interpret results at the project rather than just at the activity level.

Statisticians are often accused of looking for problems and never giving credit for well-designed work. To counter this we offer the reader a selection of good practice case studies, found on the web at www.ssc.rdg.ac.uk which includes the following:

- Good practice in well-linked studies using several methodologies, based on a project in the Philippines.
- Good practice in survey design and analysis, based on a project in India.
- Good practice in on-farm studies, based on a project in Bolivia.
- Good practice in researcher and farmer experimentation, based on a project in Zimbabwe.
- Determining the effectiveness of a proposed sample size, based on work in Bangladesh.
- Good practice in data management, based on a project in Malawi.
- Three examples of good practice in the preparation of research protocols, based on projects in West Africa, Nepal and Bangladesh.
- Developing a sampling strategy for a nationwide survey in Malawi, based on a bilateral project in Malawi.

There are also many distractions to the use of best practice, some of which are described in the next chapter.

Chapter 3

Sidetracks

3.1 Introduction

Incorporating good statistical practice into natural resource management research activities is often not just a simple matter of understanding an issue and the options available for resolving it. A researcher discussing sample size with his or her colleagues may find that the debate becomes one of the merits of different types of farmer participation. A technician trying to establish a database for raw data from a project flounders over issues of data ownership. When the results of a research project are being prepared for publication, the discussion of the relative merits of standard errors or least significant differences being attached to tables meanders into an argument about the role of drama in communicating with farmers. These discussions, and many more, will inevitably arise in any NRM research project, and that is healthy. However, they can be sidetracks from addressing the statistical issues.

In this chapter we discuss just a few of these potential sidetracks. Our aim is to show why these distractions arise and how they can interact with areas of good statistical practice. Often the statistical points remain much the same whatever the outcome of the sidetrack argument. For example, the sample size required to generate information with a given level of confidence does not depend on the reasons for farmers to participate. The desirability of attaching metadata to raw data files remains the same whoever has access to the database. Standard errors and least significant differences attached to tables or results both require a certain level of statistical sophistication of the reader. For other audiences and media we have to find different ways of expressing the uncertainty in what we know.

The issues discussed here are a sample of those that could have been included, but serve to illustrate their sidetrack nature.

3.2 The Research Team

3.2.1 Understanding each other

Natural resources projects often involve multi-disciplinary teams. The ideal researchers in this team have some knowledge of all the areas of the project, while contributing extra information on their particular speciality. However, effective functioning of these teams can be hampered by:

- Misunderstandings of each other's perceptions of the problems.
- Misunderstandings of the contribution that each component or specialist can make.
- Lack of familiarity with methods and traditions of each specialist.
- Perceptions of special interest or hidden agenda.

These can adversely affect the research if they obstruct the team from making sound decisions. For example, social science has a strong tradition of participant observation, whereby the researcher joins activities to be investigated such as meetings of a farmers' group. This method of investigation is quite alien to the soil scientist, used to taking and analysing samples according to a rigorous protocol. However, these differences can be resolved to allow good statistical practice.

First, aim to understand the reasons why different disciplines use different methods, as well as the reasons why, and contexts in which, they were developed. It will then become clear that some of the key requirements for generating useful data and defensible results are the same in both cases. An element of consistency in approach is needed, to make sure we are comparing like with like. An element of repetition is needed in both studies to understand where the common features lie. In both cases, we would like the observations to be representative, and so on. What distinguishes the approaches is often no more than the constraints of the measurement tool (for example, an auger and chemical test compared with participation in a discussion), and the nature of the variability and repeatability that we need to know about.

Second, look again at why the different scientists, and their activities, are in the project. The aim is generating integrated understanding of, and solutions to, the resource management problems. This has implications for the way in which the studies are designed and analysed. If the soil scientist and social scientist are both there because social and biophysical aspects of soil management need to be put together, then this must be built into the statistical aspects of the design. For example, the objective of understanding some of the soil fertility implications of social processes will require the two components to be conducted together, involving the same farmers. On the other hand, if the soil scientist needs to do an on-farm trial evaluating new soil management practices, and the social scientist is examining the social constraints of current practices it may be best to separate the two studies. If they are collocated then the second study may be disrupted by the novel techniques in the experiment.

3.2.2 A statistician in the team

The ideal team includes a statistician with similar characteristics to the other members: able to communicate and understand others' perspectives, while adding his/her own specialist knowledge. The statistician within the team is often able to resolve some of the differences described in the previous section, as they have no other disciplinary 'turf' to defend.

Unfortunately, it is rare to find such a statistician, and consequently the statistical aspects of a project are often undertaken by scientists within the team. This book is for these scientists, and we hope it will alert them to the wide range of decisions that need to be made on statistical issues. Often it should encourage them to look for a statistician to be involved in their team.

Statisticians are not always involved in research teams to the extent that they wish. The reasons for this include the theoretical training that many of them receive and the tendency for statistics to be seen purely as a branch of mathematics. Another reason is sometimes their lack of experience in some areas of NRM research. A statistician trained in medical research, for example, may lack the confidence to advise on an ecological survey. Scientists also often expect more wide-ranging capabilities from a statistician than they would from someone from another discipline. They are then disappointed and fail to consult the statistician again. More patience with willing statisticians will often result in better statistical inputs to a project and improved research outputs. We hope this book will be of use to those statisticians who want to be effective in NRM research projects, but have limited experience.

In the absence of a statistician, a 'statistical auxiliary' is sometimes consulted or included as a member of the research team. This is typically an economist, social scientist, agronomist, breeder, or soil scientist, who is more comfortable with statistical ideas than his or her colleagues. They are useful in their own right and can also be effective as the means of communication between the team and an external statistician.

3.2.3 Competing objectives

A team that functions well should be working towards the common project objectives, and should be evaluated according to the progress made. Scientific publication, however, has long been regarded as a valuable way of judging quality and innovation in research, and is still a primary means of evaluating performance. There are therefore pressures to publish results at the activity rather than the project level. This can sometimes distort the big picture of the research by turning it into a set of disjointed activities that take up the available time and resources, yet do not fulfil the objectives of the project as a whole.

Pressures may be evident for statisticians as their primary role is one of a service provider, often applying known methods rather than developing new statistical ideas. This work, when done well, should be acknowledged as valid and worthwhile participation in the research process, and result in sharing in authorship of reports and papers. One particular sidetrack for statisticians is their lack of statistical publications, without which they sometimes receive minimal recognition within their institute. Many statisticians, aware of this problem, seek 'nuggets' within a project that might benefit from new statistical work. Whilst this can be valuable it should not be allowed to sidetrack the project. Academic institutes need to give more recognition to the statistician's contribution to project work. Equally, statisticians should think more about publishing imaginative examples of applied work. The statistical literature would welcome this.

Similar pressures can also apply to other team members. The soil scientist may feel that the work at the watershed level does not help in publications in the soil science literature. The economist may have difficulty with the lack of hard data for what they consider a full analysis. These issues need addressing early in the project, or they will tend to fragment the work and possibly render it much less useful for the stated objectives. Effectively the individual team members are working to a different set of objectives, so nothing more can be expected. Once recognized, it is often possible to add objectives so the project as a whole remains on course, while individual aspirations are also catered for and the good practice messages of design, collection and analysis of data can be applied.

3.3 Project Management

NRM projects can be fairly complex, with multiple activities, for the reasons described in Chapter 1. This adds management challenges and, with current pressures from results-based funding, monitoring and evaluation challenges. There is a tendency to look only at the project level in this monitoring process. Deficiencies may then be missed at the lower activity level, which may be serious enough to cast doubt on the validity of the project-level results. A recent review of Natural Resources Projects (Wilson *et al.*, 2002) included the suggestion that projects could be improved by adding some kind of methodological review at the 'activity-level'.

3.4 Statistical and Research Approaches

Project team members are often obsessed with a particular type of data collection and analysis approach that they wish either to use or to avoid, instead of thinking about what would be best for a particular situation. The enthusiasm or fear is often derived largely from ignorance of other methods, from examples where they were poorly applied, or simply from the traditions of the subject.

We examine three such areas here. The first is the use of qualitative methods as opposed to quantitative, or formal, survey methods that use questionnaires. The second concerns the use of models in NRM research, and the third is on statistical traditions or schools of thought.

In this book we emphasize similarities, rather than the differences between the approaches of collecting and processing research data. This is to encourage users to recognize more easily which method they could use, and to understand the methods used by colleagues.

3.4.1 Using a qualitative approach

Qualitative survey methods have been used by social scientists for a long time. They started to gain prominence in development projects during the 1980s, primarily in response to the drawbacks of questionnaire type surveys, which were considered time-consuming, expensive, and not suitable for providing in-depth understanding of an issue. This led to a polarization in collection and analysis of information with traditional, quantitative techniques on the one hand, and qualitative methods on the other.

One reason that they are called 'qualitative approaches' is because they often give rise to open ended responses. These usually need to be coded before they can be summarized and the resulting codes provide what statisticians call qualitative data.

The terms 'qualitative' and 'quantitative' are not without potential problems. The distinction has also been made between 'contextual' and 'non-contextual' methods of data collection and between qualitative and quantitative types of data. Contextual data collection methods are those that attempt to understand poverty dimensions within the social, cultural, economic and political environment of a locality (Booth *et al.*, 1998). Examples include participatory assessments, ethnographic investigation, rapid assessments and longitudinal village studies. Non-contextual data collection methods are those that seek generalizability rather than specificity. Examples are epidemiological surveys, household and health surveys and the qualitative module of the UNDP Core Welfare Indicators Questionnaire.

Using the terms above, we deal primarily with 'non-contextual' methods of data collection in this book, because we usually wish to generalize from the particular respondents to some larger recommendation domain, a point discussed earlier in Chapter 1 (Section 1.4).

During the second half of the 1990s, attempts were made to highlight the complementarity of the two types of approach; and their pros and cons were examined along with the potential for synergy in a general development context. In natural resources research it was realized that while some research practitioners were combining methods as a matter of course, their experiences were often not documented.

The key question is something like the following: 'Given a set of objectives on the one hand, and constraints such as time, money and expertise on the other, which combinations of qualitative and quantitative approaches will be optimal?' We are thinking here of research involving both socio-economic data (e.g. livelihoods, wealth, gender) and natural scientific information (e.g. entomology, epidemiology). The question should be asked for any data collected in a research context, whether it be as part of an experiment, survey or in the context of participatory activities.

For those unfamiliar with qualitative methods, a key component is the rich set of methods for data collection used by anthropologists, qualitative researchers and others, which are potentially of great value in NRM research. These methods are sometimes called 'participatory methods', because they encourage more discussion and involvement by respondents who may be individuals, focus groups or village committees. This involvement allows views to be recorded on complex issues that are impractical with standard questionnaires.

Participatory researchers now recognize that conducting a good survey involves much more than a questionnaire. There is basically no controversy about the need for sensible sampling, whatever the method of data collection, as we discuss in Chapters 6 and 7.

There is also little difference in the statistical aspects of the analyses. The data management process may be more involved because of the open-ended responses, but any coding problems for example are similar to those of coding open-ended questions from a standard questionnaire. Other types of data collection lead to ranks, scores, or causal diagrams (Burn, 2001), and methods for their processing are available.

One key component of a questionnaire-type study is that the questions are clear and simple so all respondents answer them in the same way. Otherwise it is difficult, or impossible to make sense of the results. This is a greater challenge for participatory studies. The amended quote: 'The plural of anecdote is not evidence' states that participatory studies must ensure that there is some uniformity of approach if the results are to be generalized, and not merely become a set of interesting case studies.

3.4.2 Statistical and other models

Table 3.1. Possible models of a watershed.

	Model type	Useful for
1	Mental model describing the interaction between water, land and people.	Starting to plan a project.
2	Physical scale model of the watershed, built by villagers from clay.	Delineating land uses, identifying resource management problems and possible interventions.
3	Input-output model connecting outflow of water and sediment to rainfall.	Estimating flood risks from hydrological records and long climate records.
4	Spatial hydrological model of surface and subsurface flow.	Predicting the effect of land use change on water availability.
5	Dynamic model of household decision-making.	Predicting the effect of policy changes on land use patterns.

The term 'model' is used in many ways in research and development, and it means many different things. For example an agronomist might immediately think of crop simulation models, whereas the statistician would think of the types of statistical models discussed in Chapters 19 and 20. In Table 3.1 a few possible models of a watershed are described with some examples of the purposes for which each may be useful.

None of these models is right or wrong; each has its own purpose. Statistical models are used for much of statistical analysis. In this book, statistical models are used to describe quantitative relationships between variables, based on measured observations. These will be part of some of the models in Table 3.1. For example, infiltration is likely to be a key component of Model 4. The relationship between land use and infiltration could be determined by a survey, measuring infiltration in a sample that covers a range of land uses. The effect of change of land use on infiltration may be further measured through an experiment in which infiltration is measured before and after changes. In both studies statistical models will be used to elucidate and quantify the relationships, which can then be built into other models used to solve the real problems.

The connection between the statistical models used to analyse empirical data and the other models used in problem solving is:

- Statistical models are used to identify and estimate the relationships built into other models.
- Other models are used to integrate and interpret the results of statistical models.

These are really two different facets of the same thing. The starting point is unlikely to be either the data or a model. Instead the ideas and information in models will evolve over cycles of observation and reflection, the data collection becoming more tuned to the purpose each time.

3.4.3 Approaches to statistics

The approach to statistics that we take for granted today is relatively recent. In this book we summarize the standard approaches to descriptive and inferential statistics (which probably incorporates much of what is needed by NRM research projects). These models use what theorists describe as a frequentist approach to statistical analysis. It encompasses all the methods taught in basic statistics courses and widely used in applied science, from the t-test to regression models for unravelling complex interacting effects. The latter have recently grown enormously in scope to cope with a wide range of realistic data types and structures, as described in Chapters 19 and 20. They also include non-parametric methods, though we find those only rarely of use (Section 20.3.3). Most NRM research will benefit most from sensible use of existing statistical methods rather than the development of new methods, and we have therefore resisted the temptation to include new methods in the book for fear of sidetracking projects from their main objectives.

One topic that should be mentioned is Bayesian statistics and decision theory. In principle Bayesian methods are attractive. The scientist's prior views about a particular parameter of interest are stated, the data are collected, and then combined with the prior beliefs to give the current views. This is appealing because it formalizes part of the scientific method, quantifying how beliefs develop as you gather more evidence.

These ideas are not new and controversy in the approach to statistics has raged since the 1960s. In these early years it was easy to ignore Bayesian methods, because they were impractical for the processing of complex datasets. Nowadays almost half the research papers in statistics take a Bayesian approach. More important still is that the Bayesian approach can now be applied routinely to fit useful models to complex datasets due to the existence of customized software. However the software packages available do not yet provide the same ease of use as some other statistical software.

Bayesian ideas can also combine naturally with a formalized approach to decision-making. Data are required to support sensible decisions. Once a decision is made there are possible 'losses' or 'gains' for each stakeholder, and the 'best' decision might be defined as the one with optimal trade-off.

Within a few years we would expect Bayesian ideas to be included within the standard tools for data analysis and be used more routinely for analysis of individual components of NRM projects. We are less convinced by the further step of including them within a decision-making framework. This is mainly because we anticipate that most users will prefer to give the evidence in a less digested form, so that individual stakeholders are able to come to their own decisions in a transparent way.

However Bayesian methods will never be able to completely formalize the scientific method for issues as complex as are usual in NRM projects. Rarely can we actually formulate a whole problem in terms of a set of parameters to estimate. Also evidence is rarely just in the form of new observations to be combined with the old, in a relatively simple updating process. Evidence may be in the form of observations that suggest the whole model in which the Bayesian analysis is embedded is inappropriate, and we must therefore start with a new conceptual framework.

The implication is that researchers must keep an open mind on appropriate methods, rather than following a dogma that cannot be universally applicable. Analyses will use a wide range of approaches as we show in Part 4. There is no 'magic bullet'.

3.5 Research and Professional Ethics

Ethical concerns in research have strong interactions with good statistical practice in design, data management and analysis. Examples include:

- The need for human subjects to be included in experiments only with informed consent. Most guidelines and regulations on this have been developed in the context of medical and psychological research. However numerous examples in NRM research show that participants often do not understand, and researchers sometimes find it difficult to convey, all the experimental aspects of research and the associated risks and responsibilities.

- Confidentiality of personal data. This can be a difficult topic in NRM projects that require sharing of data. Often it means that shared data contain gaps in the information; for example, where detailed information is given about specific locations but names of farmers within the locations are omitted.
- Requirements to feed results back to people who have provided information and taken part in research.

Researchers should strive for high standards, and need to be familiar with the research ethics requirements of their own organizations and those of project partners, along with any relevant legislation in the country where they are working.

There are also ethical dimensions to many professional and working practices. Some of the more obvious dubious practices to rigorously avoid are:

- Wasting people's time, and misleading them by collecting data that is not fully used.
- Dishonesty with data. This includes falsifying data, which often occurs when field workers are paid little and supervised even less. It most commonly is a result of failing to make sure field staff understand the basis for the research and the importance of the data. Dishonesty also includes choosing observations because they support the researcher's theory.
- Hiding data or stealing data. Here we include not making data available to others, perhaps in the hope that some day the researcher will get round to the analysis. Another example is where data are taken away for analysis, whereby they are no longer available in the country where the research was undertaken.
- Dishonesty with papers and reports. Giving the correct credit for all contributing to a project output can be difficult. There are clear guidelines on what it means to be an author of a scientific paper. Simply being the project leader, or from a collaborating institution is not good enough! For products other than scientific papers there are ways of crediting all with the actual contribution made.
- Not providing reports at the conclusion of a project. As an analogy, it would be an outrage for money to be taken from a public contract to build a road and no road built. Possibly the contractor would go to jail. Even if the road was built, there would be an outcry if the contractor used only a thin layer of tarmac so its life was short. The parallel is to take public money for a research contract and then to fail to produce results, or to produce only a perfunctory report.

Many professional societies have prepared codes of ethics and researchers should be familiar with those relevant to their work. They may also like to look at a statistician's code of practice. For an example of such a code, look at the Royal Statistical Society's website www.rss.org.uk/about/conduct.html That might help explain why statisticians don't always do quite what the researchers hoped, such as provide a *p*-value on demand.

References

Booth, D., Holland, J., Hentschel, J. and Lanjouw, P. (1998) *Participation and Combined Methods in African Poverty Assessment: Renewing the Agenda.* DFID Report.
www.dfid.gov.uk/Pubs/files/sddafpov.pdf (accessed 23 February 2004).

Burn, R.W. (2001) *Quantifying and Combining Causal Diagrams*. Theme paper from DFID project R7033: 'Combining Quantitative and Qualitative Survey Work'. Natural Resources Institute, University of Greenwich and Statistical Services Centre, University of Reading.

Wilson, I.M., Abeyasekera, S., Allan, E.F. and Stern, R.D. (2002) *A Review of the Use of Biometrics by DFID RNRRS Research Projects.* Statistical Services Centre, University of Reading.
www.ssc.rdg.ac.uk/dfid/BiometricsReview.pdf (accessed 24 October 2003).

Part 2: Planning

Chapter 4

Introduction to Research Planning

4.1 Introduction

The following chapters in Part 2, concentrate on the planning of research studies, focusing on 'good statistical practice' in both the design and the execution of surveys and experiments. A recurring theme throughout this part of the book is that it is the objectives which determine the design. In this chapter we look briefly at four areas that impact on the problem of deciding the study objectives, and hence on the quality of the research activities designed.

The main message is that attention must be paid to all levels of project design, and not just to those top-level aspects needed to raise funds and to get a project started. While many researchers spend much time on general project planning, the time and effort spent on the details of each activity is often minimal. It can take a long time for projects to move from the initial concepts stage to that of implementation; researchers can then find themselves under a great deal of pressure to move quickly to show that data collection has started in the field. Hasty decisions can result in field research activities that fail to meet objectives. The other less likely, scenario is 'overkill' whereby objectives are met, but could have been met at much less cost.

4.2 Linking Activities and Picturing Strategies

Natural resources projects usually involve a set of linked research activities. It is crucial that the objectives for individual studies link together and contribute to the overall project. The results-based management (RBM) framework is sometimes helpful for this. RBM is defined on the CIDA website (www.acdi-cida.gc.ca) as: 'Results-based Management is a comprehensive, life cycle approach to management that integrates business strategy, people, processes and measurements to improve decision-making and drive change. The approach focuses on getting the right design early in a process, implementing performance measurement, learning and changing, and reporting performance'.

An outline for an RBM view of a project is shown in Fig. 4.1. A problem is first identified and then the activities that will produce outputs (changes in knowledge) are planned and implemented. These activities will include (but not be limited to) studies using surveys and experiments. The outputs will lead to outcomes (people doing things differently) and the outcomes lead to impact (a change in the status of the problem). Using this model, it should be possible to map out the activities that are needed, and see how every result from an individual activity will eventually be used. If there are gaps, or outputs that are not needed, then the study objectives are incorrect.

Fig. 4.1. Outline RBM view of a project.

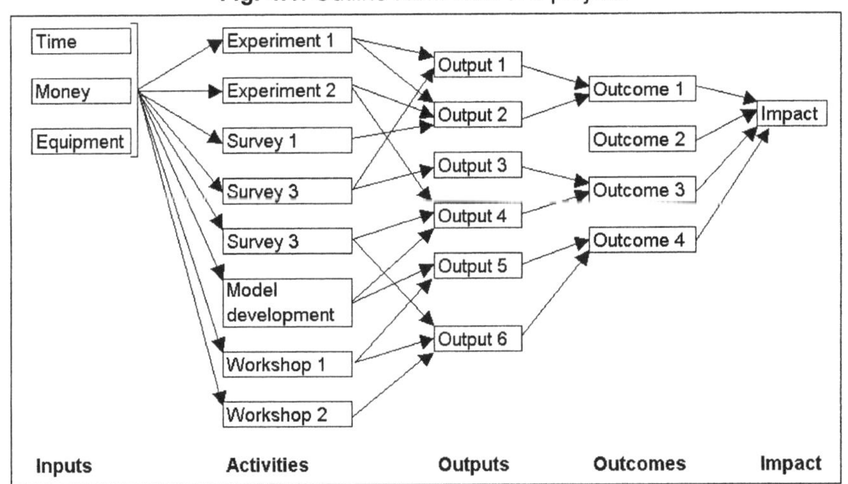

The model in Fig. 4.1 does not capture the fact that most projects have a sequential element, where the results of one activity help determine how the next will be carried out. In such situations it may be helpful to think of the process that we would like to follow step by step to implement a solution to the problem. Some of these steps cannot yet be taken because of gaps in our knowledge and understanding. Hence, we carry out research studies to fill these knowledge gaps. Once we have filled a gap, we can move to the next step in the implementation process. A simple example is shown in Fig. 4.2. Each rectangular box represents an activity, and the ellipses represent milestones in the project's progress. With this style of map of a research strategy, it should be possible to see each individual activity as a part of the whole puzzle, and design it to meet those requirements. Diagrams of the logic and structure of research strategies, such as that shown in Fig. 4.2, can be very useful for keeping all participants focused on the purpose of their research. Some project offices insist on having these clearly on display for this purpose.

Fig. 4.2. Simplified representation of the strategy for a project aiming to solve a fodder supply problem, through domestication of indigenous species.

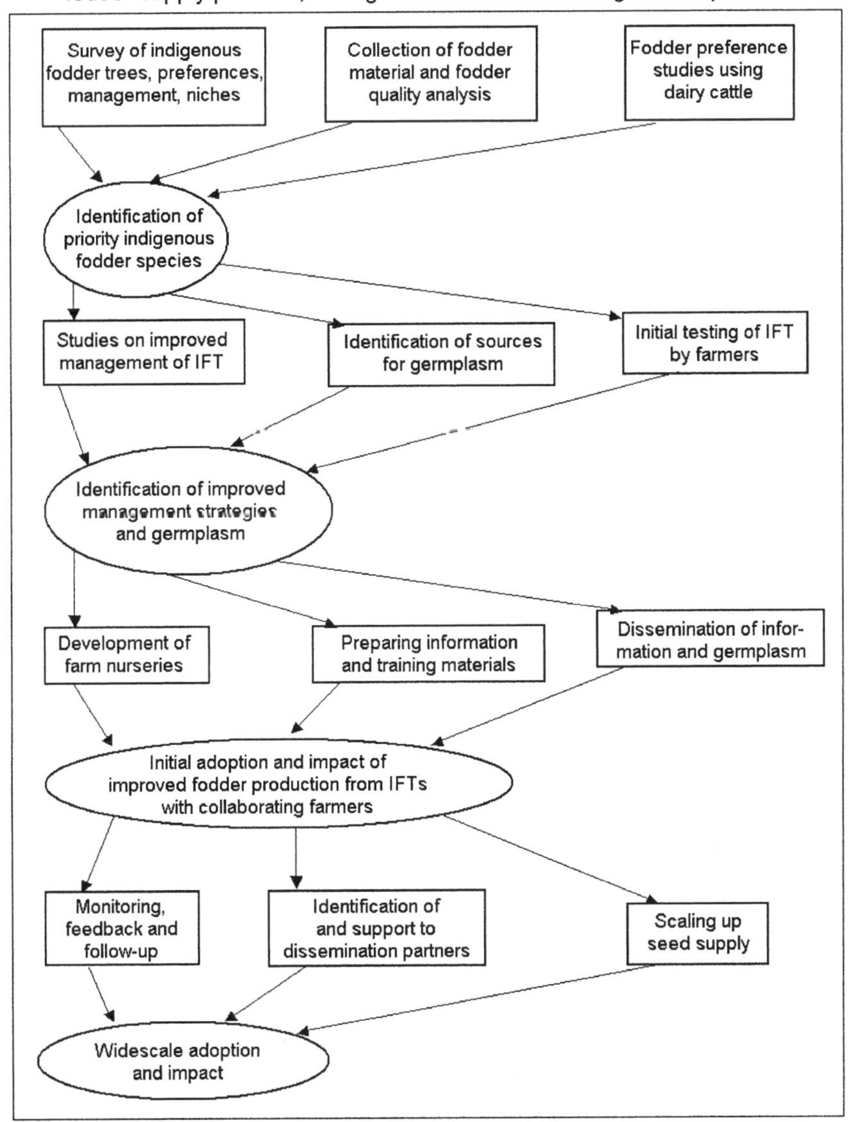

4.3 Priorities

Most projects have to go through various levels of priority setting, making reasoned choices between alternatives. For example, will the project look at all fields or only maize fields, as maize is the most important crop? Will it work on upper watersheds or wetlands as well? Should it focus on the people who can benefit most or the poorest? There are many approaches for choosing priorities, from estimation of economic surplus to systematic use of expert opinion. The methods used are usually less important than being consistent within a given project. This reduces the chance of parts of the project going off in different directions, and increases the likelihood that all those involved in the project understand the reasons for the decisions being made.

Table 4.1. Choosing factors for investigation, in a study on management of improved fallows.

	Criteria			
Factors	Effect on biological performance	How likely to be modified by farmers?	How critical for adoption?	How easy to research?
Method of land preparation	1	1	3	3
Direct seeding vs. transplanted seedling	3	2	3	3
Effect of seed inoculation	1	1	3	3
Tree population per ha	2	3	1	3
Establish with or without crops	3	3	3	3
Frequency of weeding	3	3	2	3
Length of fallow	3	1	1	3
Management of fallow: burn vs. incorporate	3	1	1	3
Remove tree stumps or not	1	1	3	2
Mixtures and sequences of crops	1	3	1	1

A systematic approach to priority setting is equally important when setting the objectives of an individual study, such as a survey or experiment. Technology development experiments, for example can have a large number of factors that could be studied. Which ones are worth investigating? Again, there are many approaches, and a simple method is illustrated in Table 4.1. In the example in Table 4.1, the problem was the management of improved fallows for soil fertility replenishment. A brainstorming meeting produced the list of possible factors to investigate. These included methods of land preparation (e.g. no-till, flat ploughing and ridging), and the crop mixtures and sequences to follow the fallow (e.g. maize monocrop, maize-beans rotations and maize-beans intercrops). Four criteria were then selected to choose between the factors to investigate. The first, 'effect on biological performance', is the extent to which each factor is likely to impact on the soil fertility of the fallow. Other criteria were whether farmers have a choice on this factor, whether it is likely to affect adoption and the ease of doing the research. Then each factor is scored, from 1 to 3, on each criterion, with a score of 3 being positive for the factor. The resulting scores were used to narrow down the number of factors investigated and to justify some of the reasoning behind the decisions.

4.4 Problem Domains and Research Locations

Defining the problem domain is a critical part of planning and objective setting. You have to be able to answer the question 'What is the geographical extent of the problem you hope to solve?' For many years, a common approach was to start investigations at one site, which was often a research institute. If success was achieved the efforts would be 'scaled up', to bring the positive results to a larger area, with disappointment when this did not work. The problem is that solutions that are optimized for one place are not always appropriate in another. One way to minimize the chance of this happening, is to begin the study by delineating the area and conditions within which you hope to find answers, and then match the research to this problem domain. The domain will usually have a geographical boundary, for example Nyanza Province of Western Kenya, or the Kenyan Lake Victoria basin less than 1500 m above sea level, or the Nyando river basin. The domain may also have other defining characteristics, such as 'Households for which farming is the main source of income', 'Villages with established farmers' groups' or 'Fields with sandy soil'. Whatever the definitions, it must be possible to identify whether or not a field, farm, or village is within the problem domain.

A key question is 'Where in the problem domain should research be carried out?' We want results to apply, with some degree of certainty, to the whole domain, but this does not mean that every part has to be included in every component study. Some cases to distinguish are:

- The study involves processes that we are confident apply anywhere in the problem domain. For example, an experiment might be done at just one site to investigate the way rooting depth modifies competition between trees and crops. While the results may be different depending on climate or soil types, a single location will be sufficient to establish the principles and understand the key processes.
- The study involves processes (biophysical, social or economic) that we expect to remain constant across the problem domain, but we are not sure. We run the study at a scattering of locations across the domain to check. An example is the disease resistance of a new variety. While we expect it to be resistant to a pathogen wherever it is grown, it is possible that there will be unexpected interactions with the farming system or some other factor. A similar example from social science would be the hypothesis that farmer's investment in soil conservation requires secure land tenure. A survey to check this would sample from the whole problem domain.
- The study involves processes that we expect to vary across the domain, or to interact with factors that vary. For example, we may think that the viability of different soil conservation technologies depends on rainfall, slope, soil

type and ground cover. The research locations have to be selected to cover suitable ranges of each of these factors. This problem is discussed further in Chapter 9.

In agronomy, these same ideas are known as 'the G×E interaction problem' (Genotype by Environment), the possible existence of which means that genotypes must be studied at several sites. In the absence of any G×E interaction, the same genotypes can be recommended across all sites. If there is G×E interaction, several sites should be studied to understand which environmental factor affects our recommendations. In NRM research, we can still use the G×E terminology, though in this broader context, we consider 'technology by environment' or 'intervention by environment' where 'environment' now includes the social, economic, and physical environment.

A related problem is deciding if a study needs to be repeated over successive seasons or years. Guidelines similar to those above could be employed, but substituting the year or season for the site. There are often good reasons for repeating studies in successive years, but doing a study over several seasons just because 'journals do not accept the results from a single study' is not one of them.

One of the main reasons for repeating a study over several years is the variation in weather patterns from one year to another. If this aspect is important, then just two or three years is a very small sample from which to draw wider conclusions. Researchers therefore need to look for ways in which the results of their studies can be 'embedded' in a much longer time frame. One possible way is to make the processing of climatic data a linked activity in the project, so that this small sample of information can be put into context. This should not be difficult, as meteorological stations are sited throughout most countries and are often located at research stations where components of many studies are undertaken. Another option that can be viable is to substitute sampling over several years by sampling several sites in the same year; locations are sampled over a wide area, giving sufficient variation in the effects of the weather in a single time period. Of course, it is important to make sure that other factors that vary in location are not going to confuse the results. The feasibility of this will depend on the exact problem.

We return to the idea of embedding studies in Chapter 6, where a similar argument is used in research studies where the participatory component is carried out at only a small number of sites.

4.5 Iterative Planning and the Activity Protocol

Defining the objectives for an individual activity, and larger parts of the study, will be an iterative process. Tentative objectives will be set and an initial attempt made to design the trial or survey. This may reveal that some objectives need modifying, perhaps because they are too ambitious, or a study to meet them may be too expensive.

It may become apparent that the original objectives will need more than just one activity; tentative plans will need rewriting to accommodate this. It is important to look again at the revised objectives to see if they can be met with realistic studies. At each step, the planned objectives must be checked to see if they still fit into the overall strategy. Often researchers continue with what they first envisaged, sometimes implementing an experiment or survey which clearly will not fully meet objectives. This is a waste of research effort and resources.

A written protocol should be prepared for every research activity. The protocol describes in detail how the study will be carried out and the information should be sufficient to allow someone to repeat the study. Appendix 1 contains an example checklist for a study protocol for an on-farm trial. This is not a pre-formatted document to complete, but a list of items that the protocol should cover. The protocol must be kept up to date, from the iterations of the design details, until completion of the study. It is usual that the design will be modified once the fieldwork starts, and these changes must be recorded. The written protocol is important for the following reasons:

- It forms the basis for discussion with other scientists and a statistician at the planning stage.
- The management process may require the detailed study plan to be approved before fieldwork starts.
- It can be shared with all who are involved in the study (e.g. collaborating scientists, technicians) to ensure that all have a common understanding of what will be done.
- It is a plan that can be followed, reducing the chance of mistakes being made in the implementation.
- In long-term studies, there may be staff turnover during the project and the written protocol provides continuity.
- It is a record of the data collection process, which can help analysis and interpretation.
- It can be archived with the data so that in the future it can be re-analysed and interpreted.

Within the protocol all terms used must be sufficiently clearly defined to allow the plan to be used. Sometimes objectives use words such as ‘diversity’, ‘poverty’, ‘eco-system health’ and ‘adoption’. Such concepts can be useful in general project planning but have to be tied down to specific and usable definitions when preparing detailed study objectives.

Chapter 5

Concepts Underlying Experiments

5.1 Introduction

An experiment is characterized by a researcher imposing change on a more or less controlled structure, and measuring the result. Experiments can give insights into cause and effect, with a measured level of uncertainty. A few years ago, much of natural resources experimentation was carried out on-station, which allowed for a high degree of control. These experiments are still important today, particularly for investigating biophysical processes. NRM research also involves investigating a much broader range of issues than can be accomplished in the typical research station setting. Experimental design ideas must still be taken into consideration at the planning stage of a project, regardless of the type of research. They are likely to be relevant even if you do not consider the research primarily 'experimental', and will certainly be relevant if using an adaptive management or participatory research approach.

This chapter describes the key concepts that are important in designing an experiment. These concepts can be applied to most types of study, but are easiest to explain when based on the design of on-station trials.

The main concepts explained here concern:

1. Identification of the objectives of the experiment in the form of specific questions.
2. Selection of treatments to provide answers to the questions.
3. Choice of experimental units and amount of replication.
4. Control of the variability between units, through systems of blocking and/or by using ancillary information collected on the units.
5. Allocation of treatments to particular units within the overall structure, involving where possible, an element of randomization.

6. Collecting data that are appropriate for the objectives of the research. These may cover the entire experimental unit, or may involve sampling within the unit.

We look, at sampling in more detail in Chapter 6 and apply the concepts explained in this chapter to NRM experiments in Chapter 9.

5.2 Specifying the Objectives

The current situation of the target problem must be reviewed before the research objectives can be identified. The aim is to identify knowledge gaps and specific aspects that need to be explored within these gaps. This should lead to a clear specification of the objectives. The importance of clear objectives cannot be emphasized enough. Objectives determine all aspects of an experiment, particularly the context in which the trial is run, the treatments to be compared, and the data to be collected. Vague objectives are of no value in determining the rest of the design. It is not unusual to see experimental objectives which say essentially 'The objective is to compare the treatments'. Such a statement cannot be used to determine the rest of the design. In practice, the process is usually iterative, whereby tentative objectives lead to a tentative experimental design, and then a re-examination of the objectives.

Experiments, which are part of applied problem solving, should always lead to a logical next step that could not have been reached without conducting the experiment. Check what will be done once the experiment is completed that cannot be done now. Then make sure the experiment will really allow that next step to be taken. The linking of studies to produce a coherent overall strategy is discussed in Chapter 4.

For example, an on-station experiment (trial) was planned as part of a strategy for introducing improved tree fallows for soil fertility restoration. The on-station trial involved looking for improved soil fertility after fallows with a range of different species. The next step was to be an on-farm evaluation of the species that produced positive results. However, 60% of the farms in the problem domain were found to be severely infested with striga. Therefore, it was decided that the trial objectives should be extended to include a study of the effectiveness of the different fallow species as a striga control. The study was then carried out in farmers' fields as a researcher-managed trial, as the research station fields did not have high levels of striga.

Most experiments have multiple objectives which have to be identified individually and collectively to ensure that they can all be met by one experiment. For example, an experiment to determine the extent to which a *Sesbania sesban* fallow can reduce striga incidence will need the fallows to be established under realistic farmer conditions. An additional objective of understanding *how* the sesbania reduces striga, may need a very different experiment. Some possible mechanisms will have to be hypothesized, and experiments devised to test them. These experiments might be in artificial conditions, such as laboratory investigations of the effects of sesbania root extracts on striga seed germination.

The objectives should also indicate how assessments will be made. Objectives such as 'To assess the effect of sesbania fallows on striga' are not sufficient. Should we measure the viable striga seed bank, the emergence during the fallow or after the fallow, growth rates and flowering, effects on associated crops or farmers' perceptions?

The assessments will also suggest the set of treatments for study. If the objective is stated as 'Assessing the effect of a sesbania fallow on subsequent maize yield', the sesbania fallow needs to be compared with something; it cannot just be observed in isolation. The question remains whether it should be compared with current fallowing practice, no fallowing, both, or something else altogether. The sesbania fallow could be compared with a continuous maize crop (i.e. no fallow) and current fallowing practice. The expectation is that the sesbania fallow would perform better than the no-fallow system, and be no worse than the current fallowing practice. Two hypotheses are of interest here: (a) sesbania fallow is superior to having no fallow; (b) there is little difference in performance between the sesbania and current fallowing practice. In this way, the objectives can be expressed as testable hypotheses, which translate directly into the analysis to be undertaken.

Objectives should not be over-simplistic or they will not add to our understanding of a problem. For example, farmers in West Africa asked a researcher for the optimal time to plant their cotton. This was stated as the objective and an experiment was conducted at the research station with six different planting dates as the treatments. After three years experimentation, the main conclusion was that the optimal date is different in each year. This was no surprise to either the researcher or the farmers.

Unfortunately, this type of objective is quite common. It is equally idealistic to look for the optimum level of fertilizer, optimal date to spray and so on. The reason that these objectives are over-simplistic is that life is rarely so simple. The optimal amount of fertilizer may depend on many other factors, such as the initial fertility, the type of soil, the date of application, the way it is applied, and so on, as well as on the farmers' objectives. A more fruitful strategy would be to develop a theory of optimal planting or fertilizer application, and design experiments to test and refine that theory.

Good experiments can only be devised when the objectives include some clear hypotheses. An example aimed to identify the key factors that determine the conditions under which sesbania fallows can be recommended. The plan was to measure the effect of the fallow on a number of farms and hope to spot the condition under which it did well. However, this necessitates some hypothesized factors, otherwise we will not know which characteristics to record, nor will we know whether the set of farms includes a suitable range of these factors. Once more, the objectives must be specific.

Thinking of the question 'What comes next?' should help researchers to conduct experiments with more insight. For example, consider the problem of searching for the suitable species to recommend for a three-year improved fallow. A direct strategy is to try the promising species, with each considered just as a treatment. An alternative is to look for the characteristics of the species that make them acceptable. Specific suggestions that suitable species should be fast growing, deep rooted legumes will lead to comparisons of species that vary in each of these factors, rather than simply working with a list of hopeful guesses. Such an experiment could lead to the two next steps. First, we could start making recommendations to farmers that might allow them to select species that are likely to perform well as fallows, and at the same time meet their other objectives. Second, this might serve as a basis for proposing, and then investigating, the mechanisms by which the fallows restore fertility.

In many NRM studies, the treatments are alternative technologies, which come in the form of a whole package of recommendations. For example, an alternative technology might be to plant early, using minimum tillage of the soil, and rotating maize with groundnuts. Objectives of comparing such packages of recommendations often turn out to be uninteresting. When important differences in performance of packages emerge, the experiment cannot show which components are responsible. The trials rarely represent alternatives that farmers will use, as they tend to adapt complex sets of recommendations to meet their own requirements. More successful approaches might focus on understanding why things work, rather than simply comparing them.

It is not always feasible to investigate the effects of change by direct experimentation. Focusing the objectives on understanding the processes involved, rather than trying to collect empirical evidence, may be more useful. As an example consider the problem of understanding the effect on overland flow and erosion of changing land management in a catchment. The direct experimental approach would define experimental units (catchments) and impose a number of treatments of land use change. The practical difficulties are obvious, and it may not even be possible to define realistic treatments. However a simple model of the processes describes surface flow as dependent on slope, roughness and infiltration, and it is roughness and infiltration that may be changed by land use. It is feasible to experiment on the effect on these of changes in the management of landscape elements. Models can then be used to interpret the results at the catchment scale.

5.3 Selection of Treatments

We have already seen that the choice of treatments depends on the objectives of the experiment. The minimum number is two, appropriate for the most straightforward of objectives. An example is assessing the impact of introducing a new extension method compared to the existing one, with treatments of the current and new practice. Many experiments will have less straightforward objectives, and less obvious choice of treatments.

Suppose the objective of an experiment is to assess the effectiveness of a pesticide spray on the vigour and yield of groundnuts. Assuming that we know the formulation of the spray and when, how and how much should be applied, there are two possible treatments: with spray or without spray.

This experiment is risky as the results are based on a single variety of groundnut, and we cannot be sure whether the effect of the spray depends on the variety. To validate the conclusions, the spray could be tested on several varieties. This introduces more choices. Suppose we decide to include four varieties of groundnut in the experiment. These could be four varieties that are popular with farmers in the domain of interest. Alternatively, we could be more strategic, with the objective of understanding the effect of spray on groundnut in general. There are numerous types of groundnut but understanding the pathogens suggests two characteristics may be important: growth form (spreading or erect) and resistance (susceptible or resistant). We choose one variety of each combination.

There are now eight treatments in our experiment, shown numbered in Table 5.1.

Table 5.1. Treatment combinations for evaluating spray on four varieties.

Growth form	Resistance	Variety	No spray	Spray
Spreading	Resistant	A	1	5
Spreading	Susceptible	B	2	6
Erect	Resistant	C	3	7
Erect	Susceptible	D	4	8

We may also want to investigate other characteristics, such as time to maturity. Table 5.2 shows the possible treatment combinations when we include this feature (long- or short-season) while still using just four varieties. There are many combinations that could have been chosen, but the key point is that each characteristic is represented by two varieties. Therefore we have two erect and two spreading varieties, two resistant and two susceptible varieties, and two long-season and two short-season varieties. We will see the same idea again in Section 6.5, when choosing sample villages.

Table 5.2. A possible treatment combination for evaluating spray on four varieties (A, B, C and D) selected from eight samples.

Season	Growth form	Resistance	Variety	No spray	Spray
Long	Spreading	Resistant			
Long	Spreading	Susceptible	A	1	2
Long	Erect	Resistant	B	3	4
Long	Erect	Susceptible			
Short	Spreading	Resistant	C	5	6
Short	Spreading	Susceptible			
Short	Erect	Resistant			
Short	Erect	Susceptible	D	7	8

5.3.1 Factorial treatment structure

The experiment shown in Table 5.1 has eight treatments defined by combinations of two factors, spray (Spray and No Spray) and variety (A, B, C and D). This is known as a factorial treatment structure, and in this example is a '2 by 4' structure (usually written as '2×4'). Each treatment is made up of a set of factor levels. For example, treatment 1 is variety A without spray, and treatment 8 is variety D with spray.

Project objectives often involve several factors. For example, these could be associated with crops (e.g. variety, fertilizer, spacing, planting date), livestock (e.g. feed type, feed supplement, time of feeding), or social factors (e.g. types of extension material, sex of head of household). Factorial treatment structure is the norm in agronomy trials. However, for problems that are unusual subjects of experimentation, factorial treatment structures are seldom used. Factorial structure is an important method of experimentation as it allows interactions to be detected and studied. We describe this below.

The spraying example above is a simplified version of an actual trial that had 16 different spray treatments. The options in the trial were whether or not to spray on any combination of four occasions. The times of possible spraying were 45, 60, 75 and 90 days after sowing. With the four varieties, the treatment structure was therefore a 2×2×2×2×4 factorial, with 64 different treatments, as shown in Table 5.3.

One objective in using this structure was to answer the question 'Is early spraying useful?' Factorial treatment structure allows interactions to be studied, answering questions like 'Is spraying in mid-season equally useful, even if there was early spraying?'

Table 5.3. Treatment combinations (64) for studying spraying (S) on up to four occasions on four varieties.

	Date				Variety			
Spray	45	60	75	90	A	B	C	D
1	-	-	-	-	1	17	33	49
2	S	-	-	-	2	18	34	50
3	-	S	-	-	3	19	35	51
4	-	-	S	-	4	20	36	52
5	-	-	-	S	5	21	37	53
6	S	S	-	-	6	22	38	54
7	S	-	S	-	7	23	39	55
8	S	-	-	S	8	24	40	56
9	-	S	S	-	9	25	41	57
10	-	S	-	S	10	26	42	58
11	-	-	S	S	11	27	43	59
12	S	S	S	-	12	28	44	60
13	S	S	-	S	13	29	45	61
14	S	-	S	S	14	30	46	62
15	-	S	S	-	15	31	47	63
16	S	S	S	S	16	32	48	64

Another way of looking at the spraying treatments is shown in Table 5.4. The first consideration is the number of times to spray, which ranges from zero to four. For a given number of sprays, we then ask 'What are the most effective times to spray?'

Table 5.4. The same information as in Table 5.3, but arranged as frequency and regime factors.

Spray		Variety			
Frequency	Regime	A	B	C	D
0	Never	1	17	33	49
1	45	2	18	34	50
1	60	3	19	35	51
1	75	4	20	36	52
1	90	5	21	37	53
2	45, 60	6	22	38	54
2	45, 75	7	23	39	55
2	45, 90	8	24	40	56
2	60, 75	9	25	41	57
2	60, 90	10	26	42	58
2	75, 90	11	27	43	59
3	Not 90	12	28	44	60
3	Not 75	13	29	45	61
3	Not 60	14	30	46	62
3	Not 45	15	31	47	63
4	All	16	32	48	64

What is the alternative to factorial structure? Consider the example above but with just two spraying factors; on day 45 and on day 60. With four varieties, this would result in 16 treatments. An alternative would be two simpler experiments, each with eight treatments, consisting of the four varieties and a single occasion when spraying is done or not. In the first experiment, we spray some of the plots on day 45. In the second experiment, on day 60 we evaluate the first spraying that occurred on day 45.

Suppose that in each small experiment we found that spraying was useful. What are our recommendations? It would probably be to propose spraying on day 45 (using the information from the first experiment), and on day 60 (from the second experiment). We would therefore be recommending a spraying pattern that we had not actually tried. The previous experiment, with the two spraying factors, might confirm that this was the sensible recommendation. On the other hand, it might show that all we need to do was spray once, and it did not matter when we sprayed. The results of several single factor experiments will be equally misleading if the objectives are to understand processes rather than generate recommendations directly.

Sometimes a factor is included in an experiment just to increase the recommendation domain. The variety factor in the above experiment is of this type. This was not a trial to select a certain variety, and the objective of the protocol did not refer directly to the variety factor. Hence, discovering which was the best variety was of no interest to the researcher in this study.

Consider this experiment without the variety factor, but instead using the single most common variety grown by the farmers. This experiment would then support the majority of the farmers who use that variety, but might not be of use to farmers who grow a different variety. However, if the variety factor is added and spans the range of varieties used by all the farmers, then two outcomes are possible. If there is no interaction between spray and variety, then the same spray recommendation can be given for all varieties grown. If there is an interaction, (say upright and spreading varieties needed a different regime), then this can be reflected in the spray recommendation. The variety factor was included in this trial because of its possible interaction with the spraying recommendation. If the researcher becomes convinced that there is never an interaction, then it could be omitted in further trials. The experiment is then more straightforward, though another factor like plant spacing, could be investigated instead. Section 4.3 looks at ways of choosing which factors could be investigated.

Interaction often forms the main focus of the experiment. Consider an irrigation trial, with a treatment factor of irrigation at two levels, one to provide plenty of water and another to stress the plants. Showing that plants prefer water to being stressed is not a new discovery! However it is still of interest to investigate which varieties, spacing levels or type of tillage are better able to withstand stress, and why.

Another key reason for encouraging factorial treatment structure is hidden replication. To explain this point we can look at the following example. In semi-arid areas of West Africa, most farmers grow millet, and in the 1990s, research emphasis concentrated on the effects of mulching with millet stalks before sowing. The mulching protects against erosion, reduces soil surface temperature, returns some nutrients to the soil and increases organic matter. However, the millet stalks are of value to the farmers as they are used for building, for fuel and as fodder.

The mulch trials had multiple objectives leading to three treatment factors:

- Application of mulch at three levels (0, 0.5, and 1.5 tonnes per hectare).
- Different ways of adding phosphorus at four levels, (none, rock-phosphorus, triple superphosphate, and a micro-dose of phosphorus) in the planting hole.
- Cropping pattern (growing millet as a sole crop or intercropping with groundnuts).

This produced a 3×4×2 factorial treatment structure, with 24 different treatments. If there were two replicates, there would be 48 plots.

Assume that the analysis of this experiment indicated an interaction between the application of mulch and cropping pattern (sole or mixed). There was also an effect of adding phosphorus, but this did not interact with the other two factors. The mean yields could then be presented as shown in Table 5.5.

Table 5.5. Structure for tables of means, from a three-factor experiment with a two-factor interaction.

a

Mulch (t/ha)	Sole	Mixed
0	X	X
0.5	X	X
1.5	X	X

b

Phosphorus			
None	Rock	TSP	Micro-dose
X	X	X	X

Table 5.5a shows the six means for the combinations of cropping pattern and mulch level. This trial was conducted on 48 plots, so each mean is the average of

eight observations. In Table 5.5b, the means for the four phosphorus levels are each an average of twelve observations. The 'hidden replication' occurs because each mean is based on eight or twelve observations, although there were only two explicit replications in this experiment (i.e. only two plots had exactly the same combination of all three factor levels).

We often find that scientists do not appreciate the value of factorial treatment structure. They sometimes devise rules, such as having a minimum of four replicates in every experiment, and these rules do not include hidden replication. The result can put unnecessarily low limits on the number of treatments in an experiment. To those who are receptive, we often find that a major improvement can be made in their design by adding another factor, and decreasing the explicit replication. This adds either to the objectives that can be included in the study, or to the domain of recommendation.

Is there a limit to the number of factors in an experiment? No, but the practical considerations of having to implement, manage, measure and interpret the experiment will impose limits. In an example from a research institute in West Africa, it became apparent that numerous small experiments were being conducted each year, involving mulch as a treatment factor. It was suggested that these experiments should be combined using factorial treatment structure. However this resulted in nine factors and over 400 treatment combinations in order to meet all the objectives. Each team member tended to emphasize the work related to their area of expertise. However, in order to ensure the overall feasibility of an experiment, team members must be willing to compromise their individual emphasis. Any study with too many objectives and agendas becomes impossible to conduct in an efficient manner. The next stage therefore was to rationalize the work into three or four studies, using the type of priority-setting discussed in Chapter 4.

When an experiment involves several factors, the treatments chosen do not have to be the full factorial set, if the objectives or practical constraints suggest otherwise. A 'near-factorial' set with perhaps an extra treatment, or one or two combinations omitted, may be appropriate in some circumstances. For example, a study of the application of different amounts of fertilizer (three non-zero levels) at different growth stages (three levels), might also include a 'no fertilizer' control, making a total of 10 treatments (3×3+1). Alternatively, in an on-farm trial looking at the effects of manuring (none, one bucket, two buckets), and sowing rate (usual density and double), the farmer may not be prepared to sow at double density without manuring.

5.3.2 Quantitative treatments

The examples above have had two types of factors. Qualitative factors such as variety, have levels that are defined by type rather than quantity. However, a factor such as an amount of fertilizer, is quantitative. This introduces some important choices when designing the trial, particularly when the objective is concerned with the response to changes in the level of a factor. To illustrate this, a curve can be estimated from the data, as shown in Fig. 5.1.

Fig. 5.1. Response (R) to increasing levels of a factor (F) and some possible designs for investigating it.

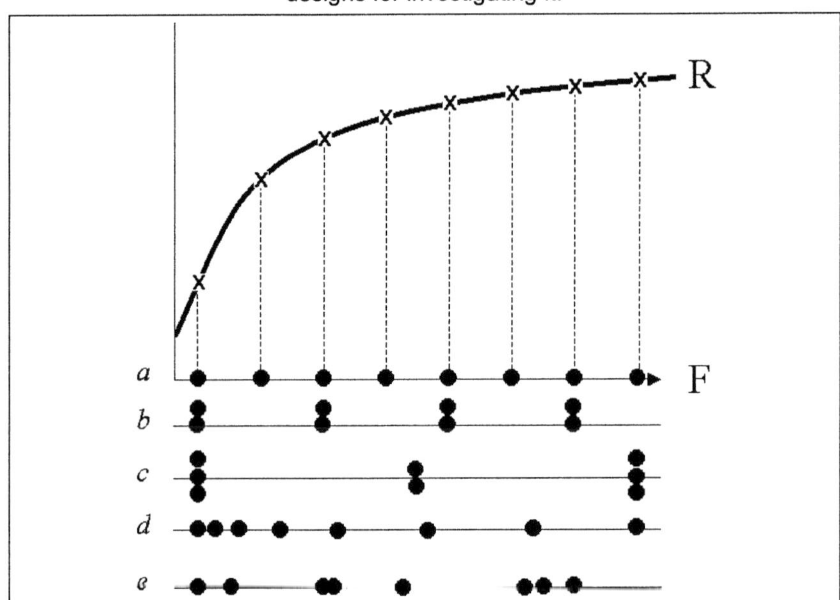

The response (R) to increasing levels of a factor (F) is shown as a smooth curve. The objective of the experiment is to find this curve. Responses are measured (×) at different values of F (•). Some of the choices available are illustrated. Design *a* has 8 equally spaced levels of F, giving eight observations from which to estimate the response. Design *b* has four different levels, but two replicates of each. Design *c* has unequal replication, and design *d* has unequally spaced levels. In design *e* the levels were chosen randomly.

The design questions are therefore, which levels of the quantitative factor should be used, and how many observations of each level are needed? To answer these we need to consider not just the objectives, but also prior information, or suspicion, concerning the response to changing levels of the factor.

Consider again the example of applying mulch to millet fields. The initial proposal was to use three levels, namely 0, 0.5 and 1.5 t/ha. There was a discussion on the levels to use, of which the main arguments were:

- Even if the trials are to be carried out on-station, the levels must be realistic. Farmers rarely have a total dry matter as high as 1.5 t/ha. The millet stalks are traditionally used for other purposes, therefore it is unreasonable to have a mulch level that is so high.
- In an on-station trial, the aim is to estimate the response curve. With high inherent variability this would be much more difficult to do if low levels of application (e.g. 0, 0.25, 0.5 t/ha) were used. The response will be clearest if it covers as much range as possible so high levels of mulch are needed.

The second argument was adopted for the on-station trial. However, the farmers added a further element as the parallel on-farm studies confirmed that they clearly understood the importance of mulch, though used it selectively on less-fertile patches of their fields. On those patches, they applied mulch at a rate that often exceeded 2 t/ha.

For quantitative factors, general guidelines on the number and spacing of factor levels follow from regression ideas because at the analysis stage we usually fit a curve to the data. The principles of fitting the curve provide an indication on the choice of levels. The key points are:

- For given resources, it is better to have fewer levels with more replicates of these levels rather than to have more levels with fewer replicates.
- Unless the expected response curve has a particularly unusual shape, there is usually little point in having more than four or five levels of the factor.
- The levels used depend on the form of the expected response curve, and do not necessarily need to be equally spaced. For example, it is better to have levels closer together where the response changes quickly. The key element is that the levels should allow us to capture the main features of the shape.
- The range of levels studied should be wide enough to fit the required curve well, and this may mean covering a range that is wider than the researcher had anticipated.

The distinction between qualitative and quantitative factors is not always clear-cut. The varieties might be chosen to be of different durations if the key issue was the season length. The variety then effectively becomes a quantitative factor. The same guidelines apply when choosing these levels. We might decide to use levels of 70, 90, 110 and 130 days duration but find that varieties with these exact durations do not exist. The design is equally good if near equivalents are used (say 72, 91, 108 and 135 days). It is the guiding principle that is important rather than strict adherence to the detailed advice.

5.3.3 Control treatments

Control treatments need to be justified through the objectives of the experiment in the same way as any other treatment. The role of a control is to act as a baseline against which other treatments can be compared. There can be one or more control treatments, and they must be chosen so that you have the right baselines for all the comparisons you want to make. For example in the groundnut example considered at the start of this section, either never spraying, or spraying on each occasion could be viewed as the control. These treatments are included in the analysis in the normal way. Control treatments should not be thought of as 'zero input', 'do nothing' or 'usual practice'. Each of these may be the right baseline against which to assess other treatments, but it might not be. Hence, phrases such as 'three treatments and a control' are confusing. If there are four treatments, just describe each of them.

Many experiments involve mixed cropping, particularly if we interpret a *mixture* broadly as including two components such as maize and beans, crop and trees, or crops with weeds. A common debate is whether each component alone should be included as control treatments. The answer is that it depends on the objectives: there are realistic objectives which require these sole components but others which do not, or only require one of them. Determining the control treatments needed by the objectives often leads to reassessing the objectives themselves.

5.4 Choosing the Units

The experimental unit is the unit of resource material to which a treatment is applied. The conclusions of the study are drawn from this unit. In an on-station crop experiment, the unit is the plot. In other instances the unit could be:

- A single animal; for example a study looking at the effect of dipping on tick prevention. An animal is dipped, and the number of ticks remaining on that animal, are counted.
- A group of animals; for example when goats are housed in pens and different feed regimes are given to each pen of goats.
- A household; for example when a change in the management of the farm, aims to improve the conditions of the household as a whole.
- A farmer's field; for example when different weed management practices are compared on different fields.
- A village; for example, when the impact of a new extension service is being assessed at the community level.
- A forest block, watershed or track of rangeland; for example, when looking at larger scale effects of management changes.

5.4.1 Type and size of units

The choice of an appropriate unit depends on practical considerations combined with the study requirements. Past experiments provide guidance to researchers on the type and size of the unit required for a given factor. For example, irrigation treatments or land preparation methods, are normally applied across larger areas of land, while factors such as crop variety or amount of fertilizer, can be sown in (or applied to) smaller areas of land. These will still produce conclusions that are relevant to real-life practices. In experiments with a social component, the impact of market access may be assessed at the village level, while use of credit, could be addressed by individual households.

Field experiments may be more precise if plots are long and narrow than if they are approximately square. This is because long narrow plots reduce the average distance between different treatments. However, management issues, such as ease of cultivation, harvesting, or type of equipment used, may dictate the shape of the plots. On particular types of land, other shapes may also be feasible, such as plots following the shape of the land contours on terraced land or other obstacles on the terrain. In a nutrition study on goats, the search for precision may indicate a preference for using individual animals as the unit, but practical issues of applying feeding regimes to pens of goats may have to be considered.

Sometimes the natural units, such as a community group or a terrace, may be of unequal size. These should be accepted, rather than trying to force all units to have the same size, though understanding the effect of size on response may become an additional objective in such a trial. However the variation, in size or other characteristics that will occur in a group of naturally occurring units, may obscure the responses of interest and decrease precision. Where practical reasons and the need for precision are in conflict the practicalities should dictate the unit, but the discussion may lead to an assessment of the feasibility of the study to achieve the stated objectives.

5.4.2 Interference

Interference amongst units occurs when what is happening on one unit influences a neighbouring unit. Generally interference should be avoided, as the data can be very difficult to interpret.

In field experiments, one way to deal with interference at the design stage is to create border areas around the plots. This often leads to square shaped plots, as this minimizes the land area committed to borders. Sometimes solutions have to be more extreme. For example, in agroforestry trials it may be necessary to dig trenches between plots to prevent root interference. A 'no-tree' control is often an end plot. This however, is in conflict with the idea of randomizing all treatments; here the practical issues should again take priority.

Where the units are people, interference may involve neighbours copying or competing with each other. A simple solution is to work with a higher level of unit (perhaps a village), although the occurrence of interference can indicate flaws in the conception of the study. For example, a series of trials with four plots on each farm, in a large on-farm study, had the objective of assessing the value of agroclimatic information at the plot level. This information included when to plant, when to apply fertilizer and so on. The main factor was 'with and without the agroclimatic advice', and the second factor was 'with and without added fertilizer'. The interference problem was predicable: the farmer often copied the agroclimatic advice on the control plots when relevant, and this interference resulted in the 'control' plot being nullified as a true control. When assessing the farmer level trials, this resulted in studies that had control and 'treated' farmers in the same villages. However, this was also nullified by the farmers, as the human factor was involved and they continued talking to and watching each other.

This type of study could possibly avoid the interference problem by incorporating the whole village into the unit, though it might be more constructive to assess the viability of the research as a whole. A more appropriate approach might use participatory methods to evaluate the significance of information that the farmers received.

5.4.3 Hierarchical structures

When there are two or more factors, the researcher must consider whether the same size of unit is necessary for assessing both factors. In the irrigation method/crop variety example, all plots could be large units, however, this is likely to be too costly or result in an experiment with too few replicates. An alternative is to use the large areas of land for the irrigation treatments, and split them into smaller plots to accommodate the different crop varieties. This is called a split-plot design. The large areas of land to which the irrigation treatments are applied are the main plots; the smaller areas are the subplots or split plots.

The split-plot experiment is a specific example of the more general hierarchical (also known as nested or multilevel) structure, in which there are different 'layers' of experimental units, and different treatment factors at different layers. Examples of hierarchy include plots within fields, farms within villages, cows within herds, beehives on trees, individuals within households, blocks within sites and sites within regions.

There is no need to make structures 'more hierarchical' purely to accommodate several factors. For instance if spraying is introduced as a third factor in the irrigation experiment above, the most appropriate design might be to have combinations of irrigation and spraying applied to the large areas, with variety being sown in the smaller plots within the areas. A split-split-plot is only necessary if practicality dictates that each factor requires a different sized unit.

Hierarchical structures have become more common in experiments with the move away from on-station work. On-farm trials introduce different levels of variability. For example, there will probably be less variability between farms in the same village, than between villages in a region; variation among cows within a herd will also be different from the variability across herds. Since understanding the variability is at the heart of analysis, it is important to recognize and understand the layers of the hierarchy in any design. There may also be layers in addition to those to which treatment factors are applied. For example, insect counts are made on quadrats within each plot, and soil nitrate is measured on cores within each quadrat.

5.5 Replication – What, Why and How Much?

Replication refers to the number of experimental units per treatment. The reasons for replication are:

- To make it possible to estimate the variability at the unit level.
- To quantify the uncertainty – that is, to attach an estimate of precision – to the results.

Increasing the replication so that averaging is done over more observations generally improves the precision.

A question frequently asked in designing experiments concerns the number of replicates that should be used. Details of sample size calculation are discussed in Section 16.9.5. The elements of the calculation depend on anticipated treatment outcomes and the variability of the resource material. Getting prior estimates is an important part of planning an experiment. Expected treatment effects should be realistic rather than optimistic, while estimates of variability are usually from published research or previous experiments on similar material.

The decision about replication also should take into account:

- The benefits of hidden replication from the factorial structure.
- Possible loss of information due to unexpected occurrences (e.g. failure of equipment, wildlife damage, floods and fire).
- The need to accommodate unforeseen objectives and emerging priorities in long-term trials.

A trade-off between precision and available resources is usually needed, and this may mean revisiting the study objectives. Simplifying the objectives can reduce the number of treatments, thereby increasing the replication of individual treatments for the same resources.

Increasing the replication is not the only way to improve the precision, and it is often the most expensive way. Two other ways to increase precision are:

- Reducing variation through good management of the experiment. Anything from the use of cloned rather than out-crossed planting material, to improved training and monitoring of technicians, can have a dramatic impact.
- Using a design that recognizes probable heterogeneity, which is discussed in relation to blocking in Section 5.7.

5.5.1 Replication, hierarchical structures and levels of variation

In hierarchical structures, it is important to be clear on the unit of replication for each treatment factor, and to ensure that there is sufficient replication for comparisons to be made.

Consider a tree species trial with five species, planted with four trees in each plot, and three plots per species. This gives a total of 60 trees in 15 plots. Growth rates are measured by recording initial and subsequent heights of individual trees. The experimental unit to compare species is the whole plot and not an individual tree. Therefore, there are three replicates per species, even though we have 12 trees and hence 12 measurements of growth for each species. The species have been allocated to whole plots of four trees, and hence differences between the species must be assessed relative to the plot-to-plot variation, not the tree-to-tree variation. The latter gives the within-plot variation while the former is the between-plot variation. Using the within-plot variation to compare species is not correct and does not give a valid estimate of the precision of a treatment comparison. It may overestimate the between-plot variation or, more commonly, it may be an underestimate because of homogeneity of the within-plot environment.

The split-plot experiment is another example with different levels of variation, but with treatments factors at both levels. In the example of the irrigation trial, there are two sources of variation:

- The variation between main plots against which irrigation treatments must be evaluated.
- The variation between split units (within main units) for assessing variety differences and the interaction between irrigation levels and varieties.

We need to ensure that the replication is suitable for the treatment comparisons of interest; particularly the treatment factors at the top level of the hierarchy (e.g. the irrigation methods in the split-plot experiment). These same ideas apply to other studies with a hierarchical structure.

5.5.2 Unreplicated experiments

The previous section emphasizes the need for replication, though this is not always possible. Consider an investigation into the effect of a new watershed management practice on the water quality of a river. The study is conducted for the benefit of the local farmers who want to assess the impact of the intervention in their community. One site is chosen for the new practice and another site located nearby (similar in all aspects), is the control site. Measurements are made at both sites before and after the new practice was implemented. This is called a 'before-after control-impact' or 'BACI' design; in this particular example there is just one site per treatment and hence a single replicate.

As it stands, this investigation addresses its objective of demonstrating that a change has occurred. However, it cannot prove that the change was caused by the new watershed management practice. This confirmation can only come from replicate sites and the random allocation of the treatments. Hence this unreplicated experiment is not one from which wider conclusions can be drawn. A land manager who would like to learn whether the intervention has a role to play across the region would have to replicate the study across several sites. The general efficacy of the intervention can then be assessed relative to the site-to-site variation. Ideally this should mean replication of both control and 'treated' sites. However, a cheaper replication option may be to have several control sites and only one or two 'treated' sites. Of course, the opportunity for replication is also limited; there are unlikely to be many sites available on which to carry out the experiment. Nevertheless, some replication is needed to generalize the results.

5.6 Choice of Sites

The choice of sites for an experiment – whether it is conducted at a research station or on farmers' fields – depends on the study objectives and prior knowledge of the resource material and the effects of treatment. It needs careful attention. It seems unlikely that the research station where you happen to have an office, or the village where you did your last fieldwork, will provide the right site for future experiments.

The sites used will be a sample from the 'recommendation domain'. The researcher should understand what the recommendation domain is, and make sure that the sites do represent the population of interest. For instance in an experiment looking at the effects of weeding practices in sandy soil, obviously we only need sites that have sandy soil. If the recommendation domain is 'all soil types' then this needs to be reflected in the sites chosen for study. Since the results from the sites will be generalized, the key issue is to ensure that the sampling process is representative. This is discussed in Chapter 6.

However, where the objectives demand specific areas, the situation is different. For instance, if new interventions for disease control or pest incidence are being investigated, it is necessary to choose sites that are known to have high levels of disease incidence or pest attack. This is necessary to observe and evaluate the effects of the interventions.

Single site experiments will be appropriate if you are sure the effects of interest will be large compared to any treatment by site interaction. They can also be appropriate when the objective is to confirm that a phenomenon occurs. For example, a trial was conducted successfully at a single site to investigate whether changes in soil depth can modify tree-crop competition, as theory predicted. Of course, if the trial failed to show the expected effect we could not conclude that it would not become apparent in a different environment.

Often the effect of treatments is expected to vary across sites. The implication for site selection depends on the role this site by treatment interaction plays in the objectives. For instance if the objective is to determine whether some weeding practices are more successful in some types of soil as opposed to others then a range of different soil types is required. We want to learn something specific about the interaction and choose sites to allow it to be estimated. If, on the other hand, the intention is to make general recommend-ations to all farmers in the region, the researcher needs to ensure that the spectrum of soil conditions in the region is covered in the sites. Here the interaction is not of primary interest, but we need the average effect of the treatment to be realistic for the region. This of course raises the question of why an interaction that is thought to exist should be deliberately ignored.

The number of sites used is also an important consideration, and is part of the decision-making on replication. A common procedure in on-station research is to do an identical experiment at three sites. Similarly, on-station experiments are often repeated for two or three years, because of differences caused primarily by climatic variation. These are simply samples of size two or three, and so do not provide much information in terms of generalizing to a wider population. In the case of climatic variation one way to generalize is to use historical records or simulated series of climatic records to help interpret the results. Of course if it turns out that the three years actually sampled very little of the expected weather variation, then the conclusions of the experiment are still limited. Similarly, analysis by means of a GIS (geographic information system) may help put the few experimental sites into the wider context of variation across the recommendation domain.

The characteristics within each site play an important role in determining the details of the experimental design. These include the identification of blocks and the size, shape and orientation of the plots. Typical sources of variability at potential sites are due to soils, topography, other physical features and previous management. It is important to take account of these sources of variation and minimize their effect when choosing areas for laying down the experiment and this may mean omitting parts of the site. Soil variability is generally the largest likely source of variation. Where resources permit, a uniformity trial or measuring soil samples may help in identifying patterns in soil variability. A map of the site and information about its past history are also useful.

5.7 Choosing the Blocks

The concept of blocking in experiments is the equivalent of stratifying in surveys. The main reason for using blocks in an experiment is to increase the precision of treatment comparisons; stratification is also used to ensure that important subgroups are represented, and for administrative convenience.

It is unusual to find that the experimental material is homogeneous, but the experimental units can often be grouped so that they are relatively homogeneous within groups. These groups are called blocks. In a field trial, it would be expected that adjacent plots, or plots with similar soil properties, would give similar yields; these would constitute blocks. In social investigations, farms that are ranked similarly in terms of wealth could be grouped together to form blocks. Blocking is the grouping of a set of experimental units into these homogeneous blocks. Researchers' knowledge of the material to be used, and experience with past experiments of a similar type, should help in choosing blocks. In experiments with field crops, fertility gradients in the field or physically defined variables such as moisture and water levels, height and slope of the land can all be candidates for blocking. In a study of people's livelihoods, village, farm size and gender of the head of the household may all be useful blocking factors.

A design that is used frequently is the randomized complete block design, in which the blocks are all the same size, and where the size is the same as the number of treatments in the experiment. Here a block contains a single replicate of each treatment. This design is so common that scientists often confuse the concepts of blocking and replication.

One reason the randomized complete block design is common, is that the analysis is straightforward and intuitively appealing. However, now that software for the analysis of other designs is widely available, researchers could look afresh at the way they block experiments. Blocks should be chosen to consist of similar experimental units and not be forced into the randomized block design structure. It is much more important to have homogeneity within blocks, than for blocks to be of equal size and to be the same size as the number of treatments.

Experiments sometimes involve more treatments than can be accommodated in homogeneous blocks. In such instances, the experiment should be carried out as an incomplete block design, using blocks that are smaller than the number of treatments. This idea is familiar to breeders where it is generally not possible to include all of the large number of genotypes typically found in a single trial within homogeneous blocks. For years, breeders have been doing experiments with large numbers (>100) of treatments in small blocks. The problem is so common in this area, that special designs have been developed for breeding work. However, the general principle is not widely understood, or used in other application areas.

With incomplete block designs, it is not possible to have all treatments in a block. Care is then needed in the allocation of treatments to the units in each block. Two further points are:

- If the material naturally suits a randomized complete block structure, then this is the most efficient design to use.
- It is not advisable to have many blocks whose size is smaller than necessary; this makes the design less efficient.

The desirability of factorial treatment structure can lead to many treatments in an experiment. Small blocks are also possible with experiments that involve several factors by sacrificing information on one or more of the higher order interactions amongst the factors. Thus, less important information is *confounded* (mixed) with the block-to-block variation, while allowing all the important information to be estimated precisely, because of the homogeneity of the small blocks.

Software is available to help in the design of incomplete blocks. These programs are most effective for design of 'regular' designs, in which (for example) all the blocks are the same size and their size is a factor of the number of treatments. The programs are very helpful for the rapid generation of designs of similar types, as needed by breeders. They are generally less useful for the generation of the one-off and irregular designs common in NRM research.

5.8 Allocating Treatment to Units

Treatments should be randomly allocated to the units in a block. The principle of randomization is important, because it guards against unidentified sources of variation that may exist in the experimental material. It can be regarded as an insurance against results being biased, due to unforeseen patterns of variation amongst the units. It also underpins the statistical analysis of the data collected from the units.

In some situations, practical considerations outweigh the statistical requirement for randomization. For example, when siting control plots in an agroforestry experiment, trenches surrounding the plots may be needed to guard against interference from trees in neighbouring plots (this was mentioned earlier in Section 5.4). If the control plots are placed in a corner of the experimental area, then trenches are needed on only two sides of the control plot, rather than on all four sides, thus limiting the effort needed in conducting the experiment. Here the researcher has taken the decision not to randomize; but if the data from the control plots are to be included in the statistical analysis, he has to feel comfortable that the data are no different from what one would expect if randomization had taken place. The burden of justifying the conclusions, through showing that there is nothing special about the corners, shifts to researchers. The unbiassedness of the comparisons is no longer guaranteed by the design.

Hence, randomize the allocation of treatments to units or blocks wherever possible.

5.9 Taking Measurements

The general questions relating to measurement to be considered, are:

- Which measurements best suit the research objectives?
- What additional measurement could be useful?
- At what scale and when will the measurements be taken?
- How will the measurements be made and data recorded?

Measurements are taken and data are recorded, for three broad purposes. The first is to give the overall context of the experiment. These are measurements such as the location of the trial, the dates of various operations, and climatic and soil characteristics. These measurements are normally recorded for the experiment as a whole, and enable users to see the context of the experiment and hence the 'domain of recommendation'. Often too little data of this type are collected, or data are collected which the researcher does not know how to use.

The second and most important reason for taking measurements is to record the variables that are determined by the objectives of the study. These are normally recorded at the plot level (i.e. the level at which the treatments were applied). They may be recorded at a lower level (e.g. the tree or animal level), when the experimental unit consists of a group of trees or animals. Sometimes too much data are collected that do not correspond to any of the stated objectives.

The third reason is to explain as much variation in the data as possible. For example, a yield trial with different species of millet may suffer from bird or insect damage or waterlogging in a few plots. Although these variables are not of direct interest in relation to the objectives of the trial, they are recorded because they may help to understand the reasons for variability in the data. In the analysis, they may be used as covariates (concomitant information) or simply to justify omitting certain plots. Failure to make recordings of this type leaves differences due to these effects as part of the unexplained variation. This might make it more difficult to detect treatment differences, i.e. to realize the objectives of the trial.

When planning the measurements to be taken, it is important to clarify whether particular measurements will be taken at the experiment level, at the unit (plot) level, or at the sub-unit (plant) level. For example, soil measurements may be taken at the experiment level to characterize the site, or within each plot to include in the treatment comparison.

Occasionally there are unplanned events that necessitate a review of the objectives of the trial. For example, if the rainfall is low in a variety trial, should irrigation be applied? This would keep the stated objectives, but make it more difficult to specify the domain of recommendation for the experimental site. An alternative would be to reconsider the objectives and perhaps to study drought tolerance instead of yields. The new objective(s) would lead to different measurements from those originally planned.

For any experiment, the measurements to be taken can be considered in three phases:

- Measurements taken before the experiment begins. These can have two purposes. The first is to assess which site is appropriate for the trial and will often involve a small survey. The second purpose is to record the initial conditions before the trial begins, for example, soil constituents, details of household, animal weights or initial plant stand.
- Measurements taken during the experiment. Possible measurements include labour use for different operations, weed weights, tree height, dry matter, quantity of food eaten by animals, animal weights, and disease incidence.
- Measurements taken at the end of the experiment. Examples are yields in crop trials, degree of damage in storage trials, wet/dry weight of fish in aquaculture experiments, germination rates in seed storage experiments, milk yields in comparison of diets in a livestock trial, or whether the farmer thinks a new intervention has been successful.

Sometimes it is not possible to measure the whole unit (plot), and a sample is taken from within each unit. For example:

- A soil sample may be taken at five locations within each plot.
- Nitrogen content may be measured for four leaves for each of three bushes in each plot.
- Grazing time may be recorded for two hours for each of four animals in each fenced area.

These are effectively small sampling exercises within each plot or unit, the aim being to use the sample to estimate the overall value for each whole plot. Sometimes, particular measurements such as soil moisture are sufficiently expensive, that it is not possible to record the information on all the plots, and a subset must be taken. In both instances, an effective sampling strategy is required (see Chapter 6 and Chapter 8).

Definition of measurements also require information on when they will be taken. Sometimes this is obvious (e.g. harvest). In other cases, we have to select sample times, a problem discussed in Section 8.5. A common problem is illustrated by the example of measuring a disease progression curve, which shows the change in the proportion of a plant stand affected by a disease over time. This starts at zero, begins to increase and then reaches a maximum. In scoring for disease, it is common to measure every ten days or two weeks, and this can result in eight to ten measurements during the season. The curve may cover a period of 30-40 days, but we cannot predict when these 30 days will start. We can either accept the extra measurements, or look for a way to make this prediction. This could involve climatic observations, because disease incidence is often triggered by climatic events, or rapid observation to check for the start of infection. Once a crop has been infected, observations times can be selected using the ideas of response curve estimation (described in Section 5.3.2), increasing the observation frequency when the infection rates are increasing rapidly.

In field experiments, measurements during the growing season can be destructive. It is best to avoid destructive sampling where possible, as it leads to several practical difficulties, such as the need for larger sized units, or altered competition within plots. Often a less precise non-destructive measurement is valuable as well as being economical, in terms of costs and time taken for measurement. For example, a disease score on a scale of 0 to 9 can be used, instead of a precise measurement of the percentage of the plant that is diseased. Such a scale could be calibrated through a limited collection of detailed, though destructive observations.

5.10 Analysis and Data Management Issues that Affect Design

Identifying the data analysis methods required for a trial, forms part of the process of design, with implications for objectives, treatments, and other aspects. There are two reasons for this. The first is the identification of the response variables that will address the study objectives, together with the specification of how they will be obtained. Are they raw data variables (e.g. crop yield), or do they need to be derived from other variables, and if so which ones? In a study of vegetation diversity, interest may lie in species richness and evenness. Which indices will be used, and from what measurements will they be derived?

The second reason is to ensure that the random variation between the experimental units in the study will be estimated with reasonable precision (as discussed in Section 5.5). As a check, it is useful to draw up the structure of the analysis of variance table, listing all anticipated sources of variation and their degrees of freedom. The example in Table 5.6 is for a randomized block design for a field experiment.

Table 5.6. Skeleton ANOVA table for randomized block design.

Source	DF
Blocks	
Treatments	
Residual	
Total	

The replication should be sufficient to provide an adequate number of degrees of freedom for estimating the variation between plots in the experiment (the 'residual' in the ANOVA table). A rule of thumb is to say that 10 degrees of freedom (DF) is a minimum, while more than 20 DF is not necessary. In a factorial experiment, the latter may indicate an excessive amount of explicit replication: the inclusion of an additional factor in the experiment might lead to a more efficient use of resources.

Constructing the skeleton ANOVA table is particularly useful when the data have a hierarchical structure: to check that there are sufficient degrees of freedom for estimating the variation at the different levels. It will also be useful when the planned analysis incorporates other effects and interactions, not only treatment effects, but also covariate effects. Statistical software can be used to provide the skeleton ANOVA and to perform a 'dummy analysis'. This checks both the degrees of freedom, and whether all the comparisons of interest can be estimated in the analysis. This is particularly useful when the planned experiment has a complex structure.

Good data management, and archiving of the current trial, can help with the planning of future research. In field experiments, it is useful to look at features of past trials and past uses of the proposed trial area. At the experiment level, information is likely to be available (somewhere!) from soil samples conducted for past trials. Knowledge of the coefficient of variation (CV) from similar trials helps to assess the number of units that are required. Information on the effectiveness of blocking schemes can help in determining the need for small block sizes.

5.11 Taking Design Seriously

Studies have often suffered due to insufficient time spent on their design. This is compared to the time spent conducting the study and analysing the data. The process of clarifying objectives, identifying treatments to address these objectives, determining replication needed and so on is usually iterative. It is only in very simple situations that there will be no cycling round these different components that make up 'design'. We are not implying that it is difficult to design an experiment; merely that it takes time to get it right. That means more work, but it is worth it! The result is an effective experiment – one that addresses the questions posed using the resource material as efficiently as possible.

There is always a balance between effort spent and the consequences of getting it wrong. Short trials that go wrong can sometimes be repeated easily and at little cost. Longer-term trials are usually more costly in terms of time and resources, and researchers cannot afford the opportunity of a repeat run. In these instances, the extra time spent on getting the design right is surely time well spent.

Chapter 6

Sampling Concepts

6.1 Introduction

The ideas on sampling presented in this chapter are relevant to both qualitative and quantitative studies. They apply to research concerning people's livelihoods and natural populations. We consider not only human informants but also the sampling of activities, terrains, crops and so on. In Chapter 7, we discuss some special aspects of studies of communities and households, and in Chapter 8 we deal with specific issues, such as sampling in space, that are of particular concern to soil scientists and ecologists.

This chapter is about the ideas that are needed to devise an intelligent sampling plan. We are concerned with basic concepts that are widely useful. Our aim is to present general principles for achieving good, defensible sampling practice, by the systematic application of common sense rather than mathematics. Sampling is usually a crucial stage when committing resources.

The topics in Sections 6.2 to 6.7 include defining the objectives, identifying the sampling units, sample size determination and stratification. These ideas are useful in their own right and also set the scene for the later sections, where we consider, in particular, how to make do with a small sample while still producing results that can be generalized to a larger population.

6.2 Concepts of Good Sampling Practice

6.2.1 Simple random sampling and objectivity: a basic idea

The first sampling paradigm introduced in quantitative research methods or statistics classes is commonly that of simple random sampling. For this is needed an accessible, enumerated list of members of the population. The members have no distinguishing features, and each has an equal chance of inclusion in the sample. It is often assumed that there is only one clear-cut objective in such idealized sampling. This is to produce a confidence interval for 'the' mean of 'the' measurement. How different this all seems when selecting sites or informants in actual research!

Simple random sampling is seldom applied in practice, but it serves to provide some 'feel' for the benefits of other schemes. Broadly, a stratified sampling scheme will provide improved estimates but increased complexity and cost. Hierarchical (cluster or multistage) sampling will usually be cheaper and easier to manage, but the estimates will be less precise for a fixed number of subjects or sites.

The main argument for random sampling is that sample membership is determined in an objective way, and not influenced by subjective decisions. In practice, there are problems if we have a non-random sample. For instance, a sample that is selected because of administrative convenience, personal preference, or a vaguely substantiated 'expert' judgement, inevitably leads to the appearance or suspicion of bias. Selection bias, whether conscious or not, is a serious failing in research.

Probability sampling is the general term for methods where sample selection is objectively-based, i.e. based on known chances of inclusion in the sample. If there is an equal probability of inclusion, it is simple random sampling, and estimating characteristics of the population is straightforward. If the probabilities of selection are not equal, then estimation is still possible provided the probabilities are known, since they are used in the calculation of quantitative summaries. In some development project settings, it is hard to ascertain the probabilities because of inadequate time frames and sampling frames (i.e. listings from which to sample, incomplete respondent compliance). It is still good practice to be as objective as possible about sample selection, even if this is sometimes difficult, as it helps to equalize the *a priori* chances that individuals are included in the sample. Likewise it is good practice to record the arguments and procedures that support the claim to representativeness.

Random sampling offers the benefit that common, but unsuspected, peculiarities in the population will be 'averaged out' in a large sample. If 30% of households are female-headed, a random sample of 100 households should have close to 30% female-headed households even if we have not controlled for this. Alternatively, we could control for it by taking fixed-sized samples separately from the male and female headed subgroups of the population (i.e. stratifying by sex). Rarer features need large random samples if their representation in the sample is to 'settle down' to the right proportion. If samples are necessarily small, a greater degree of control may be needed to ensure the sample selected is not an odd one.

6.2.2 Representing a population

Representing a population often involves dividing sampling effort across segments of the population, and thus entails relatively small samples from minor subsections of the population. The choice of how to divide the effort depends on either the size of the different segments or their importance. If you can predict that results will differ systematically from one stratum to another, but results for the whole population are of interest, then it is usually desirable to ensure the strata are represented proportionately in the sample, in order to give a fair picture of the population.

On the other hand, if particular sections of the population (say female-headed households or striga-infested fields) are important to the researchers, and may be reported separately, the population may be stratified on this basis (gender of head of household, or low/medium/high prevalence of striga). An interesting stratum can then be sampled more intensely than others, provided its results are 'scaled down' to the appropriate level by weighting procedures in an overall summary.

6.2.3 Hierarchical or multistage sampling: a central idea

Often real-life sampling involves hierarchical structures and sampling processes (e.g. selecting regions where there are major issues about water rights, identifying and sampling localities within the regions where the issues are important, defining groups with interests in the issues), and then working out suitable ways to sample and work with members of these groups. We refer to the largest units – 'regions' in our example – as primary or first-stage units; the next level – 'localities' – as secondary or second-stage, and so on.

Another example would be a study concerning an invasion of *Pinus patula* in a forest reserve in southern Malawi – the reserve is subdivided into large 'blocks' of which a few are selected, several areas per block are chosen at each of three different altitudes, samples taken from these areas and the number of pine trees counted.

In textbook terms, these are 'multistage' sampling schemes, the stages being the levels in the hierarchy. There is no implication of multiple points in time and, unless otherwise stated, the sampling gives a point-in-time snapshot.

Multistage sampling is carried out for practical reasons. One example is the absence of a good sampling frame from which to organize a probability sample (i.e. either a simple random sample or a stratified sample). Carrying out one's own census-style enumeration exercise to produce such a list is beyond the scope of most projects. A standard way out of the problem is multistage sampling which entails the development of just the essential elements of population listing. For example, at the first stage, we list the locations of districts and take a sample of districts. When we visit the administrative centres of those districts in the sample, we can ascertain the names of all the functioning government veterinarians in just those districts. When we select and visit just some of them, we can ascertain the names of all the villages they serve, and in turn, perhaps by participatory mapping techniques, a list of livestock owners in just those villages that are selected.

For results to be applicable to the wider population, any sample must have been selected objectively. In hierarchical sampling, this applies most to the ultimate sampling units such as households, and least to primary units such as large 'well known' areas like provinces. Primary units are often selected on a judgment basis; ultimate sampling units ought to be sampled in an objective way. The researcher may find that producing a well-documented sampling plan, which details the different sampling procedures at each level of the hierarchy, can help to ensure that the sample has been selected with a suitable amount of objectivity. This can then be effectively demonstrated and when necessary defended to users of the results.

6.3 Study Objectives

6.3.1 Broad objectives

By broad objectives, we mean a brief, general description of what we expect to learn from a study or a set of studies.

(a) One possible objective is to provide an overall picture of a population. For example, rice will be imported if the national production of rice, plus stocks, will be insufficient to feed the people. In such a case, the objective is to come up with a reliable figure for the total size of the forthcoming harvest. The distribution of sampling effort in a crop-cutting survey must cover and represent the entire productive system.

(b) A different type of objective is comparison. For example, an integrated pest management (IPM) strategy is to be tried in four study areas; the results are to be compared with a set of control areas. Here it is most important that the areas under the 'new' and 'normal' regimes are matched to ensure a 'fair comparison'. Effective 'coverage' of the population of land areas is less important than in (a).

Points such as the above are relevant when the overall picture or comparison is to be based on quantitative measurement such as the size of the rice harvest. The same points apply to a greater extent if qualitative approaches to data collection are being used. The report of conclusions from only one community cannot be judged to represent the range of diversity elsewhere in the country.

Note that (a) and (b) above imply different approaches to sample selection and sample size. For the objective in the rice example in (a), something close to proportional allocation of sampling effort is usually appropriate (e.g. if one region produces about one quarter of the rice crop, it should provide the same proportion of the sample). Efforts should be made in the analysis to correct for any disproportionate representation in the sample. This idea was discussed earlier, in Section 6.2.1. Producing an 'overall' figure depends on having reasonably up-to-date and accurate information about relevant features of the population. This means having either a good approximation to a sampling frame for the whole population or adequate sampling frames at each level of multistage sampling.

In contrast, in a case like the IPM intervention in (b), the sampling frame requirement is generally much less rigorous as long as the main aim is a fair comparison. For instance, the example in (b) above may invest half the effort in field data collection in the study areas, even if these represent a minute proportion of total planted area or production. At a later stage in the project, this approach could change. After a 'new' regime has shown economically important promise in some sub-areas, it may become worthwhile to delimit its range of beneficial applicability (i.e. its 'recommendation domain'), and a reasonable sampling frame (a description of the whole population) is needed.

These two examples illustrate that there is no universal statistical result to help produce an appropriate sample size or sampling pattern: appropriate sampling strategies depend on the objectives.

Providing an overall picture of the population can also be a common objective, particularly in studies of natural populations. Sometimes this relates to a single species, as in the case of estimating the size of the elephant population in an area of land set aside for conservation. In other instances, the focus is on multiple species, for example summarizing the species diversity on the western slopes of Mount Mulanje. These studies use particular sampling tools such as mark-recapture and transect sampling methods, and are discussed again in Chapter 8. The only point to add here is that the usual sampling principles still apply. For instance, if the density of animals is thought to differ in different areas (e.g. perhaps numbers of animals are higher when closer to watering holes than when further away), then data-relevant stratification needs to be defined, the population sizes need to be determined for the different strata, and the results combined.

(c) Sometimes the objective is to typify the units (e.g. households, communities, or village fishponds), and to classify them into groups. These groups can then be studied, sampled, or reported separately; or they may become recommendation domains. This is therefore a *baseline* survey, similar to a mapping exercise, and usually involves a lot of observations.

An example is where a special group is involved (e.g. 'compliant' farmers are recruited to a 'panel') and they are visited for one or several studies in a project. If the non-compliant farmers are not studied at all, this restricts the range of generality that can be claimed for any conclusions. At least we should know the proportion that are compliant, and have some idea how they compare to the others. Preparedness to adopt innovations swiftly may be higher for readily-compliant farmers, and predictions about overall adoption can be suspect if this proportion is not considered carefully.

(d) If a project encompasses several information-gathering studies in the same population, the objective may involve a relatively long-term relationship with informants. It is often best to link up the samples across studies in an organized way to enable results to be aggregated and synthesized effectively. Suppose a three-year bilateral project identifies 150 farm households as possible collaborators. The anthropologist works in depth with seven households chosen on the basis of a baseline study, led by the economist, of all 150. Their combined work leads to a division of the cooperating households into three livelihood groups with identifiable characteristics, which might be tackled in distinct ways to achieve project goals. This is a stratification of the households.

It is also useful to keep a simple population register for the 150 households, so that the selection of participants in follow-up studies takes proper account of their previous project involvements. It is important to link the information from one study to another, as it will be the basis of synthesizing project information concerning livelihoods.

6.3.2 More detailed objectives

When deciding what and how to sample, it is essential to think about what use will be made of the sample data before committing any resources. The researcher should have a plan to use and report the findings, and so the information collected must be both necessary and sufficient for the required purpose. Of course, other parts of the research strategy must also be under control as well as the sample selection! One of the inputs to sampling decisions is an understanding of the research instruments. Their qualities under the broad headings of 'accuracy' and 'stability' are important determinants of what, and how many, should be sampled.

Many studies have several general objectives, which may pull in different directions as far as sampling schemes are concerned. It is desirable to verify at each stage that the samples obtained are adequate to satisfy all important objectives. This usually involves making the general objectives more specific either by prioritizing them or in making compromises between the different demands of the fieldwork. Modularization of field studies can be useful, so that separate exercises can be adapted to fit the different objectives.

Note that a multistage sampling design introduced because there is a hierarchy of units, will involve a need to define objectives at the different levels of the hierarchy, and ultimately to prioritize these objectives. For example, consider a design with administrative units (e.g. provinces) as the primary units, and samples taken from within the provinces. The main output is a dissemination programme tailored to localized audiences. Summarizing the adoption rates for each province is an objective at the top level of the hierarchy. At the village level, lower down the hierarchy, there may be another objective, such as assessing the effectiveness of different methods of dissemination.

In committing resources we need to consider how best to collect information to address these objectives. Say the project output is an intervention package to be implemented at village level. One multistage research strategy might focus on testing the package in two villages, collecting detailed data about internal village organization at the household and individual level as it relates to the intervention package. At village level, this is a sample of size two, and it may provide little more than anecdotal or case-study evidence that the effect of the package can be replicated elsewhere. This is because too much of the information is at 'within-village' level. An alternative strategy might be to treat one village as above, but divert the other half of the effort into briefer studies in five extra villages. There will be less information at the 'within-village' level, but more knowledge across villages (i.e. at the 'between-village' level) – where it matters.

6.3.3 Can objectives be met?

Both the studies sketched out in the preceding paragraph are based on such small sample sizes that they should probably not be funded! Ideally, the specification of clear objectives should define the data and analyses expected and the worthwhile conclusions that can be anticipated, with a financially feasible and cost-effective set of activities to complete the work. The fact that intensive use of resources only permits a small study (e.g. a case study in one locality) does not prove that the very small study is capable of generalization or of yielding conclusions that will be of real value in a wider context. All researchers need to face the possibility that a proposed study may be incapable of yielding results that are fit for the intended purpose, or even for a more modest and sensible purpose.

6.4 Units

6.4.1 Conceptualizing the unit

Simple random sampling treats units as if they are potentially identical, but isolated. However, human populations are socialized, reactive and interactive. Even with single respondents to a formal survey, the unit being investigated can be individuals, their households, or their villages, etc. Some units are easily defined; individuals are one such example. Households are more changeable over time, as are some natural populations, which can grow and shrink at different times of the year. The multistage study involves different units at the different levels. Some have a natural definition, such as the farmer's maize field; others have not, such as a plot within the field where yield will be measured. There is a choice to be made in the latter case on how large the plot needs to be to give a sensible compromise between getting a good measurement and undertaking too much work.

6.4.2 Unit levels

Different 'effects' come into play at each level of a hierarchical study (e.g. the individual's educational standard, the intra-household distribution of food and the village's access to rural transport). If there are several levels in the study, much confusion stems from the failure to recognize or deal with such structure.

Studies can become overly complicated and resource-hungry if they try to encompass many effects at many levels. Sampling can be conducted at several levels, and it is important to find an economical way of learning just enough about each level and its links to the other levels. Often, one way to achieve this is by not attempting to balance a hierarchical sampling scheme. The first strategy for the village-level intervention at the end of Section 6.3.2 suggested equally detailed studies in two villages. The alternative made the study into two different modules, an in-depth study of one village and a broader study of several villages, perhaps using rather different methodologies. While the intention of the alternative strategy is to use the results of the two modules together, there may be no need to synthesize them formally if they address objectives at different levels.

6.4.3 Profiling

A complication is added to the study design when consideration is given to how things evolve through time or space. Regularly repeated observations can provide evidence of consistency or systematic change in time, especially if the same respondents or plots are revisited each time. Then the unit is a compound of person, or plot, and time. Such sampling is often expensive compared to before/after studies. However, it is essential if the time track of events is intrinsic to the study. This is the case for example in seasonal calendars or monitoring systems to capture and identify sudden changes in vegetation of a conservation area, or staple food prices in a rural community. Now each time profile constitutes one unit; for instance, one farmer's record of farming activities over one year is a unit.

As far as generalization is concerned, data collected in one year constitutes a sample of size one (as far as years are concerned). The consequent difficulty of generalizing to other years applies not only to data at the level of detail as above, but also to larger-scale attempts to deduce anything from a small sample of years. Drawing conclusions about the sustainable management of a natural resource based on a couple of seasons is one such example.

There are two ways to generalize the results to more years. The first is to 'embed' the small sample within a sample that gives data on more years. The obvious example is climatic data, where records may be available for perhaps 50 years. An analysis of these data provides a long-term perspective for the detailed study of two or three years. The second way is to ask respondents about past years. This can be difficult to do with a questionnaire; it is more likely to be part of a participatory study. These ideas are discussed in more detail in Chapter 7.

In the same way, if a participatory activity involves a gathering of the women in a village with a facilitator to thrash out a cause-and-effect diagram, the result is a single profile in the form of a diagram (i.e. one unit): an unreplicated case study. If the focus is solely on that community, it may be of no relevance to look at any other group's version of the diagram for the same issue. However, if the exercise is undertaken as research, any claim of generalizability will require more than one unit. A simple sample of units might involve several independent repeats of the same exercise in different villages. A more structured sample might compare the results from two or more facilitators working individually in a sample of matched pairs of similar villages; this may distinguish between the effects of (a) a facilitator's approach, and (b) variation between villages. With only one facilitator, (a) is being ignored. If the villages are not matched, it is hard to decide whether to attribute differences to facilitator or village. In statistical terms, the two effects are *confounded.*

6.4.4 Unequal units

Often sampling is based on treating all units as equal in importance, but land holdings, enterprises and other sorts of unit may be of varied sizes and potentials. For summative (rather than comparative) purposes, it may be important to give larger units greater weight in sampling. Deciding on the appropriate measure of size is often difficult when a variety of answers are wanted for different issues within the same study. For example, estates may be classified by number of employees, by planted area, or by production.

When a compromise size measure is used in sampling, varying weightings may be needed in analysis for different variables. Weighting in analysis may also be used to correct for unweighted sampling, for instance when weights cannot be determined until field observation. With quantitative data, there are clear-cut ways to set about using weights, but it is still important that weighting systems are not under-conceptualized. They should be used effectively, with weights reflecting, for example, village population sizes, areas given over to commercial cabbage growing, or transport costs. Using weighting in the analysis is discussed further in Section 17.2.4.

6.4.5 Qualifying units and population coverage

Recording of the sampling procedure includes giving a careful definition of the actual 'population' sampled. Often field limitations cut down what can be covered, and therefore the domain to which the research can claim to be able to generalize. For instance, when specifying a sample, only certain types of units may qualify for membership: one example of this is if compliance is a criterion. It is important to record the hit rate (i.e. the proportion of those approached who qualify and are recruited), and the types and importance of differences between those qualifying and those not. These provide evidence of what the qualifying sample truly represents.

Difficulties in accessing the target population are inevitable in many situations. However, a research sample is not necessarily worthless when it does not match the *a priori* population, but a clear description should be given of what the study has succeeded in representing. Often there are new insights worth reporting from the hit rates mentioned above, or less formally from meetings with those who did not opt to comply.

The conclusions and associated recommendation domains resulting from research must be properly supported by evidence. It is a form of scientific fraud to imply without justification that results apply to the *a priori* target population if those actually sampled are a more restricted set that may differ in kind from the rest of the target population! For example, a sample might be restricted to farmers who are quickly and easily persuaded to try a farming system innovation. If these are compliant, higher-income, male-headed households, conclusions derived with them may not be applicable to low-income, vulnerable or female-headed households.

6.5 Comparative Sampling

This procedure arises where the study sets out to compare existing situations in areas that are clearly distinct (e.g. the incidence of damage due to a particular pest in high-grown as opposed to low-grown banana plantations).

The idea behind this is that of natural stratification. The population divides naturally into segments that differ from one another, but are internally relatively homogeneous. Stratification is discussed further in Section 6.7. If internal homogeneity can be achieved, it means that a relatively small sample will serve to typify a stratum reasonably clearly, so this can lead to efficient sampling.

6.5.1 Factorial structure

In a comparative study with several possible stratification variables, a frequent objective is to check which factors define the most important differences from one stratum to another. Say we considered communities that were near to/remote from a motorable road (factor 1), where the land was relatively flat/steeply sloping (factor 2), and which had greater or lesser population pressures on land resources (factor 3). This produces 2×2×2 = 8 types.

For each factor, when we look at dichotomies such as flat land versus sloping land, we are doing the absolute minimum to consider its effect. We are acknowledging that, since there are many other complications in the real setting, we cannot investigate, describe, or come to conclusions about them all. When taking a sample, usually of a few units, it is advisable to include only cases that clearly belong to the categories, and to exclude those which are marginal or doubtful, even if only for one of the classification factors.

We might then select n communities of each of the eight types, giving $8n$ sites to investigate. The number n should not be confused with the number of factors or the number of levels per factor. It is a separate, independent choice. In the example above, if it was feasible to look at about 30 sites, we might take $n = 4$, so $8n = 32$. If we then look at the differences between remote and accessible communities, we have samples of 16 of each type, and these samples are comparable to each other, in terms of having the same mix of the other two factors. At the same time, we have comparable samples of 16 flat and 16 hilly areas of land, and so on.

The 'three studies for the price of one' benefit illustrated here is a well-known concept in designed experiments where factorial treatment structures are often advocated. However, they are just as relevant in other settings where several factors are studied simultaneously, and therefore apply here whether we are conducting a formal survey, or a much more qualitative exercise with each community. For further discussion on the advantages of factorial structures, the reader is referred to Section 5.3.1.

The objective assumed here is to compare the levels of each factor, and to decide which factors are important. We are not concerned that the eight subgroups define equal-sized subsets of the whole population, nor that population subsets are represented proportionately.

6.5.2 Small sample comparisons

The previous section assumes we are looking at a fair number of communities as primary units. What can we do if an in-depth investigation cannot be replicated that often? It still has to be conducted in a few communities selected from a complex range (i.e. 2×2×2 types in our crude example), but maybe only a handful can be looked at in depth.

As we argue with other small samples, the in-depth study has more plausibility if it is positioned relative to a larger and more representative sample. This applies equally to qualitative (e.g. PRA-type) and quantitative approaches. A relatively quick characterization might be done in each of the 32 communities, followed by an appropriate design for accompanying in-depth studies. This may be a systematically selected subset of the types of primary community.

The in-depth study might reasonably be based in four of the 32 communities chosen from the eight combinations of near/remote, flat/sloping, and high/low population pressure as a 'fractional factorial design': one choice is illustrated in Table 6.1 where each factor is included twice and appears once with each level of each other factor. Equally appropriate would be the alternative set (i.e. the shaded parts of the table). There are some important implications with regard to the analysis of such a set: these are explained in Section 16.8.

The above ideas are concerned with one level of the multistage sampling process. They make no stipulation as to how sampling aspects of the study may be structured within communities, nor of course about other aspects of research methodology. The proper specification of a hierarchical sampling plan requires descriptions of the research protocol for several different levels.

Table 6.1. Suggested sampling scheme for a small in-depth study of several factors.

	Near road		Remote from road	
	Flat	Steeply sloping	Flat	Steeply sloping
High population	include	---	---	include
Low population	---	include	include	---

Comparative observational studies have important structural elements in common with designed experiments discussed in Chapter 5. The factorial ideas above, for instance, are just one of many design ideas that apply effectively to such sampling studies. We will see later in Chapter 9 that sampling also has something to offer to experimental design.

The above is an illustration of how to sample sensibly in a small study. With sensible sampling it is thus still possible to draw some conclusions, particularly if thought is given to the factorial structure of the communities. However, there is insufficient replication of communities to estimate the between-community variability, so it would be difficult to attach statistical significance to any comparisons that can be made.

6.6 Sample Size

There is no clear-cut method of producing an answer to the question, 'How big a sample do I need?' You have to think it through in the light of the objectives, the field data collection conditions, the planned analysis and its use, and the likely behaviour of the results. There are several situation-specific aspects to this; there is no universal answer.

Statistical texts mainly discuss the case where the mean of a numerical observation is estimated from a simple random sample. This can provide some 'feel' for other situations, as indicated in Section 6.2.1. The essential component of formulae is $\sigma/\sqrt{n}$, in which σ represents the standard deviation of the quantity sampled. Thus before you can start working out a sample size to achieve a certain accuracy, you have to estimate the variability you expect in your data. We look in more detail at this idea in Section 16.8.5.

More complicated cases are the norm. Summaries from survey samples often take the form of tables, and the sample size required is then determined by the way responses spread themselves across the table cells, as well as the level of disaggregation required, for example to three-way tables. As a simple quantitative example, say you will need to look at tables of the mean value of yield per hectare, for three types of land tenure, for five cropping systems, for male and female cultivators. You are thus dividing your data into $3\times5\times2 = 30$ cells. You will need enough good data on areas, yields and their values to give reasonable estimates for all cells. If you decide that that requires seven responses per cell, you have to aim for $7\times30 = 210$ adequate responses. This is of course a net figure and the planned sample size must be a grossed-up version, which allows for those who are unavailable, unable or unwilling to participate. If that seems more than the budget will stand, think how the objectives can be made more modest. Maybe you only need accurate figures for some of the totals, not for every individual cell.

6.7 Stratification – Getting It Right

Sampling within strata, where the strata are described as 'non-overlapping groups of units', is a well-known and commonly-used sampling technique, but statistical textbooks say little about how to group the units into strata. Here we discuss some practical aspects associated with stratified sampling.

6.7.1 Effective stratification

One of the benefits of stratified sampling is that it will give more precise results, or require fewer resources, than simple random sampling when the strata are groups of relatively homogeneous units (i.e. groups of people, plots, etc.), which have a degree of similarity to one another in response terms. We call this effective stratification. The gender of the head of household or the land tenure status of farmers for instance, might bring together subsets of people who have something in common, while altitude or agroecological zone are common in the stratification of natural populations.

Sensible groupings often encompass several factors. For example, if stratifying by people's livelihoods, subdividing individuals by occupation into say farmers, fishermen and artisans may not be sufficient; they may need to be further subdivided, perhaps by stratifying by gender of the head of household.

Relatively small numbers of units (people or plots) can typify each group, so the method is economical in terms of fieldwork. Also, it is common that a report of a study will produce some results at stratum level, so it is sensible to control how many representatives of each stratum are sampled. The information base should be fit for this purpose, as well as representing the whole population.

6.7.2 Ineffective (or less effective) stratification

Populations are often divided into subgroups for administrative reasons, and results may be needed for separate subdivisions (e.g. provinces). If separate reporting is a requirement of the study then this is appropriate stratification. However, if results are required only for the region, and separate reporting is not required, it will not be the most effective stratification unless the administrative grouping happens to coincide with categories of individuals who are homogeneous in response. Similarly, if every village contains farmers, traders and artisans in vaguely similar proportions, villages will be of little relevance as an effective stratification factor if the main differences are between farmers, traders, and artisans.

The above suggests that the subsets by occupation correspond to clearly distinguished, identifiable groups, internally similar to each other but very different from group to group. In this situation, stratification by occupation is an obvious sampling tactic. However, the groups are rarely so distinct. The challenge for the researcher is to identify groupings within which there is some degree of homogeneity.

The other challenge is to balance the costs and resources of possible different choices of stratification. For instance, stratifying by province for purely administrative reasons may allow a larger sample of data to be collected at a lower cost than another more effective stratification involving fewer individuals.

6.7.3 Pre- and post-stratification

Where stratification is meaningful, it is sensible to pre-stratify where the groupings can be detected before study work commences. In some cases, the information for stratifying only becomes apparent during the study, and the cases are sorted into strata after the event. This is called post-stratification.

6.7.4 Participatory stratification

It is sometimes suggested that useful subdivisions of community members within communities can be achieved by getting them to divide into their own groups using their own criteria. This provides useful functional subdivisions for participatory work at a local level. If results are to be integrated across communities, it is important that the subgroups in different villages correspond to one another from village to village. Thus a more formal stratification may require; (i) a preliminary phase where stratification criteria are evolved with farmer participation; (ii) a reconciliation process between villages; and then (iii) the use of the compromise of 'one size fits all' stratification procedures in the stratified study. Therefore, the set of strata should probably be the set of all subsets needed anywhere, including strata that may be null in many cases (e.g. fishermen, who may only be found in coastal villages).

6.7.5 Quantile subdivision

Stratification is not natural where there is a continuous range rather than an effective classificatory factor. If there is just one clear-cut observable piece of information that is selected as the best basis to be used, a pseudo-stratification can be imposed. For example, a wealth-ranking exercise may put households into a clear order, which can be divided into three quantiles ('terciles' – i.e. the bottom, middle and top thirds), quartiles, or quintiles. This permits comparisons between groups derived from the same ranking (e.g. the top and bottom thirds of the same village). Since the rankings are relative, they may be rather difficult to use across a set of widely differing communities, some of which are more prosperous than others.

6.8 Doing One's Best with Small Samples

6.8.1 The problem

The approaches discussed in this chapter are practical when there is an adequate sample size, but what happens when a very small sample is unavoidable? For instance, the researcher wishes to have substantial and time-consuming interaction with only a few communities or households, rather than limited interaction with many communities or households, yet the sponsor wants an assurance that these will yield 'representative' and 'generalizable' results.

A valuable property of random selection is that it tends to 'balance out' various aspects of untypical results over relatively large samples. In very small samples a random selection may be obviously off-balance in important respects. Especially for primary units, the small sample will therefore most probably be chosen on a judgement basis, but note that it still cannot cover or distinguish the large range of ways in which first-stage units will vary. To choose Nepal and the Maldives as a sample of two countries on the basis that one is hilly and the other low-lying is to overlook many other features of available profile information about climate, culture, natural resources, governance and so on. The small sample of first stage units may have to be accepted as a case study with limited capacity for generalization. It is therefore best to choose primary units that are 'well known' and important in their own right.

6.8.2 Putting small samples in context

As explained in Section 6.3.2, the in-depth study has more plausibility if it is based on a larger and more representative sample. So a relatively quick study may be carried out on a larger sample of units, with the accompanying in-depth study conducted on a systematically selected subset. There are various ways of achieving this, thereby adding to the plausibility of a claim that the narrowly based in-depth work represents a wider reality. Here we offer one or two thoughts on the topic.

Ranked set sampling (discussed in Section 6.10.7) is one way to select our small sample since it helps to ensure greater coverage of the population whilst still maintaining an element of objectivity in the selection process. The advantage of this approach arises when the elements can be ranked quickly, easily, and cheaply, allowing more effort to be spent on the in-depth investigations of the finally-selected small sample.

Alternatively, if a general study (e.g. a baseline survey) has been carried out on a relatively large random sample of the population, other studies, including those with very small sample sizes, can draw from it, as long as the baseline survey members are documented efficiently so that they can be found again! These later stage samples may be taken using probability sampling methods, hence allowing them to be described as 'representing' the original large sample. Procedures based on this idea are often referred to as two-phase or double sampling. The approach can yield more precise results than an ordinary simple random sample of the same (small) size; but this and other benefits of carrying out the baseline survey need to be balanced against the cost.

Finally, if a detailed element of the research has been based on a small number of units, it may still lead to new insights and deeper understanding, which can be applied in a wider sphere. If so, the work and its conclusions can be validated by demonstrating its predictive ability. A follow-up study can be devised to test the predictions that are made by the researcher. This study should be less resource-intensive and less wide-ranging in content than the original in-depth work but perhaps covering a larger sample of respondents, so it can authenticate key predictions made in advance by the researcher.

6.9 Impact Assessment

Impact assessment is an area of natural resources research where both experimental design ideas and sampling concepts help to ensure a well-designed study. Here we discuss only sampling. We talk mainly in terms of a project, or even a project activity, where explicit examples of the statistical principles can easily be stated. Assessment of impact can be in terms of how people's livelihoods are affected or how the environment is affected; once again this component of sampling applies whatever combination of qualitative and quantitative approaches is taken to data collection.

Table 6.2. Illustration of BACI design.

Per pair of sites	Before	During	After
Intervention = X	O	X	O
Observation = O only	O		O

The skeletal form of research design that provides a truly effective demonstration that an intervention had an impact, is to have 'before' or 'baseline', and 'after' observations. These observations should be made both in a sample of sites where the intervention occurred and in comparable control sites where there was no effect due to the intervention. This is commonly known as a 'before-after-control-impact' (BACI) design.

Arguments are made against control sites on the grounds that expectations are raised unfairly, that cooperation is poor if no potential benefit is offered, and that the cost of the study is raised both by 'unproductive' time spent on controls and by any compensation given to those used in this extractive way. However, there is a logical place for using controls, and research managers must trade the practical difficulties against the benefit of reliable, generalizable conclusions.

Say the research design involved comparing two strategies, each tried in four villages. Two ranked sets are comparable with respect to the criterion used for rank setting. It is then plausible to use one set for project work (the intervention) and the other set as the control.

6.9.1 Matching sites

Matching is a different approach to choosing controls. It is more expensive than using a second ranked set, but still relevant, regardless of how the intervention set was determined. Pairwise matching requires that for each unit of the intervention set (i.e. for each village in the above example), we find a matched control that shares appropriate characteristics. The difficulties in doing this are both practical and conceptual.

To illustrate the conceptual difficulties, consider a case where the units are individual heads of household and the aim is to help ensure that beneficiaries enjoy sustainable levels of social and financial capital. If the matching is established at the outset, comparability needs to be established in respect of factors that may turn out to be important; there could be many such factors. If the matching is retrospective, as when the comparison is an afterthought, the aim must still be to compare individuals who would have been comparable at the start of the study. The results may be masked or distorted if the possible effects of the intervention are allowed to interfere with this. Of course, a targeted approach of this sort often amounts to rather subjective accessibility sampling. Its effect depends upon the perceptions of the person seeking the matches, and the comparability may therefore be compromised by overlooking matching factors or distorting effects.

Less demanding matching procedures aim to ensure that groups for comparison are similar overall. For example, the matching groups could be made similar with respect to their mean village population, or to the proportion of villages with access to a depot supplying agricultural inputs.

6.9.2 A compromise approach

A sampling-based scheme for selecting comparable controls can be used to set up several potential controls. For example, after selecting one ranked set sample of size four out of sixteen communities, there remain three which were ranked 1 but not chosen, and similarly for ranks 2, 3, and 4. It is then rational to select a comparison set comprising the 'best matching' one out of the three for each rank.

This approach makes the scope of the matching exercise manageable, the process of selecting the best match being based on the criteria most relevant in the very small set used. Of course, the original sampling limits the quality of matching that can be achieved, but the match-finding workload is much reduced. The claims to objectivity, thanks to starting with a sampling approach, are much easier to sustain than with most matching approaches.

6.10 Common (and Not So Common!) Sampling Methods

6.10.1 Simple random sampling

A simple random sample of, for example, trees selected from a forest (the forest being the population of trees), is one where each tree in the forest has the same chance of being included in the sample. The technique is only appropriate when sampling a population of homogeneous units. Natural resources tend to be subject to various identifiable sources of variability: size (DBH) of trees in a forest, for instance, will depend on the species and age of the tree, the amount of light, etc. Hence it is seldom appropriate to use simple random sampling to select a subset on which to base conclusions about the population. A more common role for simple random sampling is as a sampling technique at one of the stages in a multistage scheme.

6.10.2 Stratified random sampling

When units of the population fall into non-overlapping groups (*strata*) and simple random samples are taken from each stratum, then this is stratified random sampling. It is particularly relevant when the objective of the sampling exercise is a comparative one, or results are required for the separate strata. Stratified sampling is sometimes carried out for administrative convenience (e.g. stratifying by province). The real benefit of stratified random sampling is when the strata are groups of homogeneous units. The benefit is that, for the same amount of resources, stratified sampling yields more precise results than simple random sampling. Alternatively, for the same precision, stratified sampling requires a smaller sample than simple random sampling. The important issue for the researcher is to determine the best possible way to group his or her material, be it human or natural populations, into homogeneous strata.

6.10.3 Cluster sampling

Cluster sampling involves sampling clusters of units and then recording information on the units within the clusters. For instance, in a small region consisting of several villages, the clusters could be the villages and the households within each village could be the units. If a sample of villages is randomly selected and data collected from every household in these villages then this is an example of cluster sampling. The approach gives less precise results than simple random sampling for the equivalent number of sampled units (households in this case). However, cluster sampling is attractive for cost reasons, since it is usually cheaper to sample several units close together than the same number further apart. A cluster-sampling scheme can include more units than are possible with simple random sampling, and can still be more cost-effective even with a larger sample size.

6.10.4 Multistage sampling

Sampling which is carried out within several layers of the population is known as multistage or hierarchical sampling. An example is where a few areas in a region are selected, several villages randomly chosen from within each area, and several households per village selected at random for further investigation. The sampling procedure used at each stage depends on the situation. There is generally more variability among the units at the top level of the hierarchy, and least variability among the units at the lowest level.

6.10.5 Quota sampling

Quota sampling is a method that is used frequently, for example by market researchers, to avoid sample frame problems. It is not a random sampling method. It usually entails determining that the sample should be structured in order to control certain general characteristics. For example, a sample of 100 individuals might be required to be divided into three age ranges of 15-39, 40-64 and 65+ years, with 20, 15 and 12 males, and 20, 17 and 16 females, respectively, to match a general population profile where women survive longer. The procedure does not require a detailed sample frame and is relatively easy to carry out as long as there are not too many tightly defined categories to find. If repeat rounds of independent surveying are done, for example monitoring public opinion without following up the same individuals each time, then quota sampling is an easy way to ensure successive samples compare like with like.

Often the task of filling quotas is left to the interviewer's discretion with respect to accessibility, approachability, and compliance, as well as checking on qualifying characteristics (currently-married, main employment on farm, etc.). When subjective sampling is involved, the method is open to interviewer effects and abuses, which need to be controlled with care. The problems may be negligible if interviewing is relatively easy and well supervised. However, particularly with small samples, the subjective element of selection can be a serious worry.

6.10.6 Systematic sampling

This is the technical term for sampling at regular intervals, down a list, or in space, or over time. It is commonly used in the study of natural populations partly because the process of sampling is simpler than simple random or stratified sampling, and helps to ensure the population is more evenly covered.

When sampling over time, the regularity of systematic sampling is usually more desirable than arbitrary intervals. Sampling may need to be more intense in particular periods or areas of particular activity (e.g. to catch the peak prevalence of an epidemic). It certainly needs to be sufficiently frequent that episodes of phenomena of major interest are not missed between sampling occasions.

6.10.7 Ranked set sampling

The approach used in ranked set sampling is indicated by the simple example in Table 6.3, where we select a sample of five households from a village.

Table 6.3. Example of ranked set sampling.

Households	Ranked households
1: A B C D E	1: C A D B E
2: F G H I J	2: H I F J G
3: K L M N O	3: L K N O M
4: P Q R S T	4: P R T Q S
5: V W X Y Z	5: X Y Z W V
Selected sample	C I N Q V

1. Five random samples each of five households are drawn up (e.g. by listing the qualifying households) and selecting five at random without replacement, five times. One subsample is then picked from the set, examined by a quick method, and ranked on a key criterion, such as wealth. The household ranked 1 is taken into the final sample.
2. The process is repeated in the four other subsamples, the household ranked 2 being chosen from subsample 2 for selection to the final sample, and so on, until a round is completed after five subsamples.
3. The other households in the subsamples are generally discarded (but see below). They have served the purpose of comparators to give some improved assurance of what the ranked set sample represents.

Here a sample of size n (5) was selected from n^2 (25) households. The number of starting sets must be the same as the final sample size, but the number of units per starting set need not be. For example, if we started with 12 sets of 4 units, our final sample of size 12 could contain ranks 1 to 4 from sets 1 to 4 respectively, ranks 1 to 4 from sets 5 to 8, and ranks 1 to 4 from sets 9 to 12.

In the example in Table 6.3, the status in the entire population of the five households ranked 1, will not be identical, and *a fortiori*, the differences between those ranked 1 and those ranked 5 will not be the same. This does not affect the claims that the sample has been selected objectively and that it is representative.

Procedures based on this idea are attractive provided the process of ranking consumes relatively little effort. Efficiency gains are generally good even if honest efforts at ranking are subject to some error. The approach is explained in the context of sampling households, but its use is broader than this; for example, in vegetation surveys where detailed measurement can be expensive, though cheaper, less reliable methods allow a quick initial assessment to be made. The approach also has the advantage of ensuring greater coverage of the population in small sample cases than may be possible with simple random sampling.

The above is a conceptual description and practical safeguards need to be in place. The initial selection of units (households) must be made in an objective way, and in our example, simple random sampling was assumed appropriate. Other situations may require different sampling strategies.

A by-product of this selection process is some spare similar ranked sets. For example, the set comprising the households ranked 3, 4, 5, 1, 2 in samples 1 to 5 respectively (i.e. households D, J, M, P, and Y) should provide a comparable ranked set, because any one of these ranked sets is supposed to be representative of the original 25. One use of this second rank set could be for a later phase of recommendation testing, or as a comparator group.

6.10.8 Targeted sampling

The processes so far described are mainly concerned with ensuring that the sample selected can be justified on the basis of being representative. In some cases, the aim is to target special segments of the population, e.g. members of a geographically dispersed socio-economic subgroup or a rare species of plant. The problem is that there is not a sampling frame for the target group and we are never going to enumerate them all, so methods are based on finding target population members. A few approaches are introduced briefly below. Protocol-derived Replicated Sampling is another approach that is discussed in Chapter 7 in the context of studying vulnerable social communities.

6.10.9 General population screening

If the target population is a reasonably big fraction of the overall population, and if it is not difficult or contentious, to ascertain membership, it may be possible to run a relatively quick screening to check that individuals qualify as target population members (e.g. 'Are there any children under 16 living in the household now?'). As well as finding a sample from the target population, this method will provide an estimate of the proportion that the target population comprises of the general population, so long as careful records are kept of the numbers screened. If the target population is a small proportion of the whole, this method is likely to be uneconomical.

6.10.10 Snowball sampling

The least formal method of those we discuss is 'snowball' sampling. Its basis is that certain hard-to-reach subgroups of the population will be aware of others who belong to their own subgroup. An initial contact may then introduce the researcher to a network of further informants. The method is asserted to be suitable in tracking down drug addicts, active political dissidents, and the like. The procedure used is serendipitous, and it is seldom possible to organize replicate sampling sweeps. Thus, the results are usually somewhat anecdotal and convey little sense of how fully the subgroup was covered.

6.10.11 Adaptive sampling

This is still a relatively new method and allows the sampling intensity to be increased when one happens upon a relatively high local concentration of the target group during a geographical sweep, such as a transect sampling in a vegetation survey. It provides some estimation procedures that take account of the differing levels of sampling effort invested, and is efficient in targeting the effort. Until now this method has been developed primarily for estimating the abundance of sessile species; but it is still not as widely used as some of the other spatial sampling methods which are discussed in Chapter 8.

6.11 Putting Sampling in Context

For sample results to have a wider meaning, any sample has to be regarded as representative of the much larger population that it claims to represent. Selecting the sample in an objective, rather than subjective manner is one way of ensuring this. The purpose of this chapter has been to introduce the reader to some of the different techniques that exist, and to emphasize the importance of representativeness and objectivity. The actual approach taken will ultimately depend on the objectives of the particular investigation and any practical considerations and constraints associated with it.

Anyone with experience in questionnaire surveys will already be familiar with some, if not all, of the ideas that have been covered. However, sampling issues appear in other areas of natural resource research work. One obvious example is in the move away from the traditional survey towards participatory data collection. In addition, the practicalities of studying people's livelihoods are different from those of surveying land or natural populations. Chapters 7 and 8 deal with some of the more specific issues of sampling in these different situations.

Chapter 7

Surveys and Studies of Human Subjects

7.1 Introduction

This chapter focuses on the design of investigations of the people and communities involved in resource management. We are concerned with 'observational studies' involving humans, from which wider conclusions can be drawn. Studies of land and other natural populations are covered in the next chapter. The term 'observational studies' has been used here to encompass both traditional questionnaire surveys and participatory approaches, which have become more popular in recent years for the study of people's livelihoods.

It is in the tradition of participatory work that it aims to build up an accurate, in-depth interpretation of what is being studied and therefore tends to have limited breadth (e.g. geographical spread). Generalization may not be relevant to many participatory exercises which make very limited claims beyond local impact of the work done, e.g. empowerment of a particular community. However, for many development projects which use qualitative techniques or a mixture of qualitative and quantitative methods, the issue *is* important: how can it be demonstrated that results or findings will work beyond the immediate setting in which a project has been based?

In this chapter, we discuss aspects of the traditional survey approach, and how they can be combined with participatory methods. When the objectives of a project are to form conclusions that can be generalized over a wider population, the principles underlying sampling surveys work as effectively for qualitative as for quantitative approaches.

In the field of NRM research, the combination of qualitative ideas and the traditional survey approach are already being used, and we encourage researchers to consider the optimal combination of qualitative and quantitative methods given their objectives, time, and financial constraints.

7.2 Surveys

7.2.1 What do we mean by a survey?

A survey is a research process in which new information is collected from a sample drawn from a population, with the purpose of making inferences about the population in an objective way. Most surveys of people use questionnaires to acquire data.

Surveys attempt to gather a relatively small amount of information from a large number of respondents. This implies that informants are not well known to the members of the project team. Indeed, some advantages of the survey method such as objectivity of sampling and confidentiality for the respondents arise from the interest in the population from which informants are a sample.

The use of questionnaires can be costly; therefore, when the information required is more general, a less expensive method can provide some useful data. For example, a survey on household size may be replaced by the use of aerial photos, whereby the area of each compound is used as a proxy variable. We view such an alternative as complementary to the survey. If aerial photos are available, combining their use with a (smaller) survey of 'ground truth' might be better than devoting all the effort to a single method.

7.2.2 Strengths and weaknesses of surveys

On the plus [+] side, surveys have much to contribute, but only if well thought out. On the minus [-] side a survey can represent substantial effort within a project. Many failures, blamed on the survey method, are due primarily to attempts to take short-cuts. This usually happens where surveys are poorly conceived or managed, or where there are unrealistic expectations of what data can be collected.

[+] A well-organized survey can achieve breadth of coverage with many respondents, so that the wide range of characteristics (e.g. types of livelihoods, lifestyles, situations, ethnicities, knowledge, attitudes, practices), in the population can contribute to the overall picture.

[+] With a carefully planned sampling procedure, we can stratify to ensure proper coverage of the major characteristics.

[+] With good methodology and an adequate sample size we can expect reasonable representativeness through coverage of important factors that differentiate sections of the population, even if these were not specifically controlled in the sampling scheme.

[+] The sampling procedure, for a worthwhile survey, will involve such elements of randomness as provide an assurance of objectivity. Respondents should be selected according to a procedure which will not give 'unfair' prominence to particular groups who may be untypical, e.g. villagers who have cooperated with the researcher previously, whose livelihoods and attitudes have therefore been altered.

[+] If a survey follows qualitative work carried out on a more intensive and smaller scale, it may serve to replicate earlier findings on a wider scale. Similarly, it may delimit the range of applicability of those findings. If project results are to be of more general and lasting value, samples should be representative, adequately sized and selected objectively.

[+] Surveys frequently provide sufficient data to allow reliability to be assessed in various ways. Repeated surveys may show consistent and interpretable patterns of results over time.

[+] A few results from a survey can often be checked against respected independent sources to provide a measure of concurrent validity. If the results agree, there is basis for saying both must be measuring the right thing correctly, and through extension we can claim added plausibility for other survey items.

[+] A suitably structured survey can: (i) take account of varying sizes of units (e.g. farms); and (ii) correct for under-enumeration and to some extent non-response.

[-] Sound surveys require considerable time and effort to plan and complete; inexperienced managers sometimes do not think carefully through the tasks entailed, they may underestimate the time and resources required, especially for computerization and analysis.

[-] Unless a survey has a target length, a clear purpose, and a predetermined analysis plan, it has a tendency to grow and become unwieldy and unrewarding.

[-] Conducting a survey before the project staff has completed identifying their objectives will produce incomplete and possibly irrelevant results.

[-] Ill-phrased questions, poorly linked to objectives, lead to results that are neither digestible nor informative. Examples of this type represent a failure of survey management not of the method!

7.2.3 Some types of survey

Snapshot survey

Surveys are frequently cross-sectional (i.e. constitute a point-in-time 'snapshot'). In order to give a representative picture, a descriptive survey sample has to encompass the target population effectively. It should ideally cover well-defined and different sections of the population (strata) proportionately to their size.

Baseline survey

Sometimes a 'before' snapshot will be used as a project baseline, to be compared with an 'after' picture (e.g. for impact assessment). Fair comparison depends on collecting information on the same footing before and after, and this makes it important to keep careful records. To take account of changes through time that are not due to the project, control studies should ideally be done at the same 'before' and 'after' times, in settings that are not affected by the project (but within the domain to which the survey results are to be generalized).

Longitudinal survey

When surveying the same population twice, once before and once after a lengthy period, it often strengthens the comparison if one visits the same panel of respondents.

When looking repeatedly through time to observe features as they evolve, a long sequence of visits to the same informants may be burdensome, causing ever more perfunctory responses, or withdrawal of cooperation. A rolling programme of panel recruitment and retirement may be sensible, so no informant is troubled too often.

Comparative survey

For a comparison between subgroups (e.g. attitudes of rural and peri-urban farmers), it may be more important to ensure the comparison is of 'like with like' (e.g. in terms of the age composition of the samples of farmers, rather than to achieve full representative coverage of the population). To make the comparison effective, ensure the samples from each subgroup are big enough. Often this will mean the samples should be of equal size even if the subgroups are not equally common in the population; sampling strategies need to be sensible in relation to study objectives.

7.3 Good Survey Practice

7.3.1 Setting objectives

Objectives must be clearly specified. There should also be statements detailing how the survey is intended to contribute to project outputs and how the results will be synthesized with other information. Related to the specification of the objectives, is a detailed definition of intended survey outputs, delimiting the target population in space and time, and setting quantity, quality and time requirements on the data.

Surveys are often part of large projects with elaborately-developed perceptions of the target information. There is then a risk that one survey is expected to address multiple objectives, with contradictory demands for information quality, quantity, and timing. In such cases, linked modules may be better (e.g. a short questionnaire for a large representative sample, linked to some in-depth studies on smaller subsets). The set of studies is designed so no respondent is overburdened. This is discussed in more detail in Section 7.9.

There are surveys where the main objectives are to assess frequencies, for example how many farmers use soil conservation techniques or fertilizer, or what proportion of women practise some form of family planning. An alternative type of objective concerns relationships, for example between erosion and soil fertility, between yields and fertilizer, and between education and family size.

7.3.2 Pilot studies

When surveys fail, it is often because the pilot study phase was omitted, or given too little time or attention. Consider piloting the survey a number of times, so that changes are also pilot tested, and that staff are familiarized with different settings. Pilot studies provide many opportunities (e.g. to see how long an interview takes, to check on the viability of the questionnaire as a 'script', to assess and to contribute to the training of interviewers). They provide a framework and a small amount of data for working out which analyses will be interesting. It is also important to make sure that you are only collecting relevant data.

The main purpose of pilot testing is to learn how informants respond to the survey. Look for any clues such as sensitivities, or difficulties with willingness or ability to give information (e.g. truthfulness, remembering items, understanding concepts and words, or carrying out tasks such as mental arithmetic). Reduce the demand of information if your schedule induces fatigue, boredom, or restlessness.

7.3.3 Non-response

Survey non-response arises when a targeted respondent fails to cooperate with the survey. This may mean that he/she is an unusually busy person, is especially secretive, or has a different relationship than other people with the interviewer. If this is the case, then it is hard to work out how he/she would have responded. If there is a high proportion of non-respondents, a worrying question arises as to whether the remaining sample represents the intended population. Replacing each non-respondent with someone more compliant will not answer this question.

Survey procedures should minimize non-response by:

- Accommodating busy informants (short questionnaires, flexible times for interview, appointments made in advance).
- Winning the confidence of the reticent (transparent purpose, confidentiality guarantees).
- Careful selection and training of interviewers (knowledge of subject matter, ability to portray survey objectives honestly but positively, ability to work objectively).

When some sector of the population clearly cannot be surveyed effectively, it may be best to acknowledge this at the start and redefine the target population that the survey will claim to represent.

Often the planned sample is structured (e.g. businesses stratified by activity and size). When the achieved sample is affected by non-response, it is informative to record and consider the success rates by stratum. It may be desirable to weight actual responses to reflect intended stratum sample sizes. This is always uncomfortable because the greatest weighting up is required in strata with the highest levels of non-response.

Eliminating non-response is often impossible, but ignoring substantial levels of non-response and reporting survey results as if the achieved sample were what was planned, should be evaluated as seriously inadequate performance. Often the effects of non-response have to be inferred from informal reports by interviewers, and clues from outside the survey. A good survey will have a data quality report, where data collection and handling procedures are carefully reported and appraised post hoc; this should include realistic guidance as to how non-response may have, for example biased the results, or led to underestimates of standard errors.

7.3.4 Questionnaire design

If one had to choose a single indicator of a successful survey, it would be the questionnaire as this is the main instrument for data collection. A good questionnaire does not guarantee a useful survey, but unless the questionnaire is well designed, the survey is unlikely to be of much value.

The structure of the questionnaire will depend on many factors, including whether the survey is postal or by interview, whether the population is a general one or a specific group (e.g. managers, heads of households, women). However, some general guidelines can be given.

There is increasing evidence that a thoughtful introduction can be very important to establish a good rapport. For example, it dispels any suspicion that the interviewer works for the government tax department and it introduces the themes and purpose of the survey. The introduction can also develop the respondents' mindset (e.g. refreshing past events, or recalling particular situations that will be questioned in the survey). Practice during training, and effective supervision, should ensure interviewers are reliable, and cover the correct topics.

Transparency of intent should be established in the introduction and also by following clear lines of questioning (e.g. sections on household demographics, land tenure, crops, livestock). Within sections, it may be useful to follow a regular sequence of question types (e.g. facts, practices, knowledge, attitudes and beliefs).

All questions to be included must be consistent with the objectives of the survey. It is often at the planning stage of the questionnaire that there is a realization that the objectives have not been precisely specified.

Constructing an effective questionnaire is a time-consuming process. Researchers inexperienced in questionnaire design should recognize that it is easy to construct a questionnaire, though more difficult to construct one that is effective. To avoid rambling or obscure questions, put some issues and words in: (i) a preamble; (ii) lists of permitted answers; or (iii) reiteration, confirmation, and extension of the first response.

If questions demand recall, should checklists be given to help recollection? Partial lists may bias the response pattern.

How many alternatives should be given for attitude questions? Often there are five, ranging from 'strongly agree' to 'strongly disagree', unless there is a wish to deny the respondent the lazy choice of a mid-point, which sometimes has no meaning. Careful thought is needed to 'profile' attitudes in a way that ensures meaningful results. Informants should participate directly in deciding the importance of profile elements.

Open questions, which allow for free expression, require disciplined data collection and may be difficult to summarize.

If translators are employed who are inexperienced in survey design, they may not appreciate the precision required in the wording of a question, with completion instructions, and units of measurement. Take care to notice any formally correct translations in which the emphasis in dialect or culture are inappropriate to the respondent.

There is information available from past studies to help with constructive approaches to many problems of questionnaire design. Seek advice of those with relevant experience.

7.3.5 Quality in survey data entry

Data management is explained in Part 3, and the general guidelines for survey data entry are the same. Particular features of survey data need special consideration, and careful thought must be given, for example to multiple dichotomy and multiple response questions, and ranking data.

Even if the survey is not an in-depth case study, the human informant, often in conjunction with an interviewer, should have provided data that make up a consistent individual profile. Survey datasets can be large and the data entry can be monotonous. It is worthwhile to set up data entry screens that simulate the appearance of the questionnaire. Checks can be built in to avoid entries in the wrong field, to catch out-of-range numbers due to incorrect typing, and to match 'skip' instructions in the survey schedule. These procedures speed up data entry, and improve quality compared to the amateur approach of typing into a spreadsheet and assuming the numbers are accurate.

To ensure accuracy, it is best to re-enter the data independently in a program with a verification or double entry facility. This should produce two files, whereby subtracting one set of data from the other produces zeros where the data matches. Even quite rudimentary software can be used to yield such verification.

It is worth setting aside three times as much time as required for single data entry, to undertake data entry, verification, and the reconciliation of any values that disagree in the two versions of the data file. Relative to the total cost of the survey, time spent on data quality verification is well spent.

Coding involves developing a code system for open-form responses. This is not usually completed until most questionnaires have been scrutinized. Codes are given to common written responses, though additional codes may be added during survey examination as particular responses are found to be more frequent. This will then involve revisiting early forms to add these codes. Achieving consistency in coding requires concentrated thought and effort.

7.4 Study Management

A properly organized survey requires carefully planned managerial procedures. Some technical aspects of study management are discussed elsewhere in this chapter, but they are dependent on being developed and used effectively as indicated below.

Resourcing review

Include a clear commitment of the necessary manpower, equipment, facilities, material and 'political' support to ensure the exercise can be carried through.

Time budgeting

Allow for preparatory activity, pilot testing, seasonal delays, holidays, absences, equipment breakdowns, error correction, and contingencies. Use activity analysis that identifies who does what, and when (e.g. Gantt charts or critical path analysis).

Training plan for staff

Include manuals of survey procedures, assessment materials, and interviewers' and supervisors' instructions. Provide training for interviewers and supervisors for the interviewing process, and organize the necessary staff to be sent on appropriate courses, or for trainers or consultants to be brought in.

Schemes for dissemination and utilization of results

Consider several variant outputs, possibly including: (i) feedback to contributors of local data relevant to them; (ii) discussion materials for a workshop; (iii) an executive summary for policy makers; (iv) reports; and (v) a computerized archive for those who will use the data in future.

7.5 Participatory Research

7.5.1 What do we mean by participatory research?

During the 1980s and 1990s, questionnaire surveys became less popular in development programmes, because they were thought to be too time consuming and expensive, and were not providing in-depth understanding of issues relating to people and their livelihoods. At the same time, participatory methods became increasingly popular, especially for relatively quick and informal studies. In Section 3.4.1, we give an overview of the methods and their evolution; but nowhere in the book do we go into detail about participatory methods and their tools. There are plenty of other books dedicated to that. Here we merely remind the reader what they are.

Participatory methods encourage more discussion and involvement by the respondents who may be individuals, focus groups, or village committees. The tools used for data collection are a rich set, familiar to anthropologists and other qualitative researchers. They include visual tools such as matrices, diagrams, maps, and time lines, all of which can generate useful insights into a community and its dynamics. These tools allow information to be collected on complex issues; an example might be people's reactions to the building of a large dam, where communities were dispersed and relocated, and livelihoods changed. The appeal of participatory methods is that they offer much greater insight than can be obtained from questionnaires. A downside is that the methods are resource-intensive, and work cannot be done on a large scale unless carried out over a long period of time. Another aspect is that for useful results the approach requires good-quality facilitation.

Participatory investigations often give rise to open-ended responses, and one way of summarizing such data is via coding. Statisticians refer to these codes as qualitative data; hence the term 'qualitative approach' is sometimes used for participatory work. However participatory work and qualitative work are not synonymous; the data collected from participatory investigations can be qualitative, quantitative or a combination of both.

This book is about good statistical practice in NRM research, and it is in the context of research that we will discuss participatory methods. We are concerned with situations that generate data from which conclusions can be more broadly drawn and not those studies whose sole objective is one of empowerment.

7.5.2 'The plural of anecdote is not data'

Participatory methods provide an alternative to formal surveys in terms of concentrating efforts. The survey can be regarded as 'broad and shallow' (i.e. large number of respondents providing not very detailed information) as opposed to participatory approaches which are 'rich and detailed' (i.e. a few 'respondents' or groups of respondents providing in-depth information). If, however, the claim is to be made that participatory work is worthwhile because the results will be widely applicable, the issue of generalizability of the research should be addressed seriously. Often projects are vulnerable to the criticism that they are 'just another case study' and do little to provide knowledge rather than hearsay about the wider population.

One major purpose of the following sections is to bring some sampling survey principles and practices from the statistical into the participatory arena, so that results from studies can be claimed as representative of a population or a clearly defined part of it. We also hope that the reader will see the parallels between surveys and participatory studies. For instance both need: (i) sensible sampling; and (ii) data collection activities that will produce reliable results. Other issues discussed earlier, such as (iii) clear specification of objectives; (iv) good study management (Section 7.3.2); (v) the benefits of pilot testing; and (vi) ensuring that there is good rapport and transparency of intent when data are being collected, are all good practice issues as important in participatory work as they are in surveys.

7.6 Doing One's Best with Small Samples

7.6.1 Study design compromises

The fact that participatory exercises tend to concentrate effort within rather than across communities means that often the researcher only has resources for studying a few communities, and there is therefore a trade-off in terms of effort between and within communities.

If we select twenty communities instead of two, for the same total effort, the depth attainable in any one community becomes much less and the study loses some richness of detail about human variety, social structure and decision-making within communities.

If we select two communities instead of twenty our sample of communities is of size two! If our two communities turn out to be very different from one another, or have individual features that we suspect are not repeated in most other places, the basis of any generalization is extremely weak and obviously suspect. Even if the two produce rather similar results, generalization is still extremely weak, though less obviously so, as we have almost no primary information about between-community variability.

Roughly, one can think of:

[total effort] = [effort per community] × [number of communities].

If one factor on the right hand side increases, the other must decrease for a resource fixed in terms of effort, or project budget. Including an extra community often entails a substantial overhead in terms of travel, introductions and organization, so that:

[effort per community] = [productive effort + overhead].

If this means that:

[total overhead] = [overhead per community] × [number of communities],

there is a natural incentive to keep down the number of communities, to reduce the budget. This seems efficient, but there needs to be a deliberate, well argued case that an appropriate balance has been struck between this argument about overhead and the argument about having too little information about between-community variability.

7.6.2 Phased sampling

By drawing on general principles of selecting units by sensible statistical sampling, an in-depth study carried out in a few sites can be underpinned more effectively. One way is via phased sampling, or 'sequencing'. This is only a formalization of what people do naturally. As a first phase a relatively large number of communities – perhaps tens or hundreds – may be looked at quickly, say by reviewing existing information about them. Alternatively it could be done by primary data collection of some easily accessible, but arguably relevant, baseline information on tens (though probably not on hundreds) of them. Either route can lead to quantitative data such as community sizes or to qualitative classifications or rankings.

Sampling of communities for a more intensive study can then be guided by the information gathered by the phase one data, but only when the two phases are linked. Suppose that phase one gives us an idea of the pattern of variability of a quantity, X, which could be the wealth index, size of the community, or distance to a secondary school. We may then be able to argue that a small sample of communities can be chosen for intensive investigation in phase two, and yet be representative of a larger population as far as X is concerned. This argument is simplified by referring to only one variable (X); in reality a selection of key items of phase one data is used to profile communities, and the argument is that the communities chosen for phase two are reasonably representative with respect to this profile.

The above process can work if the first phase and subsequent work are qualitative, quantitative, or combined. It is apt to use simple quantitative data followed by qualitative work because the first phase is intended to be less costly and less time consuming than the subsequent phase. The above does not limit the second-phase methodological choices made by discipline specialists working in development settings. What is suggested is just a systematic approach to selection of settings, which offers some strengthening of the foundations on which the work is based.

A second sequencing idea can be added to strengthen the claims of in-depth research carried out within in a few communities. The knowledge gained should lead to recommendations, which can be tested in communities other than those impacted by the main research work. These can be sampled from the set considered in the first phase, possibly as a 'replicate' of the set chosen for intensive phase two work. Once again, this applies whether the work is qualitative or quantitative in the various phases.

7.7 Site Selection in Participatory Research

7.7.1 Structure: hierarchy and strata

Given that an investigation requires the study of 'a few communities that have to be studied quite intensively', researchers often focus on the question of 'how?' this should be done, and delve immediately into the methodologies to be used within communities. However, just as important is the way the respondents are chosen. Sensible selection depends on recognizing that the population of interest has a 'structure'. This structure can best be described in two ways.

The first is that the study units have a hierarchical or multilevel, aspect: for example individuals within a household, households within a village (or community), and villages within areas. The selection of primary sampling units, which are often communities, is likely to be a hierarchy of connected decisions, and needs to be thought out carefully.

This hierarchical aspect of the population can be thought of as a vertical element in its structure. The second structural aspect can be described as having a horizontal dimension. By this, we mean that there are features at the different levels of the hierarchy that can also have a bearing on people's livelihoods, points of view etc. For instance, whether the head of household is male or female might influence a decision on whether to keep or sell some produce; villages in different agroecological areas may use different tillage methods, and so on.

These two ideas are not new in survey work; their existence is the reason why multistage sampling and stratified sampling have been used to select respondents in questionnaire designs.

We illustrate how these ideas fit into participatory work with an example in the next subsections. A researcher is concerned with a socio-economic aspect of a set of potential innovations, which might be taken up by farmers in poor rural communities in a part of Malawi. Let us assume that within this region there are five Extension Planning Areas (EPAs). Given the holistic nature of the approach, a few communities have to be studied quite intensively. The innovations are not equally suitable for everyone, because of variations in resources.

7.7.2 Choosing a sample

The real purpose of random sampling as the basis of statistical generalization is objectivity in the choice of units (communities in our example) to be sampled. However, a small sample that is random may be unrepresentative. The claim to objectivity is achieved if the target communities for phase two are selected based on reasonable, clearly stated criteria using phase one information.

Some phase one information demonstrates that selected communities qualify for inclusion in the study (e.g. maize growing is the primary form of staple food production); other items indicate that they are 'within normal limits' (e.g. not too untypical in any of a set of characteristics). A few key items of information show that the selected communities represent the spread in the set (e.g. different quality of access to urban markets). Representation can be of differing sorts. For example, we might choose two sites with easy access and two with difficult access. A different strategy could be via systematic sampling where we select four samples taken 1/5, 2/5, 3/5, and 4/5 of the way through a ranked list, omitting the very untypical extremes. The case studies can then be seen or built on a more solid evidential foundation in terms of what they represent.

Another approach would be to use ranked set sampling, as described in Section 6.10.7. For instance, in our Malawi example, if we required five communities for sampling from the region, we could first stratify by EPA, select five qualifying communities per area and then use ranked set sampling to select a sample of five communities, one from each area. The example in Chapter 6 is almost identical to this, except that we selected five villages and then households from within the villages. The selected sample should span the range of characteristics of communities in the region, and also include an element of randomness.

If we also wanted to investigate male and female heads of households separately, we could stratify by this trait, and randomly select farmers from both groups within the villages.

7.7.3 Participatory tools as an aid in sampling

One of the advantages of multistage sampling is that it only needs a sampling frame for each stage of the sampling, and not one for the study as a whole. In our example, selection of the primary sampling units (the communities) has already been discussed, as was the distinction at the farmer level of male- and female-headed households. But, what if instead of gender, we were interested in whether some aspect of food security or wealth ranking had a bearing on uptake of the innovations? This type of information about individual farmers or households is not readily available on any kind of 'list', and compiling such a list using a traditional census style approach would be time consuming. It might also not allow for the fact that community sizes can swell and shrink according to seasonality, food crises, etc. On the other hand, community mapping can offer an easy and reliable way of constructing a sampling frame from which the sample of households may be drawn; particularly when the 'community' has a clear definition, as in our example (Fig. 7.1).

Fig. 7.1. Example of a community map.

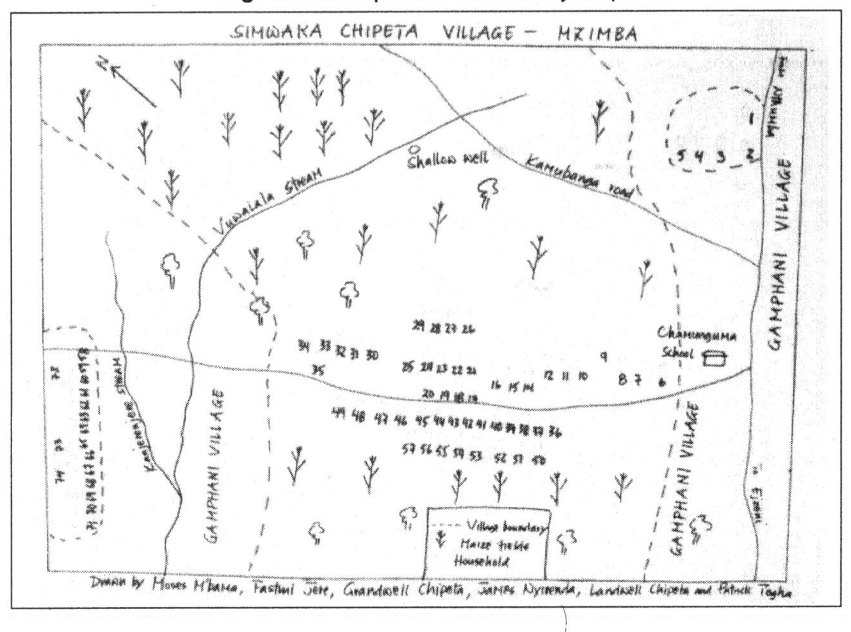

By involving a few informed members of the community, it is possible to get a clear description of the community. The maps provide the total number of households in the community, and the location of each household. With some careful questioning of the community members, it is also possible to acquire information about the wealth, ranking, or food security status of the different households. With this information, the researcher can identify groups of households for stratification, know the size of each stratum, and thus organize a sensible stratified sampling of households in the community.

Here we are not advocating that participatory exercises should always use ranked set sampling, followed by stratification using community mapping. Instead, the above example illustrates how standard sampling techniques and participatory tools can be used together. The choice of tools and techniques will depend on the particular situation. Indeed, in our example if we had a wealth ranking that we believed in, then taking a systematic sample through the ranks might be better than stratification. Equally, if there were no sensible strata, the households could be chosen randomly.

7.8 Data Collection in Participatory Research

Replication of sites is necessary for generalizability of results in participatory studies, where one of the main aims is research (rather than empowerment). The data collected at the sites are often the outcome of focus group discussions and this information needs to be pulled together into a coherent summary for generalizability to be possible. Broadly speaking this means that the data collected from the sites need to have some elements that are common to all the sites. This can be achieved by simple adaptations of everyday PRA tools.

7.8.1 Standardization

Using PRA tools in the same way, at all sites, and collecting the same information with the same level of detail, is essential for combining results across sites. If the same tools are not used then the types of information collected may be different and hence cannot be combined. Even if the same types of information are produced, there may be some bias resulting from the particular tools used.

This message may sound strong to the qualitative researcher. It is necessary only for the elements of the investigation that require synthesizing to address research objectives. PRA exercises often search wider and deeper than this, and there is no reason why greater flexibility cannot be employed for the other information collected for other objectives. In fact, we invite this greater flexibility for the other objectives; achieving these should not be compromised by having to meet the narrower research objectives.

One way to ensure that tools are used in the same way when required, is via a field manual that sets out clearly the practice(s) to be employed. The manual should be in line with project objectives, and will almost certainly need testing. Preventing facilitator bias can be ensured by training the field teams.

This whole process of standardization should be participatory. Different stakeholders could be involved in testing and selecting the approaches to be used, and in developing the field manual. This way the end product addresses everyone's requirements and concerns. It is similar to designing a questionnaire that addresses study objectives without being contentious. The difference is that standardization in participatory studies involves a greater choice of tools and techniques and more interested parties, and takes more time and effort to finalize. The researcher should not be discouraged, but should perhaps make sure that enough time is allowed for this part of the work.

7.8.2 Comparability

In answer to the question 'Is this a better or worse year than you would have expected?' it is difficult to combine information from two sites that use different years (e.g. where one uses 1999 as their reference and the other 2001). More appropriate would be to establish a reference time point and ascertain the information relative to that time point. Using the same definitions, indicators, time periods at all sites, and using absolute rather than local reference points, are ways of making information comparable across sites. Without this comparability, it is not possible to combine results.

For similar reasons, when faced with making preference statements, for instance when communities are asked to assess the cooking characteristics of four types of rice, or to make assessments of different types of welfare transfer, it is preferable to use a clearly defined scoring system as opposed to ranking.

Table 7.1. Using scores as opposed to ranks.

Using scores 7-point scale: 7=excellent, 6=very good, ... 2=very poor, 1=terrible, wouldn't use					*Using ranks* 1=worst, ... 4=best				
Variety	A	B	C	D	Variety	A	B	C	D
Community 1	3	4	2	1	Community 1	3	4	2	1
Community 2	4	7	3	2	Community 2	3	4	2	1
Community 3	6	7	3	2	Community 3	3	4	2	1
Community 4	5	6	4	1	Community 4	3	4	2	1
Average score	4.5	6	2.5	1.5	Average rank	3	4	2	1

The reasons are purely to do with combining data. As the example in Table 7.1 shows, ranking varieties of rice gives no indication of the extent of differences between two adjacently ranked varieties, either within a community or across communities. On the other hand, with a well-defined scoring system, the scores have a meaning that is consistent in all sites and the difference between scores is consistent across sites. The average scores also have some meaning and the differences between them can be interpreted in a quantitative way. Furthermore, statistical analysis methods are fairly well developed for dealing with score data, as opposed to ranks, when explaining patterns of response in terms of site characteristics.

7.9 Combination of Studies

We first illustrate these ideas by reference to a conventional survey, which may be formal or semi structured. The wider application of the approach follows.

7.9.1 Segments of one survey

A single survey is often a combination of segments, each comprising some questions in a questionnaire, or themes in a semi structured investigation. Having performed one round of data collection, there is a tendency to treat the survey instrument as a unitary entity. Every respondent answers every question, and especially where the survey has been allowed to become large, the exercise is too cumbersome. Often this is blamed on the survey method, in the same way that a bad workman blames his tools.

Fig. 7.2. Segmenting surveys into themes and modules.

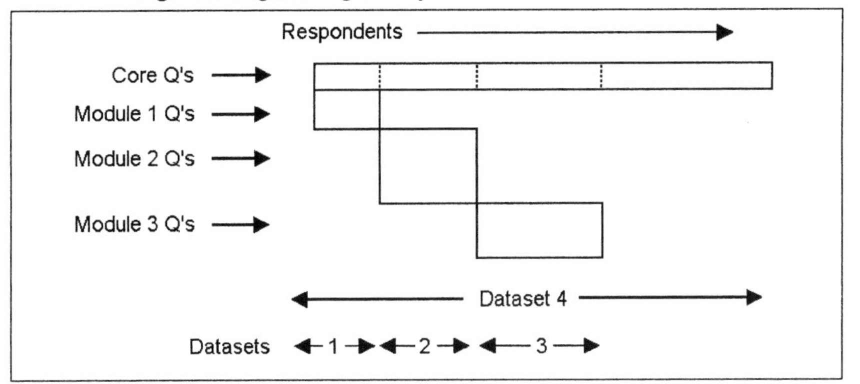

Segmenting a survey into carefully constructed sections can save unnecessary effort. A core set of questions is usually crucial (e.g. to establish a baseline for evaluation or impact assessment purposes). This set should be kept small to be manageable. Questions relevant to individual researchers within the survey may not need or justify such a large sample of respondents. Themes can be set up as modules, to be answered by only a subset of the respondents (for example in one community, or by a structured subsample of the respondents within each community). If there are perhaps, three such modules, the survey can consist of a relatively large sample responding to the core questionnaire, and three subsamples, each responding to one of the modules. This can be shown diagrammatically as in Fig. 7.2, where all respondents contribute to analyses of the core questionnaire (dataset 4), while analyses of module 1 questions, and their inter-relationships with core questions, are restricted to the relevant subsets of respondents (i.e. dataset 1). The modules are deliberately shown as different sizes. There is no reason they have to be of equal length, or have the same number of respondents. There is no time dimension implied here.

In the above, we consider different audiences for different questions. Warning bells should ring if researchers start to take responses to module 1 from one set of people, and to 'compare' them with responses to module 2 from different people. Modules should be clearly defined so that this does not happen.

7.9.2 Phasing as modularization

Segmenting divides a large survey into modules, but can also link a series of separate studies together. In phased research (discussed earlier in Section 7.6.2), the idea was of a broad but shallow baseline study (perhaps quantitative), which led to a second phase with a smaller sample of locations, for a more in-depth study.

Figure 7.3 represents the example in Fig. 7.2, whereby the 'tabletop' indicates the wide but shallow phase one study and the 'legs' represent narrower, but deeper follow-up work in a few communities. This applies regardless of whether the phases involve collecting qualitative or quantitative data, or a mixture.

Fig. 7.3. Showing the extent of a phase one study, with four phase two studies.

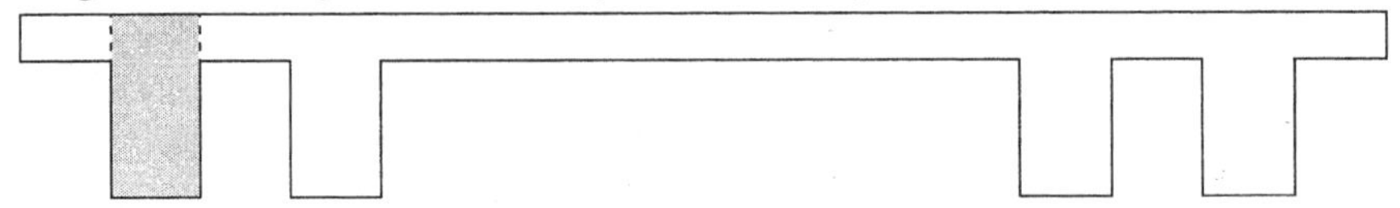

If the data collected in phase 1 can be connected directly to some from phase 2, then the analysis after phase 2 can incorporate some of the earlier information, and this is shown diagrammatically in the shaded 'leftmost leg' in Fig. 7.3. Such 'read-through' of data becomes more powerful when it adds a time dimension or historical perspective to the later-phase data, as is required in impact assessments. It also serves the purpose of setting the small case study and its phase 2 information (in the 'table leg') in context. The shaded part of the tabletop can be considered relative to the unshaded to show how typical or otherwise the case study setting is.

7.9.3 Project activities as modules

The same style of thinking applies if we consider a development project, where samples of farmers are involved in a series of studies. These studies are likely to be conceived at different points of time, and by different members of a multi-disciplinary team. However, they may still benefit from a planned read-through of information, which allows triangulation of results, time trends, cause and effect, or impact to be traced, as the recorded history gets longer. This is not inconsistent with spreading the research burden over several subsets of collaborating rural households. The horizontal range in Fig. 7.4 corresponds to a listing of respondents, the vertical shows a succession of studies.

The separate interventions, processes, or studies within a project sequence are of greatest interest to the individual researchers for whom they were conducted. These correspond to a 'horizontal' view of the relevant sections of Fig. 7.4. Looking at the project information from the householders' point of view the information that relates to them and their livelihoods is a 'vertical' view. Two groups of respondents, who have been involved in a particular set of project exercises or interventions, are shown diagrammatically by two shaded areas in the diagram below. If a project sets out to focus on the livelihoods of the collaborating communities, it ought to be of interest to look 'down' a set of studies, at the overall information profile of the groups of respondents. As a result, the programme of studies should be designed, and the data organized, so that such 'vertical' analysis can be done sensibly for those aspects of interest, rather than a snapshot view.

Fig. 7.4. A sequence of studies.

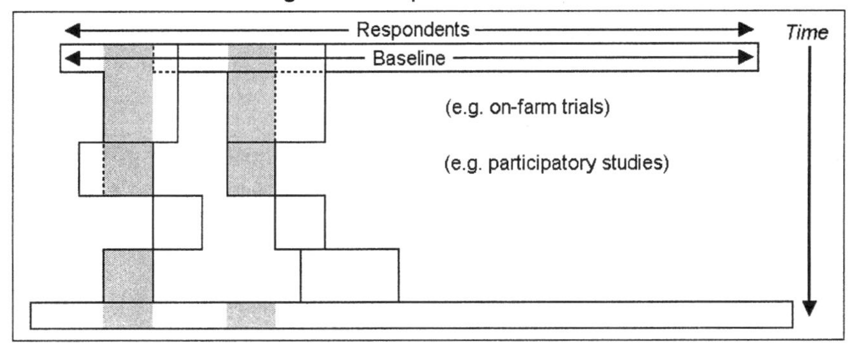

7.10 Targeted Sampling

In Section 6.4, we introduced the problem that the target population may not be a readily ascertainable population, but a subset whose size and boundaries are rather ill defined. One example might be people who are engaged in activities that carry a social stigma. Here we offer another approach, in addition to the methods discussed in Chapter 6. It is primarily intended for those interested in targeted sampling. The combination of ideas, and the suggestion to use it in the development setting, makes it novel in the sense of being untried at the time of writing this book. It needs further development through practical application.

7.10.1 Protocol-derived replicated sampling

The notion of replicated sampling is highly adaptable as a basis of valid statistical inference about a wider population. The original idea concerned a standard quantitative survey, probably with a complication such as multistage structure. If this could be organized as a set of replicates, (miniature surveys, each with an identical structure), then an estimate of key measures could be derived individually. That set of estimates can then be treated as a simple random sample of data. The replicate-to-replicate standard error would incorporate the whole set of complexities within the stages of each miniature survey, and provides an easy measure of precision of the final answer.

Here we need to combine that idea with two other notions. The first is the idea of developing a prescriptive sampling protocol to be used in the field as a means of systematic targeting (i.e. for particular households). The protocol prescribes in detail how to reach qualifying households. As an example, suppose our target comprises 'vulnerable, female-headed, rural households' in a particular region. This involves sorting out all necessary procedural details. One element might concern interviewing key informants at the primary unit level (e.g. NGO regional officers), by presenting them with a list of twelve areas within the region and asking them to indicate two areas where they are confident there is a high number of target households. There would be numerous procedural steps at several hierarchical levels. In the preceding example, the use of key informants is just an example; it is not an intrinsic part of every protocol.

Samples are often derived in some such manner: they get at qualifying respondents cost-effectively, but the method usually carries overtones of subjectivity, and of inexplicit individual preference on the part of the selector. The protocol is intended to address these difficulties. Its development is a substantial process involving consultation, some triangulation, and pilot-testing of its practicability. It is thus a specially developed field guide which fits regional circumstances and study objectives, incorporating e.g. anthropological findings, local knowledge, and safeguards against possible problems. The protocol is a fully defined set of procedures such that any one of a class of competent, trained fieldworkers could deliver a targeted sample with essentially interchangeable characteristics.

The second added notion is that if the protocol development involves appropriate consultation, brainstorming and consensus building, then the protocol can serve to define the *de facto* target population being reached. Developers of the protocol can effectively sign up to: (i) accepting a term such as 'vulnerable, female-headed rural households' as the title of the population who are likely to be sampled; and to (ii) accepting that the population so sampled is a valid object of study, and a valid target for the development of innovation(s) under consideration in the particular locale for which the protocol is valid.

Repeated application of the procedure would then produce our equivalent of replicate samples. These carry some statistical properties, provided that: (i) the sampling is regulated as described above; and (ii) the information collection exercise within any given replicate is standardized. When the procedure is replicated, it is necessary that a common core of results should at least be collected in the same form, and recorded using the same conventions in each replicate. It is for these results that we can make statistical claims.

For example, suppose we record the proportion (x) of respondents within a replicate sample, who felt their households were excluded from benefits generated by a Farmers' Research Committee in their community. The set of x values from a set of replicate samples from different sites now have the properties of a statistical sample, from the protocol-defined population. Even though the protocol itself encompassed various (possibly complicated) selection processes, we can for example, produce a simple confidence interval for the general proportion that felt excluded.

The important general principle which follows is that if we can summarize more complicated conclusions (qualitative or quantitative) instead of a single number x from each replicate, then we can treat the set as representing, or generalizing, the protocol-defined population. There are interesting ways forward, but the practical development and uptake of such a notion poses 'adaptive research' challenges, if the concept is put to use in the more complex settings of qualitative work in developing countries.

Chapter 8

Surveying Land and Natural Populations

8.1 Introduction

Many natural resource management problems require surveys of land (and water) and what is on and in it. Examples include:

- Land use and productivity.
- Plant, animal and microbial abundance and diversity.
- Soil properties and problems.

The objectives may be:

1. Estimating the current status of land, to answer such questions as 'How much of the district is eroded?' or 'What is the carbon storage of this forest?' The objectives will have to be made more specific, for example defining 'eroded' and setting the boundary of the study area.
2. Comparing the status of different locations or strata, such as different soil types, land tenure classes, or times since forest conversion. Which stratum a location falls in may not be known until it is visited for measurement. This has implications for design of the sample.
3. Monitoring changes over time. Important sampling questions concern re-sampling the same points or moving to new ones on each occasion.
4. Looking for relationships between variables, such as erosion and soil microbial diversity.

These objectives all require sampling in space. The sampling problem may well arise as part of an experiment. In field experiments, we take samples of soil (because we cannot measure all the soil in a plot), or plants (because we want to take destructive measurements during the experiment). We also have to sample in large-scale experiments, for example in trials to measure the effect of forest management on plant diversity, or the effect of watershed management on soil erosion.

An example

The problem is initially simply stated: to estimate the average level of P (phosphorus) in the topsoil of a 1 ha field. First, we must confirm the specific objectives:

- What is 'topsoil'? Is it defined by depth or by chemical, physical, or biological properties?
- Will this be a one-off measurement, or will it be repeated to detect change?
- Is the average needed on its own, or the average plus an estimate of its standard error? The standard error of the average for that field (that is, the precision with which we have measured the average) may not be needed if the field is a single unit in a study. The variation between fields, not the precision of the average of the mean of one field, will be relevant for assessing treatment effects or other field-to-field differences.
- Is the average level really the only objective, or is the variation in the field also of interest, perhaps to help explain variation in crop growth, or to stratify the field for further work?
- What measurement of P will be used? What are the constraints and requirements (for example, soil volume or sample freshness) of this measurement?

The next step is to determine the sampling unit. Typically, soil samples are taken with corers or tubes of a fixed diameter. If only a few grams are needed for chemical analysis, then it is tempting to use a very small sample. However, such samples may be highly variable and may be biased, perhaps against larger particle sizes. Guidelines exist on suitable sampling unit sizes for different measurements, such as 3 cm diameter for nematodes, 6 cm for mesofauna and 25 cm for earthworms. Also, think through the possible distinction between sampling units and measurement units. It is common practice to take many field samples of soil, bulk and mix them, and subsample for chemical analysis. This has implications for the type of variation that can be studied. In particular, variation between subsamples taken from the bulk tells you nothing about variation in the field, only about lab measurement error and deficiencies in the mixing process.

We must then determine the population or sampling frame. If we are sampling the 1 ha field, should some border areas be excluded? Are there other parts of the field that should be avoided? Should you accept a sampling point if it falls on an 'odd' spot, such as a rock or a termite mound? If not, such spots are not part of the population. Such problems are avoided by defining them at the start.

The sample size and sampling scheme must now be chosen. Simple random sampling has the usual properties of providing unbiased estimates of the mean and its standard error, although this may not be the most appropriate sampling method in every case. In many contexts, a systematic sample, perhaps on a regular grid (see Section 8.2), makes more sense.

8.2 Sampling in Space

The 'sampling frame' is the list of the population from which a sample can be drawn. When sampling spatial locations, this list will often be implicit; a 'list' of all points is the area. If the area is rectangular, then a random location can be defined by choosing a random x and y coordinate. If it is an irregular shape, a random x and y coordinate is chosen, but the point is accepted only if it falls within the area of interest. Thus, there is nothing new in sampling in space.

However, the problem is not always as simple. Suppose you wish to find the average field size in a district by sampling fields and measuring their area. Choosing a random location and measuring the area of the field in which it falls can give a biased estimate; a random point is more likely to fall in a large field than a small one. The problem is that the sampling unit should now be the field, not the location, and there may not be a frame or list of fields to select from. It may be possible to construct a list, for example from aerial photographs. A similar problem arises if we try to select a random tree in a forest by selecting a point at random. A tree is more likely to be selected if it 'represents' a larger area. Therefore, an individual tree in the thick of the forest is less likely to be selected than one in a clearing.

Just as when sampling people, there are obvious dangers with trying to select sample locations by 'arbitrary choice'. All sorts of unquantifiable biases enter into the selection, though there are situations when little else is possible.

When sampling in space the usual benefits of stratification apply. Thus if we know the field has two distinct regions, for example upper and lower slope, or black and red soil, then it is worth ensuring that the sample covers both parts adequately by taking them as strata. However, even when no strata can be clearly identified before measurement takes place, we know that most natural spatial variables have irregular but smoothly varying, patchy distributions. A spatial picture of soil P will probably look like a contour map with hills and valleys. They will be irregular in size, shape, and position, but there will probably not be many cliffs and chasms in this P landscape. That means that sample points which are close together will tend to have similar values of P. Therefore, the information concerning the entire field should come from samples that are well spread out, rather than having sample points that are close together. The most evenly spread sample, is one that falls on a regular grid, theoretically a triangular grid; however, a rectangular grid is easier to handle in most cases.

There are two important consequences of using systematic, grid-based spatial sampling. If either of these is likely to be important for your problem, then get advice before using such a scheme.

1. Estimating the standard error of the mean cannot be done by the 'usual' formula that is appropriate for random samples. The standard error of the mean of a systematic sample requires knowledge of the correlation structure of the quantity being measured. This is roughly equivalent to knowing the typical size and shape of the hills and valleys in the 'P landscape' above. This will not normally be known beforehand, and cannot be well estimated from the regular grid of sample points. However, as explained earlier, we do not always need a standard error, if it is variation at a higher level (say between fields), which is important.
2. If the variation in the field is regular (rather than random), then a sampling grid that coincides with this regular pattern will give biased results. This is rarely a problem, because it will normally be obvious. For example, you would naturally avoid taking all your samples for P measurement along the centre of the alley between rows of trees.

8.3 Multiscale Measurement

One frequently quoted characteristic of NRM research is that it requires a multiscale approach. This means problems should be studied at various scales, for example the plot, farm, and catchment scale, and not just at one of these. However, there are only a few measurements that can actually be taken at different scales; two common examples also illustrate the general issues.

Biodiversity is usually assessed in terms of species richness (the number of species present), and evenness (the relative frequencies of those species). These can be measured by taking a sample of quadrats, and counting the species present in each quadrat. The number of species encountered will depend on:

- The size of the quadrats.
- The number of quadrats.
- The size of the area being sampled.

It is not simply that different size samples give different answers but estimate the same thing. The actual quality being estimated depends on the quadrat size and sample size in a way that is not true for the P-measurement problem. For example, you are likely to observe more species in a sample of 100 quadrats than if you had just taken 10. In the same way it is likely that more species will be observed with quadrats of 10 m^2 than 1 m^2, or in a 10 ha study area than a 1 ha study area. Most studies are comparative, so it is important to make sure that measurements in different situations can be compared. There are ways of adjusting estimates for sample size, but it is simpler to use the same measurement protocol in each case so that such adjustments are not needed.

Soil lost to erosion is also scale dependent, and can be measured at different scales. The erosion on a small plot is measured by installing some sort of barrier around the plot with a catchment device at the lowest point. When it rains, the runoff and the soil emerging from the plot are measured. The measurement can be done for larger plots, eventually taking whole catchments and measuring the sediment load in the out-flowing stream. It has been known for a long time that the soil lost from a 200 m^2 plot is usually less than ten times that from a 20 m^2 plot, and estimating the loss from a whole catchment by using measurements taken from small plots, gives answers which are huge over-estimates. The reason is simply that most of the soil leaving a small plot does not leave the catchment, but is deposited somewhere further down the slope. The sampling implications are clear; if we wish to know the erosion of a catchment, we must measure at the catchment scale. An alternative is to model the whole erosion-deposition process. If you want to make comparisons perhaps of land use, then it may be feasible to use the same small plot design for each land use. However, interpretation will depend on the assumption that land use does not change the scaling rule.

8.4 Replication May Be Difficult

In studies that are not experiments, it is not always clear what constitutes a replicate, or what the implications are for arranging samples in different ways. In an example, two areas of a forest are sampled to show the extent of herbaceous cover after forest clearance; one is sampled 10 years after it was cleared, the other 25 years after clearance. Herbaceous cover is measured by defining a sample area, perhaps a quadrat, and estimating the percentage cover by noting the number of points on a 1×1 m grid that have herbaceous plant cover.

The first measurement scheme proposed involved a single 20×40 m quadrat in each part of the forest (Fig. 8.1a). This was criticized as being 'unreplicated'. It certainly looks odd, with all the measurement effort concentrated in a small area. Figure 8.1b shows an improvement, namely eight 10×10 m plots in each area of forest. The total measured area is the same, but the samples are more spread out. If they were located randomly, we can estimate the standard error of the mean ground cover in both areas. We still have only a single replicate of each forest type (as only one forest was sampled) however, and so conclusions are limited to demonstrating a difference that *may* be due to time since clearance. Rather stronger conclusions could be made from the situation in Fig. 8.1c, in which there are several forest areas of each age. Of course we still cannot prove cause and effect, and certainly need to know the reasons why some plots were cleared 25 years ago and others only 10 years ago.

Where there are just two areas, one more scheme that may be of interest is in Fig. 8.1d. The idea is that parts of the forest that are close together may be more similar in many aspects than those further apart. Therefore, we should get the best estimate of the difference between 10 and 25 years after forest clearance by sampling plots along their common border. The precision of this comparison can be improved still further by pairing neighbouring plots; the pairs linked in the diagram. This design can be criticized for putting all the observations along the border, which may well be atypical. Therefore, more quadrats are sampled in the rest of the area to confirm the results. The outcome is a compound sampling approach with a sound justification.

We need to recognize that this is a difficult subject to study. Even in an ideal situation with experiments extending over 25 years, a trial where we have randomly chosen the areas shown in Fig. 8.1c, might still be difficult to analyse in a way that leads to clear conclusions. Although we now have a spatially random sample, we have just used a single fixed set of years. An increased percentage cover in the 25-year observations could be due to a number of factors: it might be because of the length of time, or the fact that some of the early years were warmer, with none of the 10-year samples benefiting from the warmer years. As in all research, consideration of processes and mechanisms will help in understanding observations made in this uncontrolled situation.

Fig. 8.1. Schematic plans of a forest area.

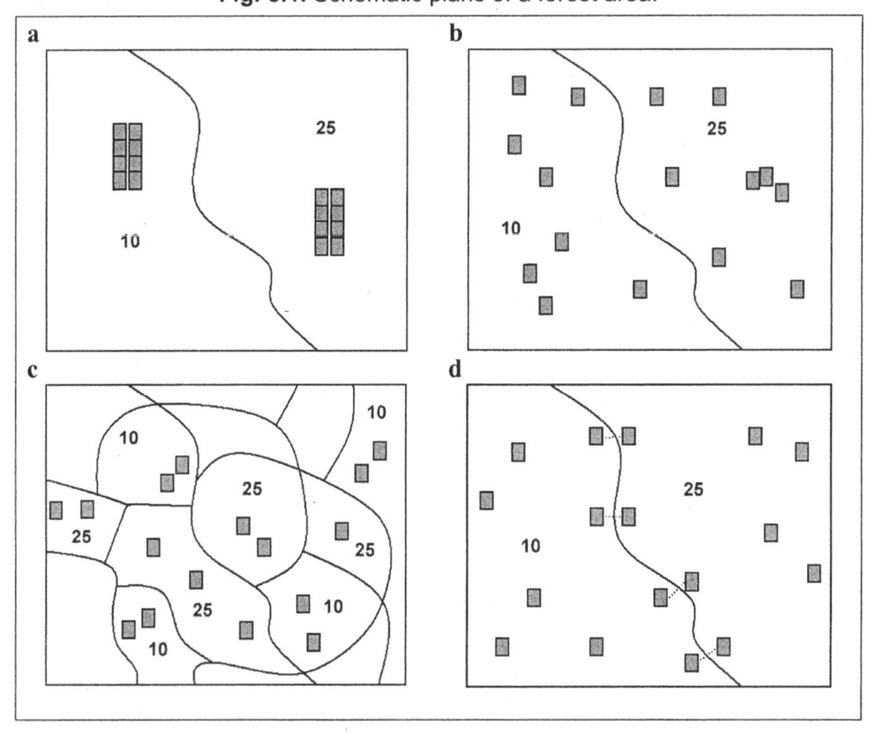

8.5 Time and Repeated Sampling

Many problems require sampling in time. There are similarities in the problems of sampling in space and time, so we discuss them together in this chapter. Objectives for studies over time include:

- Monitoring resource management activities.

- Detecting changes, particularly as a result of changing management or environments.
- Characterization of the variability in time, for example to understand risks from extremes.
- Understanding the dynamics of systems.

In each case, the design of a study will include taking decisions about when the measurements should be made; these choices interact with other decisions such as which units or locations will be measured.

Fig. 8.2. Water-use by a tree measured every hour over 5 days, plotted against time of day.

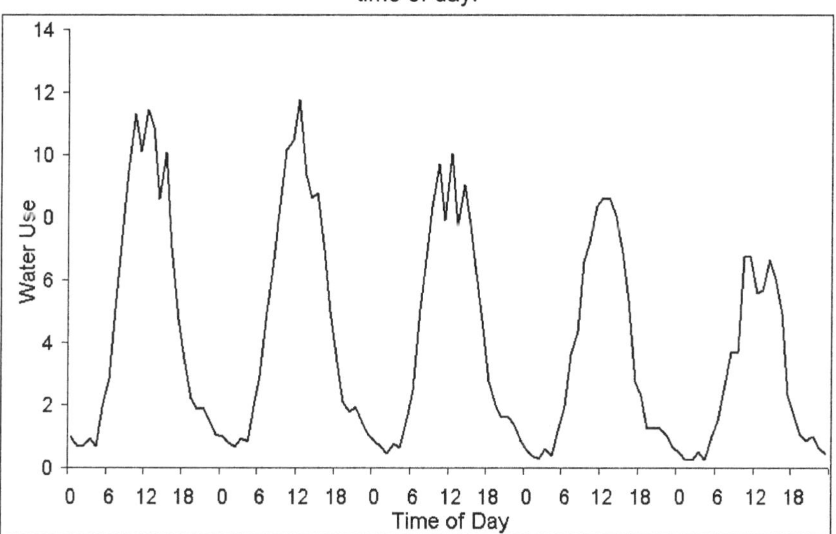

An example is illustrated in Fig. 8.2, which shows the water-use by a tree (measured continuously and recorded every hour over 5 days). If you could not record every hour, but only at a sample of points, which should they be? This depends on the detailed objectives, as well as on the practical constraints. How often is it feasible to record the water uptake by the tree? Can we record during the night? The actual trace in Fig. 8.2 shows a number of features that can be detected with varying effectiveness by using different time sampling schemes. For example, there is a decreasing trend in peak water-use over the five days, perhaps due to the soil drying out. If this trend is the main interest, then measurement should be concentrated around the middle of the day when water-use is highest. If the shape of the diurnal pattern is important, then we need to spread measurements out over a 24-hour time period. If the interest is in the high frequency variation from hour to hour, perhaps to relate it to changing weather, then it will be most effective to record during times when it is most evident. More decisions have to be made when the objectives involve the comparison of several trees, when we may have the option to measure all the trees at the same, or at different times.

It is interesting that the advantages of systematic sampling in time, rather than at randomly selected times, seem obvious to most investigators; though this is less obvious for systematic sampling in space. It is clear that when the process being measured changes smoothly with time, there is no point in taking observations too close together; a new observation will be almost identical to the previous one and therefore add little new information. Many processes measured over time will show some cyclical patterns (e.g. annual or daily, as in Fig. 8.2), and as with spatial sampling, we must consider this when choosing measurement points. If we want a good estimate of the overall average for the results given in Fig. 8.2, then measuring daily at midday is clearly flawed. However, regular measurement can give a good estimate: measuring every six hours starting at midnight, may give the desired results. Increasing the frequency would decrease the bias. Alternatively, if we already knew something about the shape of the diurnal variation, we could use that to produce a good estimate from just a few observations per day.

Continuous or very high frequency recording of observations is increasingly possible with automatic instruments and data loggers. However, having the technology to record observations continuously does not mean that this information will necessarily increase the quality of our results. It has already been pointed out that there is no reason to make frequent measurement of something that varies slowly. In the case of tree water-use, hourly measurements show variation that may be of interest, but measuring every minute or second may not be beneficial; particularly if we want to measure the overall mean, look at the trend on a scale of more than one day, or understand the diurnal variation. However, the possibility of high frequency recording does increase the range of phenomena that can be investigated, such as the short term response to a stimulus.

When the objective is to measure a trend over time, which is expected to have a relatively simple shape, then the advice on when to measure is much the same as that on levels of a treatment factor for the estimation of a response curve (Section 5.3). It is rarely helpful to have more than five time points; they should be closer together when the response is changing fastest and equally spaced in the absence of any other information.

As in other areas of research design, some common practices do not stand up to careful thought. In trials aimed at comparing the growth rates of different tree species, it was common to measure the trees every month for the first few years, and then after every year. The monthly measurement often turned out to be far too frequent to be useful, with roughly linear growth over many months. Hence, there was no more information in the series of measurements than in the first and last ones. A probing discussion of the objectives revealed that some were not explicit and imposed measurement time requirements. It had been hypothesized that some species would show greater differences between wet and dry season growth rates than others, but this would decrease with time. The requirement then was to measure trees before and after each season, not each month or year.

Most of the examples above integrate sampling in time with another sampling problem. For example, in the tree species trial, we have not only to choose measurement times but also choose which trees to measure. Many spatial measurement and sampling problems require repeated measurements in order to detect and describe change over time. In all these, the sampling scheme will depend on the objectives. Often there is a choice between repeatedly measuring the same spatial units, or choosing a new sampling point at each measurement time. For example, consider the problem of measuring soil P described in Section 8.1. If we wish to detect changes in soil P over time, should we take measurements in the same sample of locations on each occasion? The advantages of using the same locations are:

- An expected increase in precision of the estimate of change. Typically, an unusually high P spot will stay high and a low P spot will stay low. So we get the best estimate of change by keeping the same selection of high and low spots and measuring the change (that may be small relative to the spot to spot variation).
- A sampling scheme only has to be planned and mapped once.

However there are also some potential disadvantages:

- Any bias or deficiency, for whatever reason, in the initial sample remains for the rest of the study.
- There are often practical problems in identifying the same sample units on each occasion.
- Repeated measurement of the same units may introduce biases due to disturbance or familiarity.

These considerations parallel those in the repeated sampling of human subjects, in which designs vary from using the same small sample of people repeatedly (sometimes referred to as a panel study), to independent samples being taken on each occasion, to deliberately avoiding people measured in early samples because of the possibility that they will simply repeat earlier answers or otherwise be influenced by repeatedly meeting a researcher.

8.6 Special Methods

As in other areas covered by this book, a wide range of special sampling methods have been developed to cope with particular, or specialist problems. Any one of them could be the subject of a whole book. As usual, the best good practice advice is to look at how the problem has been tackled by others, look in specialist sources, and seek specialist advice. Here we mention three examples of sampling methods developed to cope with specific problems.

Think of the problem of monitoring the population of an important but rare forest tree species, such as *Ocotea usambarensis*. We need an efficient way of sampling the forest to locate, count, and measure the ocotea trees. We could apply the 'usual' method of defining a quadrat, choosing a simple random or stratified sample of quadrats, followed by counting the ocotea trees in each quadrat. This will give unbiased estimates of the numbers and sizes of ocotea. However, this method is inefficient. Since the species is rare, many of the quadrats will not contain any ocotea, but of course that is not known until we get to a quadrat and examine it. If we know that the distribution of the species in space tends to be clustered, an adaptive sampling scheme can be used. This starts with a random sample of quadrats, but when a quadrat with ocotea is found, additional adjacent quadrats are selected. This increases the chance of finding quadrats that contain ocotea. Of course the estimates of such things as number or density of the trees will have to allow for the adaptive strategy used. The way in which this is done will depend on the details of the adaptive sampling scheme.

As a second example, consider transect sampling. This is often used for estimating the numbers of a mobile population, for example of small birds. A quadrat approach is not effective for studying bird populations: by the time you have located a quadrat for sampling, the birds will have been disturbed and flown off. Instead, a transect is selected, which consists of a predefined route through the study area. As you travel along this route, each sighting of the species of interest is recorded. In using the data to estimate numbers of birds, some assumptions have to be made: for example, using the assumption that each sighting is of a different individual. We also have to make some assumptions about the area that the transect actually samples; how far off the route can a bird be detected? This will depend on, among other things, the nature of the vegetation and terrain, the behaviour of the birds and the skill of the observers. However, if we also record data on distance from the route at which sightings are made it is possible to estimate this 'distance function' and use it in compiling an estimate of the total population of the study area.

The third example is also used for studying animal populations in which is it possible or necessary to capture animals not just spot them such as fish, insects, or small nocturnal mammals. Suppose we set 20 traps and capture five animals. This tells us nothing about the population size, except that it is at least five. If we mark those five and release them, and at a later stage set 20 traps again, we can observe how many of the newly trapped animals are marked. If we have recaptured many of the same animals, we might suppose the population is rather small; if we always capture the same individuals, maybe there are no others. If most of those captured are not marked, it is evidence that the population is large. Again, use of the data requires assumptions; for example, that the population is closed or that marking does not affect the chance of capture. However, more sophisticated schemes can be found that allow some assumptions to be checked, and to modify the models and estimation procedures if necessary. This topic of 'capture-recapture' (or 'mark-recapture') is again one on which you can find books, software and experts!

Chapter 9

Planning Effective Experiments

9.1 Introduction

Farmer-participatory research trials have rapidly gained popularity in the past few years, with due consideration being given to the knowledge, problems and priorities of farming families. The move towards participatory on-farm research, means that many researchers such as breeders and agronomists, who have been trained in techniques of on-station research, are now under pressure to move on-farm. It is therefore not surprising, that the design of on-farm trials, often reflect or copy methods used in on-station experiments. Conditions for on-farm trials are typically less controlled than those for research institute fields, and this means that more consideration must be given to these designs. In addition, the range of experimental objectives is broadened to include such things as predicting adoption potential and learning from farmer innovations. The range of stakeholders with a direct interest in the trial is also increased, and care must be taken to ensure that the trial will fully meet their various objectives.

Experimental ideas are important in a wider range of NRM studies than those described as 'on-farm trials'. For example, a study investigating alternative methods of mobilizing collective action in resource management may use communities as the 'experimental units' and the alternative methods as the treatments to compare.

This chapter originated as a guide concentrating on aspects of the design and analysis of on-farm trials that are different from on-station research. It was concerned primarily with experiments where the farmer has considerable involvement and not situations where the farmer's only participation is by providing land. It did not include experiments that are really just demonstrations – the assumption is that you, as a researcher, have objectives of learning something. To this base we have added material illustrating how the ideas are applied to other resource management contexts.

We do not discuss the relative merits of different methods or reasons for farmer and stakeholder involvement in the research process, though each will have their own objectives. The critical points for sound design and analysis of a study are concerned with generating the information needed in the most efficient way; it does not depend on why the information is required, or by whom. If the objectives are acceptable, the next step is to design the trial.

9.2 Types of On-farm Experiments

Various ways of characterizing experiments with farmers have been described. For planning the trials, useful schemes are those related to the objectives and the extent to which farmers are directly involved in the design, management and assessment. For example, objectives related to biophysical performance of a technology often require a high degree of researcher control to minimize variation that may obscure results. On the other hand, if the objectives require farmers' assessments of alternative options, the farmers must be key players in designing and managing the trial.

In trials that are designed and managed by researchers, the trials are carried out on farmers' fields that are effectively 'borrowed' by the research team, becoming a temporary extension of the research institute. This type of trial is important because it provides well controlled conditions under which small scale (plot level) biophysical processes can be studied. Bringing the research institute in effect on to farms can broaden the range of soils, pests and diseases that are encountered, and encourages interaction with the farmers. The design of such trials is broadly the same as for on-station trials, so the information given in Chapter 5 is useful. However, there are still important questions of site selection that are often ignored in on-station trials.

From a statistical perspective, a key point in many NRM trials is that their design is often based on ideas that are normally associated with the design of both a survey and an experiment. Ideas of sampling and stratification are important when selecting sites, farms and plots. In addition, we collect data at different levels, such as the plot level (as in an experiment) and the farmer level (as in a survey). These concepts are often new to scientists who are used to on-station trials.

9.3 Specifying Objectives

The initial stimulus for organizing experiments on farmers' land was to broaden the range of validity of conclusions beyond the narrow confines of a research institute setting. This is still a valid reason for conducting on-farm trials, but it is now recognized that farmer participation is important, and that successful programmes must incorporate farmers' abilities to experiment and innovate.

As with any scientific investigation, it is crucial to clearly specify the objectives of the study. Time must be allowed for this phase and the objectives need to be re-assessed during the planning of the trial, to see whether they need to be revised. This is particularly challenging in on-farm trials, where researcher and farmer are now working together, often with extension staff and NGOs. It is important that the objectives are clearly identified from all perspectives.

Trials should be designed to resolve specific research questions, and researchers need to be impartial to the perception that donors expect to see the words 'on-farm' and 'participatory' before they will consider supplying funds. Usually a careful assessment of the gaps in the current knowledge will show that a series of initiatives is needed. These may include surveys and a number of trials, some on-station, and others that are on-farm, possibly some that are managed by the researcher, and others managed by the farmer.

There are obvious pressures to include multiple objectives in each study, and this makes sense for logistical and efficiency reasons, as well as for scientific ones. For example, if variations in biophysical and economic response to options are studied in the same trial, it may be possible to estimate the economic implications of biophysical effects.

However, the multiple objectives in the same trial need to be mutually consistent. Objectives relating to feasibility, adoptability and profitability of different technologies often imply different levels of farmer participation and hence may be better considered in separate studies.

Be aware of the possible confusion caused by having multiple studies going on in the same villages. Participants may not understand the reasons for the differing approaches in different experiments. For example, in one pair of studies the first experiment looked at pest resistance of new varieties, and the experimental design involves adding fertilizer to these new varieties. The farmer was provided with the inputs of seed and fertilizer. A second study aimed to assess the farmers' reaction to these same new varieties. In this second trial, no inputs other than the seed were provided. The result was confusion and jealousy in the village. Why did some farmers get fertilizer and not others? Farmer interaction between the two trials can affect the results (e.g. following the advice of fertilizer input in the first trial for the second trial, therefore skewing the results). These trials probably need to be relocated to avoid unintended interactions between them.

In defining the objectives of a trial, it is important to ensure that there remains some element of genuine research (i.e. some hypotheses to be tested). If the major objective is to encourage adoption of a new technology, then this may be important extension work but it is not research and so has different criteria for effective design. However, even the most development-orientated activity (including extension) often has a learning objective (i.e. some research); clarifying this objective will help ensure that the study meets it.

It is not adequate to state hypotheses as statistical null hypotheses. A hypothesis such as 'farmers do not vary in their views on the new varieties' may possibly be considered at some stage in statistical analysis, but is completely inadequate for planning the trial. The full hypothesis is unlikely to be stated in a simple sentence, but a more useful formulation of the hypothesis may be: 'Acceptability of the new varieties depends on wealth status of the farmers, because only the more wealthy can afford the higher levels of management needed by the new varieties'. The hypothesis suggests that the variation is by wealth status, which has two important consequences for design; we have to measure the wealth status of the farmers, and have to make sure the trial includes suitable variation in that wealth status. Giving a reason for the adoption of the new varieties adds value to the hypothesis, as it suggests action that can be taken if the hypothesis is confirmed.

While compiling the objectives and planning the design of the trial, it is also important to specify the domain to which you want the results to apply. This domain may be defined geographically (e.g. 'farms between 1000 m and 1500 m on the southern slopes of Mt Kenya'), but may also have other conditions such as 'households for which farming is the main source of income', or 'women farmers who already belong to an active women's group'.

9.4 Choice of Farms and Villages

The selection of farms with which to conduct trials must be closely related to the objectives of the research, and to the recommendation domain for which the results are intended. The ideas in Section 5.6 on site selection are relevant in this context. The large variation that generally exists between farms means they must be selected with care to ensure that the conditions are correct for the specific trial and that the conclusions will apply to the right farms – those determined by the problem domain and objectives of the trial. An initial survey is valuable in identifying how farms may be grouped, for example according to their socio-economic characteristics and environmental conditions. Decisions then have to be made whether research results will be relevant to all groupings or only to a subset of such farming groups. A representative (ideally random) sample of farms is then selected from the relevant group(s) of farms.

A multistage sampling scheme is often used, with village as the primary units and farming household as secondary units. Chapter 6 gives more guidance on this part of the design. The sample of farmers must be large enough for a valid analysis when split into different groups (e.g. by soil type, tenants and owners, access to credit). Where resources seriously limit the number of farms in the study, the objectives of the study may have to be re-examined. For example, for a new topic, the first year may become a pilot study from which ideas and objectives are refined for the following year's research.

Random sampling, though the ideal way to sample, can rarely be achieved in practice. A randomly selected farmer may have no interest in taking part in an experiment! More importantly, the participatory principle used in many projects means that trial farmers and households will always be volunteers in some way. Furthermore, there are good reasons for working primarily with groups, rather than individuals. In such cases, the groups are the experimental units. The aim is then to select groups objectively and to have sufficient groups for the objectives of the study.

In some studies, we still wish to work with farmers or households as the units, rather than the groups. Then it is important to work with the groups to explain why a 'representative' set of farmers should be involved in the trial. Whatever process of selecting farmers is used, ensure that sufficient farmers are included that meet the requirements of the objectives. Modify the objectives if necessary. It is sometimes useful to add some purposely selected farms, which might even be included under a different type of arrangement (e.g. a few farms added with a researcher designed trial to complement the self-selected farmers with farmer designed trials).

When selecting villages, consideration must be given to the length of time that a village remains associated with a research institute. Repeated work with the same villages, or work in the same village by different organizations is easy, but such villages may become less representative of the region.

When selecting farms, any restriction of the sampling scheme (e.g. including only the 'good' farmers), will restrict the recommendation domain in the same way. One justification for using 'good' farmers is that they set an example for their neighbours. Here the argument illustrates the problem of mixing the objectives between research and demonstration and perhaps not being clear in the research objectives and the design requirements to meet them.

The same considerations apply when the experimental units are larger units such as villages, watersheds, or forests, rather than farms and plots on farms; however, the constraints to applying them may be serious. For example, the objectives may require experimental units of upper watersheds of about 5 km^2. In reaching the stage of experimentation the project has been working in three such watersheds, and so it is natural to include these in the experiment, and not feasible to include any others. If these will not allow the objectives to be met (e.g. these watersheds are no longer 'typical' or representative of the recommendation domain, or they are too few) then the objectives must be reassessed. A common strategy is simply to continue and ignore (or hide) the limitations at the end of the study, or assume that some clever statistical analysis will remove them. This is foolish.

9.5 Choice of Treatments and Units

As with all other aspects of design, the choice of treatments and experimental units depends on the objectives of the study. This is considered in conjunction with the practical constraints and opportunities.

9.5.1 Choice of treatments

The same concepts of treatment structure are needed in participatory on-farm as in on-station trials:

- Treatments may be unstructured (e.g. genotypes).
- There may be a need for one or more control (or baseline) treatments.
- Factorial treatment structure remains important.
- The number and levels of quantitative factors have to be determined.

Below we concentrate on two points for which guidelines may be different for on-farm studies.

In participatory experiments, the farmers may choose some of the treatments themselves. For example, varieties or management options may be chosen from a village level nursery and demonstrations, or from open days at a research institute. This is sometimes done on a group basis, to arrive at a consensus for the trial. Alternatively, it may be done on an individual basis, with a design that then has some treatments that differ from farm to farm. It may also result in some farms having more treatments than others. This flexibility need not be suppressed for any 'statistical' reasons: the extent to which it is permitted will depend on the objectives of the trial. If the main objectives relate to the biophysical comparison of some fixed alternatives these obviously have to be present, though farmers can add further plots as they wish. Other objectives might imply more, or less freedom for individual farmers.

The second topic is that of the control treatments. These must be justified as treatments in their own right by the comparisons that have to be made. A control of 'usual practice' or 'current practice' is often called for. In a trial involving several farmers (or communities or sites), the current practice is likely to vary. Hence, the question arises as to whether a common 'control' is needed (perhaps reached through consensus, or as an average or typical practice), or whether the variation between farms is actually needed to meet the objectives. If the aim of the trial includes estimating the average change in moving from the current to a new practice, the appropriate baseline will be one that varies between farms. Other objectives will need a baseline defined in a consistent way across farms, and hence multiple objectives may need several different 'control' treatments. An added complication is that farmers' management of a 'usual practice' control, may actually be modified by the presence of the experiment, the experimenter and the 'new' treatments. This can only be confirmed through discussion with the farmer and observation on the remainder of the farm, or in neighbouring fields.

In participatory trials, farmers may wish to use the concept of 'controls' for their evaluation, in a way that is broader than researchers are familiar with. For example, in a soil fertility experiment, farmers may request fertilizer on the control, on the grounds that the new technology should be as effective as fertilizer. Similarly, the enhanced treatment may be planted on a poor part of the field, because it should bring the yield to the level of the rest of the field. Farmers often choose to use the usual or expected performance, determined qualitatively from their experience, as their basis for comparisons. The extent to which such suggestions are accepted, depends on the agreed objectives of the research. For some objectives anything is 'acceptable'; perhaps because we actually want to learn about farmer innovation. It is important to understand the comparisons farmers are making in order that researchers can interpret the results correctly. Hence, the measurements have to include information from farmers on why they put each treatment where they did, the comparisons they are making to assess the treatments, and the reasons why these comparisons are important.

9.5.2 How many treatments?

There are no rules as to the number of treatments used. The minimum sometimes appears to be one; for example, when a single new technology is introduced and is compared to the performance of a previous technology, or when farmers' expectation of a new technology is of interest. Here there is an implicit second treatment of 'usual practice', though this is only implemented mentally by the farmer. Normally there are at least two treatments being compared.

There is no maximum number of the treatments that can be applied, and we certainly do not agree with the frequent statement that four treatments is a maximum. This statement may be related to the general view that many on-station trials have between eight to ten treatments, and that participatory on-farm experiments should be simpler than on-station trials. There may be cases when simplicity is needed, but some effective participatory research involving trials with many more treatments has been done, for example by participatory breeding programmes.

Having many treatments per farm usually implies complexity of the design and this may lead to partial failure of the trial. Many resource management studies involve relatively large parts of a farm or field. For example, agroforestry based soil fertility options may need plots of 1000 m^2 or more to realistically assess (compared to maybe 10 m^2 for a new crop variety), and farmers are unlikely to risk many plots of this size in an experiment. As a second example, a multifactorial experiment that aims to study interactions between say four management options, is often not well suited for an on-farm trial because confusion and difficulty following the protocol can lead to data that is hard to interpret. Where there is no simple solution, the design team should reassess the objectives. They could also consider splitting a complex study into simpler related experiments that differ in their level of farmer participation.

There are two distinct numbers to consider. The first is the total number of treatments being compared in the trial; the second is the number being compared on each farm. These do not have to be the same, as is explained below.

9.5.3 How many treatments per farm?

Questions which experimenters often ask include, 'What do I do when there are more treatments to be investigated than there are plots in each farm?' and 'What if some farms have more plots available than others?' These are practical realities in on-farm experimentation. The trials should not be restricted only to those farmers that have a certain amount of land available to set aside to trials, nor should the number of treatments be reduced due to the land size available on the smallest farm. Care in the allocation of treatments within farms at the design stage, can ensure a successful experiment is carried out in such circumstances.

The ideas of incomplete blocks that we looked at in Section 5.7, become relevant. For example, if there are five treatments to compare, but only two or three can be carried out on each farm, a good solution may be to make sure that the pairs and triples of treatments are chosen so that each pairwise comparison occurs equally often on the same farm. In the plan in Table 9.1, the five treatments are A, B, C, D and E. Two farms can each accommodate three treatments, the other four farms can accommodate only two treatments. There are unequal numbers of replicates, but each treatment pair occurs once on a farm. In real studies, such a neat solution may not be feasible, but arrangements that come close can always be found.

Table 9.1. Allocation of treatments to farms.

	Treatment				
Farm	A	B	C	D	E
1	✓	✓	✓		
2			✓	✓	✓
3	✓			✓	
4	✓				✓
5		✓		✓	
6		✓			✓

The intended methods of measurement may also influence the choice of treatments on each farm. If each is to be compared with a control (e.g. on a scale 'worse', 'similar', 'better') then the control should be included on every farm.

9.5.4 Replication and resources

The key reasons for replicating in experiments are the same whatever the nature of the trial. They are to be able to estimate the precision of treatment comparisons, and to increase the precision of these comparisons. So the rules for determining the number of replicates should be straightforward. Just work out how many are needed for the required precision (as described in Section 16.9.5). Unfortunately, the real world of on-farm and other NRM experiments is rarely that simple.

The first difficulty applies to most experiments. Calculating the replication needed depends on knowing the level of random or uncontrolled variation that can be expected. However with on-farm trials there is often little relevant previous experience on which to base guesses of expected variation. Some on-station trials are done year after year under very similar conditions yet most on-farm trials are novel in many ways.

The second difficulty arises from the layout hierarchy typical in on-farm trials. We rarely simply have plots, perhaps arranged into homogeneous blocks. Instead there will be a clear hierarchy of the research material, with plots to which treatments are applied, then maybe fields, farms, villages or watersheds or landscape positions, then district or perhaps collaborative partner. At which level do we need replication?

The simplistic argument is this. The treatments are compared on different plots, so it is random variation between plots that is important in estimating and controlling the precision of treatment comparisons. Hence we must make sure we have sufficient plot-to-plot replication. Within-plot 'replication' (or sub-sampling) is irrelevant, and it should not matter where the replicates fall in terms of the hierarchy (replicates in one village are as good as replicates in another).

This argument oversimplifies the situation for several reasons:

1. A further reason for replicating, beyond estimating and controlling precision, is to increase the 'support' for conclusions by conducting the study over a wider range of conditions. This, combined with practical problems of sampling and managing the study, is one of the reasons for the hierarchical layout (see Section 6.2.3). It therefore makes sense to include sufficient replication to 'cover' the domain of interest. A recurring difficulty in many research and development projects is the 'scaling-up' of results obtained in limited areas or conditions to a larger domain of interest. Some of these difficulties can be avoided by covering the larger domain in initial studies.
2. Within a study, we often need to make comparisons at different levels in the hierarchy. For example, an on-farm study of soil management for steeply sloping land required the comparison of different management options at the plot level in an experiment. It also required comparison of farmer attitudes at different positions on the hill (the lower gentle slopes, steep middle slopes and flatter rocky hill tops). Hence, the study not only needs sufficient replication of plots, but also replication of farms for each stratum of the hillside. This is similar to the split plot experiment introduced in Chapter 5, where we have treatments at two levels and need to ensure replication at each level.
3. Most on-farm experimenters are interested in the interaction between treatments and other factors, and this has replication implications.

The last point is probably most important for a number of reasons. Information and recommendations concerning other options (e.g. soil management) often depend on other factors. Soil management is studied at different landscape positions because it seems likely that different options will be appropriate in these different situations. There is a hypothesis of management-by-position interaction. We therefore need sufficient replication to get the required precision not just for the treatment main effect (difference between two management options averaged across all landscape positions), but also for comparing two treatments within a landscape position, and comparing the same two treatments on two different landscape positions. The latter is a between-farm comparison. The variation at this level is likely to be high, and hence a relatively large number of replicates are required.

It is commonly stated that on-farm experiments are highly variable, with large differences in the size of treatment effects on different farms. This is true if farms are considered as a single population. However, much of this variation can usually be explained as interaction between treatments and other factors. The farms are not homogeneous but different in many social, economic and bio-physical aspects. The high farm-to-farm variation in treatment effect is simply a reflection of these other factors. Designs that allow this to be identified and estimated are the most useful on-farm trials. A conclusion of the type 'Option A is on average more productive than option B, but there is a lot of variation' is very weak compared with a conclusion such as 'On steep or eroding slopes A is the better option. On shallow rocky soil B is better. In the valley bottom there is no difference between A or B, but C is useful if farmers have sufficient labour'. Some of these interactions will be hypothesized before the trial starts. Others will become apparent during the trial or at the analysis. Designing with this in mind usually means that the number of replicates will increase.

One of the misleading messages of some statistics books (and some statisticians) is that farm-by-treatment interaction can only be detected if there are within-farm replicates, that is, several plots of each treatment on each farm. The background to this recommendation comes from considering the standard analysis of variance (which we looked at briefly in Section 5.10). Suppose we have 20 farms and three treatments, with a single replicate on each farm. The 'usual' ANOVA table is shown in Table 9.2.

Table 9.2. Outline of the usual ANOVA table.

Source	DF
Farms	19
Treatments	2
Farm x treatment	38
Total	59

The farm-by-treatment variation is used as the 'error' when assessing treatment differences; there is no error term for assessing the farm-by-treatment interaction itself. Some within-farm replication would add a within-farm error term that could be used when looking at the farm-by-treatment interaction.

However this analysis is based on a misunderstanding. Rarely is the experimenter (or farmer) interested in whether option A is 'significantly better' than B on James Kariuki's farm, but B is better than A on Stella Karanja's farm. Interest is rather on whether A is better than B on farms of one type (perhaps based on soil type, gender of farmer or landscape position) with B being superior on farms of a different type. This requires showing that there is consistency in the response across farms of the same type, and within-farm variation is irrelevant. If there are two types, then the ANOVA would be as shown in Table 9.3 below.

This form of analysis requires sufficient replication of each type of farm, and in general, a study needs sufficient numbers of farms to be able to look at the interaction between treatment and several farm-related factors.

If a good reason for estimating within-farm variation exists, then it is not necessary to replicate all treatments on all farms. For example, in the above problem an average within-farm variance could be estimated by replicating just one treatment on each farm.

Table 9.3. Outline of the more useful ANOVA table.

Source	DF
Type	1
Farms within type	18
Treatments	2
Type x treatment	2
Remaining farm x treatment	36
Total	59

9.5.5 Plot size

It is often assumed that the plot size should be larger for on-farm than for on-station trials. This is the result of on-farm trials in the past usually having been at the validation stage of the research within a 'technology transfer' model. There is no justification on statistical grounds for preferring large plots in on-farm trials. The most efficient use of a given area is normally achieved with more small plots, rather than fewer larger plots. This is provided that: (i) the plot is large enough to respond to the treatments in a realistic way; and (ii) lateral interference does not mean that large border areas have to be left between plots. This latter point can be important for example, in areas with nutrient or water movement between plots, or where plots have trees with extensive root systems. Often there is a balance between the preference of farmer and researcher for larger plots, on the basis of realism, ease of treatment application, and the statistical benefit of improved precision from more, smaller plots.

The cases for realism when seeking farmer opinion regarding treatments, or when comparing treatments with regard to labour requirements, are examples of compelling reasons for using large plots. However, the case for large plots should be made in relation to the objectives of the experiment; merely stating that on-farm experiments require large plots is not enough.

When the experimental units are not plots in a field, there are other considerations for determining size, but the same general rules apply. The units must be large enough to:

- Respond realistically to the treatments.
- Be measured or assessed realistically.
- Not be unduly influenced by unintended external factors such as movement across the plot boundary.

9.5.6 Plot layout

Layout of plots within each farm will be guided mainly by perceived or known variation in that farm. This variation can be characterized by looking at the existing vegetation and management, by rapid soil assessment or from topography. However, the best source of information concerning heterogeneity is often the farmers' knowledge. Their understanding of the variation in their fields should be used to determine the location of the plots and any blocking scheme, and perhaps to avoid using particular patches of the field.

It is important to ensure that farmers and researchers are using the same criteria to define suitable locations. Researchers normally strive for homogeneity, while farmers may have particular parts of their field where they would like to try some treatments. For example, they may feel that addition of crop residues is most appropriate on degraded patches. Where large sections of the field are degraded, this can be accommodated within the design by putting all treatments on this type of land. Otherwise, the extent to which farmers should influence layout will depend on the objectives of the trial. If farmers' opinions are of paramount importance, then the loss of randomness in the allocation of treatments to plots is of minor concern. The important sampling is at a higher level, namely in the choice of farmers. On the other hand, if a comparison of yields is an important part of the trial, then it is important to allocate treatments 'fairly' (i.e. with some element of randomness) to the plots. In such a case, use of the degraded patches could be in addition to a replicate of the treatment on ordinary parts of the field.

Many practical considerations need to be taken into account when considering block and plot layout. In an on-station trial, for instance, a split-plot experiment may be carried out because it is convenient to plant one large area at one time, while the application of different levels of fertilizer can more easily be added to smaller areas. In on-farm trials, these considerations may still apply. Another important practical aspect is the interview process. For instance if a farmer is to give an assessment of different varieties given that fertility is a secondary factor, then it may be convenient if the varieties are grouped together.

If the whole farm is the experimental unit then the design problems and questions to consider are just the same, but at a level higher in the layout hierarchy. Thus, we look for groups of farms that are homogeneous and might form a block, with different treatments allocated to farms within that block. The problem is just the same if there is only one experimental unit per farm, even if this does not encompass the whole farm. For example, the unit may be a fish-pond or an individual animal (e.g. a cow), of which any farmer has at most one, or the unit may be an orchard or woodlot that cannot be subdivided for the experiment.

Crossover designs are useful in animal and human experiments in which there is high subject-to-subject variability. They may also be used when only a single unit per farm is available. The idea is that several treatments (such as dietary supplements) are compared in succession on the same subject. The experimental unit becomes a 'subject for a fixed period of time'.

In principle, the same idea may be useful in NRM studies that have limited numbers of large units, such as villages, watersheds, or forests. However, it is hard to think of examples on this scale in which we can measure the effect of treatments in a short time and where the effect of a treatment in one period does not interfere with assessment of the next. This interference is called a 'carry-over effect'. It is the effect that an earlier experimental treatment has on the response to a later one. Carry-over can be adjusted for if it is simple, but that is unlikely to be the case with the large scale units in NRM studies.

9.6 Measurements

In participatory trials, we can distinguish between three types of measurement:

1. Primary response variables, determined by the objectives of the trial. Often these are biophysical responses that show production and environmental effects.
2. Measurements of characterization variables that 'set the scene' for the research, and are used to relate the results from the study to the whole domain of interest.
3. Measurements taken to help understand the variation in responses. These may be planned before the start of the trial or may be collected as a result of observations made during the trial.

Any of these measurements can be taken at various levels in the hierarchy (e.g. plot, farm or higher) and any may involve measurement instruments such as rules and balances, or questionnaires and discussions.

In on-farm trials, there is often too great an emphasis on plot level measurement or biophysical responses, because the implicit assumption is that the methods of analysis will be the same as for on-station trials. While these data may still be of interest, we suggest that more attention be given to the collection of measurement types 2 and 3 above. The main reason for devoting time to the concomitant information is that we need to try to understand the causes of as much of the variability as possible. In on-station trials the plots may be smaller, and are likely to be more homogeneous. In on-farm trials there may be more variation within a farm than would be found on-station, and there will also be variation between farms. In on-station trials there is a consistent management structure, whereas in on-farm trials there can be large differences in management practice between the farms. As a general guide, if a factor is not controlled then it should be measured, both at the plot level and at the farm level; this is the case if it is of direct interest or if it might explain some of the variation in the data.

In general, the objectives of the trial determine what is to be measured. Thus, the direct and concomitant measurements to be taken are normally decided at the planning stage. Often too much data is collected that is never analysed. Our encouragement to measure potential concomitant variables is not intended as support for the measurement of all possible data, just in case they may be useful.

In some trials, where farmers have chosen where they will apply particular treatments, there may be little reason for measuring yields. What is needed instead, are the farmers' reasons for choosing particular plots and their reactions at the end of the season. In less extreme situations there may still be little reason to devote much time to the detailed measurements of yield components. A quick assessment of yields using 'number of bundles', plus some idea of harvest index, will often be sufficient.

9.7 Other Considerations

In these chapters on the design of a trial, we have emphasized points of good statistical practice. There is however much more to a good design than statistical issues, and some are mentioned in this section. The points below apply to all the designs, but particularly to those in this chapter.

In Chapter 4, we explained the importance of a protocol for each activity, and included a checklist for producing one in Appendix 1. The protocol should include information on management as well as on statistical aspects. The protocol should be amended as the design is revised, both during the planning phase and during implementation.

Sometimes practical or ethical issues may appear to be in conflict with aspects of good statistical practice. Such problems are sometimes imaginary, and can result from a misunderstanding of the statistical principles. When they are real, then the practical aspects should win. It is crucial that the design is one for which all stakeholders are comfortable in collecting the data. For example, in Section 9.4 we mentioned some of the practical problems of choosing farmers at random. Insisting on randomness may cause jealousies and other problems in the village that far outweigh any possible value added to the research.

In long term studies that last for several seasons, you may find some defects in the design, or some possible improvements after the first few seasons. Extreme responses are either to refuse change (i.e. on the grounds that the study must be repeated for multiple seasons), or to change so much that the study is worthless. It may be possible to plan for changes in the design by building some redundancy into the first year. For example, an on-farm trial could start with five plots and four treatments, with a key treatment applied to two of the plots in the first season. In the second season, this fifth plot is used for a new treatment chosen by the farmer.

Unexpected design faults or shortcomings should be rectified, though this should be done with care. It is important that each phase of the trial is as informative as possible, but that some aspects remain constant every year so that comparisons can be made. If a revised design is necessary, the reasons for the changes must be noted in the research protocol and the objectives may also need to be modified.

Part 3: Data Management

Chapter 10

Data Management Issues and Problems

10.1 Introduction

Research projects usually involve the collection of a large volume of data. The data have then to be processed and analysed, with results and summaries being prepared for publication and for incorporation into the next phase or activity of the project. For this sequence to proceed smoothly, the project requires a well-defined system of data management.

The main stages of the data management process in a research project are:

- The raw data are entered into the computer, and checked.
- The data are then organized into an appropriate form for analysis, often in several different ways, depending on the analysis.
- The data are, or should be, archived, so that they remain available throughout subsequent phases of a project, and for use at a later time.

Good data management is not something that will look after itself, or evolve if left long enough; projects and institutions need to have a well-defined data management strategy. In this part of the book we consider both data management practice and strategy. We look at the role of spreadsheets for data entry and management (Chapter 11), the advantages of using a database management package (Chapter 12), and some reasons for having an effective data management strategy (Chapter 13). We start here, in this chapter, by giving some guidelines on the components of a data management system.

It is useful to compare research and financial data management. Most organizations use a financial accounting system to keep track of money. Their financial systems are designed to ensure, among other things, that:

- Each transaction is correct and legitimate (i.e. quality is high).
- Records are kept so the information can be found and understood, long after the transactions were completed (documenting, archiving and auditing).
- Invoices, payments, summary accounts and so on, can be prepared quickly and accurately (efficient data processing).

Accountants are trained for years to become proficient at financial data management. In contrast, researchers are usually expected to do a similar job with no training at all.

The following typical scenario illustrates the need for attention to data management. An organization runs a research project that involves several researchers. The accumulated data are entered and stored in several spreadsheet files. Every researcher takes copies of some, but not all of these files. When they notice errors in the data during analysis, they edit their own copy of the data, with the intention of correcting the master copy later on. Of course, the master copy is never updated. In general meetings, and during breaks, the project team often discusses archiving the data when they have time. Unfortunately, no one is given the task of archiving the project materials and no time or funds are available for this activity, so it is never done. At the end of the project, the report is written and distributed, but nothing further is done with the data. The researchers move on to other projects.

Two years later, a protocol is prepared for a follow-up project. Staff now have to spend many weeks trying to put together the information from the project, including working out which are the most up-to-date files. Some files cannot be found, while others cannot be fully understood. It is obvious how much better and easier this task would have been, had the data been properly managed in the first place. A few minutes given to archiving throughout the project and perhaps a day at the end to archive all the materials would have saved many weeks of work later.

10.2 What We Mean by 'Data'

At the simplest level, data are the values recorded in the field books, record books or data-logging devices, which are to be entered into the computer and then analysed. An example of a simple dataset – a table of rows and columns – is shown in Table 10.1.

Table 10.1. A simple dataset.

Plot	Replicate	Treatment	Flower	Total weight	Head weight	Grain weight
101	1	4	26	25.2	6.6	1.7
102	1	2	28	32.7	8.8	2.4
...	...	...	...	...	...	...
...	...	...	...	...	...	...
416	4	8	26	19.7	4.9	5.3

The information in this table is certainly needed for the analysis, but it is incomplete. Additional information in the protocol, which gives details of, for example, the treatments, the type of design, the field plan and the units used for measurements, is also needed for both the analysis and the archive. Such information is sometimes called metadata, but whatever name is used, it should be considered as an integral part of – and just as important as – the data in the table.

We are now in a multimedia world so photographs and maps can be considered as part of the data, as can reports, talks and other presentational material. However, in this book we concentrate on the problem of managing raw field observations and the associated documentation.

Roughly, one can regard the task of managing raw data in a research activity as *simple* if all the data to be computerized have been collected on a single type of unit, e.g. plots or animals. The task is *complex* where data have been collected from a number of different units or levels. For example, in an on-farm study there will often be interview data at the farm level and response measurements at the plot, animal or tree level.

Table 10.2a. Data on each site.

Site number	Site name	Country	Latitude	Longitude	Altitude (metres)	Soil type	...
1	Dori	Benin	10.654	2.813	200	C	...
2	Gaya	Niger	12.109	4.171	175	D	...
...	...	...	...	...	...	...	...
46	Mara	Niger	12.576	2.543	140	D	...

Table 10.2b. Data on each experiment at a site.

Site number	Experiment number	Year	Planting date	Stress	Pest problem	...
1	1	1997	12 June	mild	minor	...
1	2	1997	16 June	none	none	...
1	3	1998	2 July	none	none	...
2	1	1997	19 June	severe	major	...
...	...	...	...	...	...	...

Table 10.2c. Data on each variety used in the project.

Variety code	Variety name	Origin	Type	...
12	OFT1226	Mali	erect	...
14	PLO2279	Togo	spreading	...
...	...	...	...	...

Table 10.2d. Yield data from each of the sites.

Site number	Experiment number	Variety code	Yield	...	...
1	1	6	4.1	...	...
1	1	14	2.9	...	...
...	...	...	...	...	...

The complexity of data management will vary across different parts of a project. An example is a regional project, which includes a variety trial at each site, where the data are to be entered into a computer at the location where they are collected. In such a project, the set of varieties is often not identical at all sites. Then the data entry at each site is simple, the data forming a single rectangle, as in the example above. However, the regional coordinating office might need further information. Table 10.2 shows such an example.

The coordinating office could now use a relational database management system (DBMS) to combine the information from the different data tables and so provide overall management of the data across sites.

Where the data management tasks are complex, a database management package is often used. This enables all the information to be stored in a structured way. In Chapter 12, we compare the use of a spreadsheet and a database package for managing research data.

10.3 Software for Handling Data

The different types of software used for data management include the following:

- Database (DBMS) packages, e.g. ACCESS, ORACLE
- Statistics packages, e.g. GENSTAT, MSTAT, SAS, SPSS
- Spreadsheet packages, e.g. EXCEL, LOTUS-123
- Word processors, e.g. WORD, WORDPERFECT; or text editors, e.g. EDIT.

Database, statistics and spreadsheet packages have overlapping facilities for data management. All handle the 'rectangles' of data, shown in the previous section. In these rectangles, each row refers to a case or record, such as a household, an animal or a plot, while each column refers to a measurement or variable, such as the treatment code or the yield. Broadly, database packages are very good at manipulating (e.g. sorting, selecting, counting) many records or rows. They are also able to handle hierarchical data structures, such as observational data collected at both a farm and a field (crop) level, where farmers have more than one field. Statistics packages are designed primarily to process the measurements, i.e. they have powerful tools for operating on the values within the variables or columns of data. Spreadsheets do something of everything – though with limitations.

In general:

- Transfer of data between packages is now simple enough that the same package need not be used for all stages of the work.
- The data entry task should conceptually be separated from the task of analysis. This will help in deciding what software is needed for data keying, for checking purposes, for managing the data archive and for the analysis.
- Database management software should be used more than at present. Many research projects involve data management tasks that are sufficiently complex to justify the use of a relational database package such as ACCESS.

- Spreadsheet packages are ostensibly the simplest type of package to use. They are often chosen automatically for data entry because they are readily available, familiar to many researchers and flexible; but their very flexibility means that they permit poor data entry and management. They should therefore be used with great care. Users should apply the same rigour and discipline that is required when working with more structured data entry software.
- More consideration should be given to alternative software for data entry. Until recently, the alternatives have been harder to learn than spreadsheets, but this is changing.
- If a package with no special facilities for data checking is used for the data entry, a clear specification should be made of how the data checking will be done.
- A statistics package, in addition to a spreadsheet should normally be used for the analysis. We consider this issue further in Chapter 14.

10.4 Database Structure

As noted in Section 10.2, the task of data management may be simple or complex. In database terms, this distinction corresponds to whether the data tables are flat or structured (i.e. linked together in various ways). The database structure is flat if all the data exist at a single level and can be held in one table or rectangle. Familiar examples are an address list or a list of references.

Research projects usually require several linked tables to hold all the data. For instance, a regional study may produce a flat file for storing information about each site, such as the average rainfall, maximum temperature, and location of the site. Here the rows in the data file are the sites, while the columns provide different pieces of information about each site (as shown in the example in Section 10.2).

A second flat file is used to store the information on each unit within the site. These units may be the focus groups in a participatory study, the households in a survey or the animals or plots in an experiment. Rows of this file correspond to a particular unit, while columns give information about each unit such as the plot yields or the views of a focus group about a technology.

Many studies include repeat sampling of the same unit. In such cases, another flat file is needed to store the repeated information for the units. This 'repeatedness' could relate, for example, to several samples taken within the same plot, or to measurements made on the same unit but on different occasions. In our flat file a row now corresponds to a single sample or date for a unit. The first two columns in the table would usually be an identification code for the unit and the sample or date, while other columns would hold the measurements, e.g. moisture content of soil samples taken from crop plots, or milk yield collected over a period of time from one-cow farms.

When these three flat files are considered together, they form a hierarchical structure, illustrated in Fig. 10.1.

Fig. 10.1. Example showing how files are related.

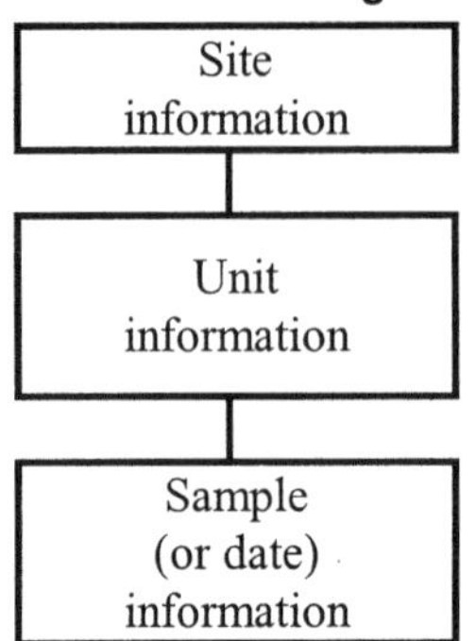

One record exists for each site.
SITE is the unique identifier.

One record exists for each unit within the site and includes a location reference for the unit.
SITE + UNIT makes up the unique identifier.

One record exists for each sample. SITE + UNIT + SAMPLE (or SITE + UNIT + DATE) makes up the unique identifier.

In this case, the site information is at the top level of this structure, the unit information is at the next level, while information collected for each sample is at the lowest level of the hierarchy. The information at different levels of the hierarchy is linked by means of *key variables* (or key fields). The key variable is a unique field or combination of fields that can be used to identify a particular record.

The key field values at one level of the structure link a record to the record (or records) at another level with the same values in the corresponding key fields. These links or relationships between the database tables define an essential aspect of the database structure. The facility to store the database structure is what makes a DBMS important for researchers. It is considered in more detail in Section 12.3.

10.5 Qualitative Data

Qualitative data may seem harder to organize than data from other types of study. We may have a transcript from each focus group, a village map, or a scored causal diagram. How are these to be managed and perhaps prepared for the analysis?

We begin with some parallels to show that all activities may have the same types of complexity in their data. Take transcripts of 30 focus-group discussions that last for one hour and compare them with 30 soil cores, each of one metre depth. From the cores, we wish to understand the chemical and physical properties of the soil, at each different depth, in relation to the growth and yield of the crops. From the transcripts we wish to understand different problems of obtaining credit within a region of the country. We may be interested, for instance, to know why some villagers can join credit groups and others cannot, and ultimately what the different problems are in relation to obtaining credit.

As another parallel, we can think of a village map or a scored causal diagram (which is another sort of map), as being like a photograph of a quadrat made by an ecologist in a study of plant diversity.

Where the data are complex, we must differentiate between the input of the raw data, and the coding of those aspects of the data that can then be analysed. For example, the maps or photos may simply be scanned, and the same can apply to hand-written transcripts. The coding is a much harder process.

In some studies, there is a small amount of qualitative information. A sample survey will often have a few open-ended questions. Commonly a survey ends with a question such as 'Please add any further points that you feel may help us to improve our service'. In such cases, often only a few respondents write anything, but those who do can provide very helpful insights.

The way qualitative data are entered will partly depend on how they are to be analysed. For example, if you intend simply to quote examples of interesting comments, then the hand-written texts can be scanned (as photos) and perhaps an extra field added to the data in the data file to indicate that this questionnaire included a text response. Alternatively, the text responses could be typed as part of the data. This could help if you wish to use software to extract codes to report on the frequency of different suggestions.

One difference between the soil core and the focus-group discussion is often the proportion of the time for the analysis that is to be devoted to the within-unit component. The soil scientist will have measurements of chemical and physical properties of the soil from different depths in the 30 cores, but the data are primarily used as a sample from the whole area of study. So the analysis is mainly at what is called the between-core level, and hence the within-core information can be dealt with quickly.

In contrast, many qualitative researchers are prepared to spend considerable time on the within-unit analysis, because their main objectives may correspond to an understanding of why some villagers can join credit groups, and others cannot. In such cases, the data will have to be typed, or scanned so they can be read by software. There are many packages that support inspection for patterns in the texts. These patterns can form the basis for a coding process, and the resulting codes may then be analysed. Pointers to relevant software packages can be found on the website caqdas.soc.surrey.ac.uk (Computer Assisted Qualitative Data Analysis Software – a networking project at the University of Surrey).

Often the richness of within-unit information in qualitative data means that the coding process becomes part of the analysis. We give an example in Section 15.2. This means that the coding must not be routine, and left to a computer, or junior staff. In such cases, the data management task (as opposed to the analysis) is primarily to organize the original data. As with other data, these transcripts, or maps etc., would eventually be made available, allowing others to examine the coding process and suggest alternatives.

The increased use of participatory approaches for research is generating a considerable amount of qualitative data. Although Participatory Rural Appraisal/ Participatory Learning and Action (PRA/PLA) tools can help in extracting numerical information, the use of these tools also produces a range of non-numerical outputs such as pictures, maps, diagrams and notes. One way to organize this type of information is through the use of a simple web page that allows browsing of computer files that contain the qualitative data. These files can be pictures (often in jpeg or gif format) from scanning notes, drawings or pictures produced in the field, photographs, typed transcriptions of discussions, or any other file that the researcher might decide to use to store the information. The web page does not need to be accessible on the Internet but can sit locally on the researcher's computer.

Data management for this type of information means providing easy access, secure storage, and a complete set of relevant documents. This is more concerned with sensible archiving of the information than with processing the data. The gain for the researcher comes from having a live archive, which can then be used in the research process to bring the non-numerical information into the analysis. The processing of this type of information is further discussed in Section 15.2.

Some information from the study on sustainable agriculture in Malawi (see Sections 15.1 and 15.2) is available from the website www.ssc.rdg.ac.uk The relevant web page contains links to two types of document:

- PRA documents: Facilitators' transcripts of the results of PRA discussions in 30 villages including resource maps, institutional diagrams, transects, trend analysis and pairwise matrices.
- Debriefing documents: Standard forms that were used by the research team to systematize the information after each PRA activity.

This web-page-based archive allows the user of the information easy access to all the documents for any village at the click of a mouse.

10.6 Designing a Data Entry System

In designing a suitable system for data entry, consideration must be given to the structure and types of data, and the measurement units. These aspects apply generally, even to data that have already been captured electronically from a measuring instrument, and so do not need re-entering on a computer keyboard.

10.6.1 Understand the structure of the data

Few projects generate simple data; most have a complex structure with several 'flat files' or data tables that must be linked in a clearly defined way, as outlined in the Section 10.4. It is essential that both the flat file components and the links are fully specified to ensure that the information meets the database requirements of completeness, integrity and minimum redundancy or duplication of information. Modern, relational database software makes this task fairly easy. Spreadsheet software does not. While there are facilities in spreadsheets for linking tables they are not easy to use and do not contain the built-in checks found within database systems.

10.6.2 Identify the types of information being collected

Try to foresee the full range of different types of data that will be collected, e.g. plot data may consist of crop yield from all plants in the plot, number of plants with pods for harvest, total pod weight and number of dead plants. Build facilities in the data collection sheet for recording all such information. Often data will be collected from the same plot on a number of sampling occasions. Dates of such records must be kept, with space available on the recording sheet for notes about the plot or farm at that specific time. Such secondary information will be valuable later, at the data analysis stage, to understand any curious aspects of the data.

10.6.3 Specify the measurement units and precision

Ensure that the system includes the units of measurement used for all quantitative variables. Changes in measuring instruments, or in field and research staff, or methods of data collection, may bring about changes in measurement units. Consideration must be given at an early stage of the database design to allow for such changes to be incorporated into the data recording system.

Specify clearly the precision (number of decimal places) to which all measurements are to be recorded. The number of significant digits should match the real precision of the measuring instruments or recording devices.

10.7 Data Entry and Checking

We consider the data that are collected in field books or survey sheets. First, we discuss the overall strategies that can be adopted for data keying and checking, and then give separate guidelines on these two aspects.

10.7.1 Strategy for data entry and checking

When planning a strategy for data entry, distinguish clearly between the activities concerned with data entry, data checking, data management and those of data analysis. The ultimate aim should be a fully documented archive of checked, correct, reliable data that can be subjected to scientific scrutiny without raising any doubts in the minds of subsequent researchers.

The process of data entry will often involve a skilled person who designs the system, while other more junior staff, e.g. trained data entry operators or field staff, carry out the actual keying. Checking is done both at the time of keying and afterwards. If the project is small, then the same person may plan the system, key the data and do the checking; but it is still useful to have a clear strategy for the activities.

When planning the system, aim to make the data entry stage as simple as possible. For example, in a replicated experiment it should never be necessary to type variety names or long treatment codes for each plot. A single letter or number is usually sufficient. The data entry system can either insert the full code, or the full names may be available in a separate look-up file, as illustrated in Table 10.2. Simplifying the keying process will speed the task, make it less tedious and less error-prone. Ideally, the data entry screen will closely resemble the paper form from which the data are being copied. The field data collection forms should be designed with the data entry in mind.

The logical checks (such as making sure the sum of the weight of plant parts is close to the total plant weight, or that the maximum and minimum values are reasonable) are best done by trained staff who understand the nature of the data. Usually this phase involves preliminary analyses, plotting and so on.

The data entry and checking steps are best designed at the same time. The way the data checking is undertaken will, however, depend on who is entering the data. Non-skilled staff should be expected to key exactly what they see on the data sheets or field books, and the logical checks (e.g. checks to rule out pregnant males, or to test whether the minimum temperature is greater than the maximum) should be done by scientifically-trained staff after the (double) entry is complete. In that way, reasoned and consistent decisions can be made about what to do. If scientists are keying the data themselves, then the entry and full data checking can proceed together.

10.7.2 Guidelines for data entry

The recommendations may be summarized as 'Make sure that the data are entered promptly, simply and completely'.

- The data should be entered in their raw form, i.e. directly from the original recording sheets or field notebooks, whenever possible. They are therefore entered in the same order that they were collected. The copying out or transcription of data prior to keying should be kept to an absolute minimum.
- All the data should be entered. Entering just the important variables, so they can be analysed quickly, limits the possibilities for checking, which can make use of relationships between variables. Often when short cuts are attempted, the full data entry has to re-start from the beginning or, more commonly, the remaining variables are never entered.
- No hand calculations should be done prior to data entry. Software can be used to transform data into the appropriate units for checking and analysis, e.g. grams fresh weight per plot to kilograms dry weight per hectare, or to take averages of replicated readings, etc.
- One of the variables entered should give a unique record number.
- Enter the geographic reference of the data where possible. For example, in field experiments, the position of each plot should be included. This enables

data, and residuals during analysis, to be tabulated or plotted in their field positions. This will be useful for checking purposes as well as for some insightful analyses. Where plots are regularly spaced with no gaps, the position can be derived from the plot number. Otherwise, two extra columns are keyed, giving the coordinates.

- The data should be entered promptly, as soon as possible after data collection. For example, where measurements are made through the season, they should normally be entered as they are made. This speeds up the whole process, because the data entry task at the end of the trial or survey is then not so large and daunting. It also helps the checking, because some checks can indicate unusually large changes from the previous value, and odd values can then be verified immediately. Feedback of any problems that are noticed to field data collectors can help maintain the data quality.

10.7.3 Guidelines for data checking

The objective is that the data to be analysed should be of as high a quality as possible. Therefore the process of data checking begins at the data collection stage and continues until, and during, the analysis.

Checks when the data are collected

- Data should be collected and recorded carefully. Consider what checks can be incorporated into the data collection routine. For example, the best and worst animals could have a one-line report to verify and perhaps explain their exceptional nature. This will also confirm that they were not written in error.
- Consider collecting some additional variables specifically to help the checking process. For example, in a bean trial, the number of plants with pods that are harvested could serve as a check of the yield values. It may be relatively inexpensive to take digital photos to record the status of each unit. Where this is not possible, recording the state of the unit, e.g. on a scale from 1 to 9, can be worthwhile.

Checks while the data are being entered

- If possible, use software for data entry that has some facilities for data checking.
- Recognize that ignoring the data entry guidelines given above may be counter-productive for data checking. For example, changing the order of the data, transforming yields to kg/ha or calculating and entering only the means from duplicate readings can all lead to errors in copying or calculation. It also makes it more difficult to check the computerized records against the original records.

- Do not trust reading or visually comparing the computerized data with the original records. Though often used, it is not a reliable method of finding key-entry errors.
- Consider using double entry, where a different person does the second keying. This does not take much longer than visual comparison and is a far better form of validation. Modern data entry software has facilities for a system of double entry with immediate or subsequent comparison of values.
- Build in further checks if your software allows. The simplest are range checks, but other logical checks can also be used. For example, for a particular crop, grain weight might always be less than half of head weight.

Checks after entry

- Transforming the data may help the checking process. It may be easier to see whether values are odd if they are transformed into familiar units, such as kg/ha.
- The initial analyses are a continuation of the checking process and should include a first look at summaries of the data. Useful things to produce at this stage are:
    - Extreme values, in particular the minimum and maximum observations.
    - Boxplots, to compare groups of data and highlight outliers.
    - Scatterplots, especially if you use separate colours for each treatment.
    - Tables of the data in treatment order.
- With experimental data, the initial ANOVA should also be considered as part of the checking process. In fact, with any reasonably large dataset, not just experimental data, it is difficult to do all the checking without taking into account the structure of the data; a value that is odd for one treatment, or one group of individuals, may be acceptable for another. So make use of software for the analysis that allows you to display the residuals easily and in a variety of ways.

10.8 Organizing the Data for Analysis

We have recommended that the data be entered in their raw form. Hence, the first step in the data organization or management stage often involves calculations to re-structure the data into the appropriate form for analysis. This can either be performed in the software used for the data entry, or in the statistics package that will be used for the analysis. We recommend:

- A record should be kept of all changes to the data. This record becomes part of the database, and is kept in the audit trail. Many packages allow data to be transformed and re-arranged visually, but still generate a corresponding file that records the transformations.
- There should be a single master copy of the data. This is a standard principle of data management, to preserve data integrity.

The master copy will increase in size as data accrue. Even after basic data entry is complete, errors will be detected, and should of course be corrected in the master copy. It is therefore something that changes through the course of data entry, data management and analysis. Not only should this process be documented, but also a consistent version-numbering system should be evolved and utilized by all data analysts and other users.

In all but simple projects the master copy will normally be stored using a DBMS. Only some of the data tables will be altered by data changes. For example, Mr A the anthropologist may not immediately be concerned with changes to experimental records made by Ms B the biologist, but should be up-to-date with additions to the site list agreed by Dr C the chief. Keeping track of and communicating changes to the master copy of the data should be a project management activity, like budgetary management.

Usually analyses and reports will be based on extracts from the master copy of the data. When final outputs for presentation or publication are being produced, it is important these are correct, consistent and complete in that they are all based on the final version of the master copy of the data. Interim analyses will have been based on interim data, and to avoid confusion and inconsistency, analysis datasets, file-names and outputs should include a record of the master copy version number from which they were derived.

10.9 Analysis

From the data management perspective, the analysis simply takes the raw data and produces summaries. The process can be viewed in two stages. The first is the production of results to enable the research team to understand their data. The second is the preparation of key summaries to be presented to others in reports and seminars. The software used for the analysis should satisfy the requirements for both stages.

- It should include tabular and graphical capabilities to facilitate exploratory investigations of the data. One use of these is to continue the checking process and hence ensure that the summaries presented are meaningful.
- The facilities for analysis should permit the presentation of the results in a form that assists the research team in their interpretation of the data.
- The software should permit the results to be displayed in a way that closely approximates to the tables, graphs and other summaries that will be included in any reports.

The software will often be a mixture of a spreadsheet and a statistics package, as we explain in more detail in Chapter 14.

10.10 Audit Trail

An audit trail is a complete record of changes to the data and decisions made about the data and the analysis, rather like a logbook. In fact, for data management it is the equivalent of the old-fashioned notion of a scientist's notebook, which is as relevant today as ever. A well-maintained audit trail, logbook or notebook greatly eases the subsequent tasks of writing reports on the data and of answering data queries.

It is important to record everything you do at the time you do it, as recollections are always poor at a later stage. For example, when errors are found during checking and when changes are made to the master copy of the data, a note should be made in the audit trail. Keep notes also on the analyses you do, including the preliminary ones done for checking purposes, writing down the names of all files created. Every entry in the logbook should be dated and initialled.

10.10.1 Analogy with financial auditing

The idea of the audit trail comes from accountants auditing financial data. One tool they use is to choose a transaction and follow it through the financial system. For example, they note that a payment has been made to X for travel expenses. They check back to ensure that the payment was approved by the budget office responsible for that account, that the claim was approved by the supervisor, that a reasonable receipt was produced, and that the travel was authorized and in line with the travel policy. Auditors do not look at all transactions, just a sample. But that is enough to identify whether parts of the financial system are not working well.

With research data the idea is the same. Processing data involves many steps, between recording the observations in the field and reporting a summary result in a scientific or review paper to farmers, or the World Bank. Following the trail of this process will reveal where errors could occur and hence misleading results could be reported.

10.10.2 Following the trail

There are two directions in which you can follow the trail. Going forwards, choose some raw observations from a dataset. Follow them through to find what was done to them. They will have been transformed, summarized, and combined with other information to contribute eventually to conclusions in one or more reports. Your records should make it possible to follow this process in the way it was done.

Going backwards, you start with a result, perhaps the key numbers in a paper that give the adoption rates for a new technology among different groups of farmers. Find the numbers from which the results were derived, then their source, leading back to the original field observations.

Using either method you are likely to be reviewing both content and process.

Auditing the process includes:

- Ways in which data have been checked and errors identified and corrected.
- Ease with which analyses can be repeated.
- Extent to which repetitive work was automated, or done by hand.

Auditing the content includes:

- Ways in which comments on the data, made in the field, have been used.
- Reasons for data selection and omission.
- Sources of supplementary data used in the analysis, for example calibrations, prices, or population sizes.

10.10.3 Preparing for the audit

In the future perhaps external auditing of research data will be routine, in the same way that it is for financial auditing. Nevertheless, the main reason currently is for the benefit of the research team, both during their current work, and to enable a usable archive to be left for future work. The following examples illustrate the sort of information that will help an audit:

- If an error is found that cannot be corrected, then do not simply delete it from the database. Instead, mark it as missing and add a comment.
- Add comments to data files describing how they were checked and edited. If a spreadsheet is used, these comments can be on a further sheet.
- Try to link, rather than copy data. Copying means that the connection between the original data and that used for the analysis is broken. If you do copy, then make a note of this action.
- Document all decisions about data selection. Why were some farms ignored, or only three of the six districts analysed?
- Document the source of all additional information. For example, where do prices, population sizes, adoption rates, moisture contents come from?
- Keep copies of all programs that do the statistical or other analyses.
- Document the programs. If you use a menu-driven click-and-point approach all the time, then keep and document the log file that is produced by most software. If no usable log file is produced, then prepare your own description.

10.11 Backing Up

It is essential to develop a system for regularly making copies, or back-ups, of your data and command files. Omitting to do so invariably results in important parts of the research data being lost. Project managers should establish a documented routine for making safe copies of the data, and should insist that all members of the research team follow the routine.

There are several types of hardware you can use for back-ups. The most common are CDs, diskettes and tapes. CDs and tapes have the advantage of higher storage capacity than diskettes. Whatever hardware you use, it is advisable to have at least two sets of back-up media and to use the sets alternately. Obviously, the back-up media must be stored in a safe environment.

10.12 Archiving

The data and programs from a research project must be archived in such a way that they are safe and can be accessed by a subsequent user. The media used for the archive might be diskettes, tapes, or CDs, similar to the ones used for back-ups.

Although the copying of data to the archive comes at the end of the project, the way the information will be transferred to the archive should be planned from the outset. Careful planning will help to promote a consistent directory structure and naming convention for computer files, and will also encourage the recording of all steps in the project (see Section 10.10).

The archive is more than a permanent storage place for the files used for the analysis. It must give access to all the information from the project. During the project's operational phase, the information is partly in the computer, partly on paper and other media, such as photographs, and partly in the minds of the research team. The archive need not all be computerized, but it must include all the relevant, non-ephemeral information that is in the minds of the research team. Where data cannot be archived electronically, the sources of information, i.e. the metadata, should still be recorded in the archive.

In the absence of a proper archiving scheme, the usual outcome is that the researchers leave the project, carrying with them the only copy of their part of the data, and hoping that the analysis and write-up will be continued later. Eventually the hope dwindles and the datasets become effectively lost to further research. To avoid this outcome, we believe that: (i) at least one full copy of the archive should be left locally; and (ii) the final report should detail the structure of the archive and the steps taken to ensure its safekeeping.

There is a distinction between what might be called a file-level archive and a data-level archive. For example, many organizations insist that any staff who departs must deposit all their data before their final salary is paid. The member of staff therefore provides a set of CDs with their files and then goes. Of course, no one is in any position to check that these are really all the files, nor that the files can be read. If the individual is conscientious there will be a catalogue that explains the purpose of each file, but usually this has to be inferred from the file name.

This is what we would call a file-level archive. However, it is unlikely to be usable. If there are multiple copies of the same dataset, how can we find which is the master copy? If we do find the master file, is there sufficient information, the metadata again, so that the data can be understood?

Of course a simple managerial solution would be just to insist that the custodian of the files, perhaps the librarian or the data processing manager, does an audit of the CDs that are supplied. No one can leave until they have deposited a usable archive. While simple to specify this would be impossible to implement, and many staff could never leave! Instead, the organization needs a data management strategy, with archiving as one aspect.

10.13 In Conclusion

In this chapter's introduction, we gave an example of a project in which different scientists worked on subsets of the data, with little thought of the problems that their inadequate data management would cause in the future. We said that a few minutes routinely devoted to data management issues, would have made a lot of difference.

In Chapter 13, we address the problem of organizing a data management strategy in more general terms. Here we conclude with the point-of-view of you as an individual, in the middle of a research project. You are poised with data, ready for processing. What could you do in your few minutes on data management? We will assume a bottom-up approach, namely you will not try initially to change the organizational, or even your project leader's way of working.

For your task, the two sections are those on the metadata (Section 10.2), and the audit trail (Section 10.10). They are linked, because the audit trail is simply a way of documenting information about what has happened to the data, and this constitutes one part of the metadata.

A key document is the research protocol, because this gives the objectives for the work that you are doing. It too is a crucial part of the metadata, so check that this document is easily available, and perhaps link it to the data files. Is there a later version of the document that revised the objectives, once you had started the study? Now look at the data file you are using. Perhaps it is in a spreadsheet file, as a workbook containing many sheets. How do those sheets relate to the data as collected? In Section 10.10 we said that an audit trail is like a diary and should be updated at the time that you perform the operations. But doing it now rather than later is much better than no diary at all. So prepare a new sheet in the workbook, or use a notebook to start with and review part of your data.

This may take more than a few minutes on the first day or two, most likely because you are catching up and learning how to manage your data in the way that best suits yourself. However, we would be surprised if you did not soon recognize that these steps are helpful for your data analysis and worth the time you are spending. Attention to data management is unlikely to extend the time devoted to the analysis, because the data have now been handled more efficiently. In any case, even if the data management part of your project does take slightly longer, the benefits are still likely to outweigh the time you spend.

What would take longer is a review of the software you are using for your data processing and analysis. Perhaps some of the issues of data management would be resolved more easily if you used your software in a different way, or if different software were used. The ease-of-use of modern software makes it much easier to consider changes that would have been very difficult a few years ago. These and related issues are addressed in the next three chapters.

Chapter 11

Use of Spreadsheet Packages

11.1 Introduction

In Chapter 10 we noted that spreadsheets are commonly used for data entry because they are familiar, in widespread use and very flexible. However, we also stated, as a warning: 'their very flexibility means they permit poor data entry and management. They should therefore be used with great care. Users should apply the same rigour and discipline that is required when working with more structured data entry software.'

Here we explain what we mean by 'great care' and 'rigour and discipline' and show how a spreadsheet package can be used effectively for data entry. In the illustrations we have used Microsoft EXCEL as an example of a spreadsheet package, but our strategy applies equally to other spreadsheet packages.

The next section begins with a simple example that shows a set of data that has been poorly entered, and we discuss the problems that can arise because of this. In Section 11.3, we consider features in EXCEL that facilitate simple and reliable data entry. The principle is this: the simpler the data entry process, the more reliable will be the data that are entered.

Section 11.4 describes how to organize the data entry process. We emphasize the need to set up the worksheet to include all the metadata, i.e. all associated background information relating to the data. This background information includes where the data came from, when they were collected, what the data values represent, and so on.

Simplifying the data entry process and recognizing the importance of the metadata has implications for the task of organizing the data entry system within the spreadsheet. The organizing phase now takes a little longer, but this extra effort is justified. It is important to separate the task of organizing the spreadsheet for data entry from the actual entry of the data.

Finally, we cover a number of other issues, including validation checks after data entry and the entry of more complicated datasets.

11.2 Common Problems with Data Entry

Figure 11.1 shows a typical dataset that has been entered in EXCEL. This is a simple set of data that can be entered without difficulty in a spreadsheet.

Fig. 11.1. A typical dataset in spreadsheet style.

	A	B	C	D	E	F	G	H
1	block	plotwb	species	rcd	height	branch	Crown_0	Crown_90
2	1	1	A.polycantha	12.7, 13.3	438	23	673	730
3	1	2	A.indica	15.1	415	17	374	354
4	1	3	A.nilotica	11.1	350	20	268	375
5	1	4	Albizia lebeck	21.1	553	17	700	620
6	1	5	Control					
7	2	1	A.indica	15.1	470	19	420	395
8	2	2	control					
9	2	3	Albizia lebeck	12	300	12	394	322
10	2	4	A.polycantha	DEAD				
11	2	5	A.nilotica	10.1	343	22	420	401
12	2	1	A.nilotica	10	330	23	443	402
13	3	2	A.polycantha					
14	3	3	Control					
15	3	4	A.indica	11	410	21	415	440
16	3	5	A.polycantha	14.25	635	23	852	880
17	4	1	Control					
18	4	2	A.nilotica	12.5	373	23	602	500
19	4	3	A.polyantha	25.8	630	25	920	750
20	4	4	A.indica	18.5	404	22	420	370
21	4	5	Albizia lebeck	198	465	10	352	340

The data were entered by a clerk, who, as instructed, typed what was written on the recording sheet in the field. However, this has led to errors (ringed in Fig. 11.1). For example, two of the names under the *species* heading have been typed slightly differently from the names for the same species used elsewhere in the same column. In the column headed *rcd*, row 2 has two measurements entered, while in row 10, instead of a numerical value, the cell reports that the plant is dead. Such entries will cause problems when the data are transferred to a statistics package for analysis.

Most of these errors can be avoided if some thought is given to the layout of the data in the spreadsheet before data collection in the field commences. This activity is, in fact, the responsibility of the researcher, not of the data entry clerk.

There are other deficiencies if these data are all that are to be computerized. For instance, there is no indication as to where the data came from, when the data were collected or what they represent. There is no record of the units of measurement used. Such information is highly relevant and should be included with the data in the spreadsheet. This is especially true if the dataset is to be integrated with datasets from other studies, or is to be passed to someone else for analysis. The entry of the metadata is discussed in Section 11.4.

You may think that this example is just a caricature and that real data would never be entered so poorly. Figure 11.2 shows some of the data from Fig. 11.1 entered in a different way.

Fig. 11.2. An alternative way to enter data.

	A	B	C	D	E	F
1	Crown 0					
2			Block			
3	Species	1	2	3	4	Mean
4	A.polycantha	673		852	920	815
5	A.indica	374	420	415	420	407
6	A.nilotica	268	420	443	602	433
7	Albizia lebeck	700	394		352	482
8	Species Mean	504	411	570	574	
9						
10	Crown 90					
11			Block			
12	Species	1	2	3	4	Mean
13	A.polycantha	730		880	750	787
14	A.indica	354	395	440	370	390
15	A.nilotica	375	401	402	500	420
16	Albizia lebeck	620	322		340	427
17	Species Mean	520	373	574	490	

The layout in this example is even worse. It confuses data entry and analysis.

The display of the data as shown in Fig. 11.2 is sometimes convenient as a prelude to the analysis. In Section 11.5 we show that with EXCEL it is easy to enter the data properly and then display the values as shown in Fig. 11.2.

You should not necessarily expect the software to help in recommending good procedures for data entry. For example, if you decided to use EXCEL for analysis of variance, EXCEL would expect the data in the form of Fig. 11.2. You would then discover that the spreadsheet's facilities for ANOVA are limited. The next step might be to transfer the data to a statistics package; but you will not find a good statistics package that receives the data in the tabulated form shown in Fig. 11.2. They all expect it in the rectangular shape shown in Fig. 11.1.

Thus, as already stated at the beginning of Section 11.1, a spreadsheet gives total flexibility to enter your data as you wish, but no guide as to how to enter the data well.

11.3 Facilitating the Data Entry Process

11.3.1 Introduction

It is possible to enter data in such a way as to eliminate the errors and discrepancies illustrated in Fig. 11.1. If the guidelines given in the following sections were applied, the data would look like Fig. 11.3. This type of layout, already described as rectangular, is also known as *list format*.

Fig. 11.3. Worksheet after data entry guidelines have been used.

	A	B	C	D	E	F	G	H	I
1	block	plotwb	plot	species	rcd	height	branch	Crown_0	Crown_90
2	1	1	101	A.polycantha	13.0	438	23	673	730
3	1	2	102	A.indica	15.1	415	17	374	354
4	1	3	103	A.nilotica	11.1	350	20	268	375
5	1	4	104	Albizia lebeck	21.1	553	17	700	620
6	1	5	105	Control					
7	2	1	201	A.indica	15.1	470	19	420	395
8	2	2	202	Control					
9	2	3	203	Albizia lebeck	12.0	300	12	394	322
10	2	4	204	A.polycantha					
11	2	5	205	A.nilotica	10.1	343	22	420	401
12	3	1	301	A.nilotica	10.0	330	23	443	402
13	3	2	302	A.polycantha					
14	3	3	303	Control					
15	3	4	304	A.indica	11.0	410	21	415	440
16	3	5	305	A.polycantha	14.3	635	23	852	880
17	4	1	401	Control					
18	4	2	402	A.nilotica	12.5	373	23	602	500
19	4	3	403	A.polycantha	25.8	630	25	920	750
20	4	4	404	A.indica	18.5	404	22	420	370
21	4	5	405	Albizia lebeck	19.8	465	10	352	340

It is sensible to spend a little time thinking about the data before rushing into using the spreadsheet. In this example, the data in columns A-D would have been determined at the planning stage of the experiment and will be the same for all the measured variables. They could have been entered into a worksheet before any measurements were taken. A paper printout of Fig. 11.4, which contains additional metadata (see Section 11.4), could be used in the field to collect the data.

Fig. 11.4. Paper data collection sheet.

	A	B	C	D	E	F	G	H	I	J
1	Study code	TS695								
2	Study Title	Evaluation of rotational woodlands and fodder for soil fertility								
3	Site									
4	Scientist									
5	Project	4.1								
6	Project title	Systems evaluation and dissemination, developing choices for farmers								
7	Objectives	Assessment of tree growth								
8	Design	Random complete block								
9	Date									
10	Trait-title					Root collar diameter	Tree height	No. of branches	Crown cover diameter	
11	Units					(cm)	(cm)	()	(cm)	
12	Orientation								0 degrees	90 degrees
13	Trait-name	block	plotwb	plot	species	rcd	height	branch	Crown_0	Crown_90
14		1	1	101	A.polycantha					
15		1	2	102	A.indica					
16		1	3	103	A.nilotica					
17		1	4	104	Albizia lebeck					
18		1	5	105	Control					
19		2	1	201	A.indica					
20		2	2	202	Control					

If you compare Fig. 11.1 with Fig. 11.3, you will see that there is an extra column named *plot*. It is useful to have a column that uniquely defines each row of data. In this example, after the columns called *block* and *plotwb* (plot within block) were entered, *plot* was calculated as follows: plot=100*block+plotwb (see in Fig. 11.5).

Fig. 11.5. Calculating the plot column.

C2 = =A2*100+B2

	A	B	C	D	
1	block	plotwb	plot	species	rcd
2	1	1	101	A.polycantha	
3	1	2	102	A.indica	

When the data have been collected, you may decide to format the columns so that the data are displayed, for example, to two decimal places for heights or lengths, or as integers for data collected as counted items. You may wish to add range checks, so that typing data outside the minimum and maximum values is made impossible. You can add comments to cells, for example, to indicate an animal has died or a tree has been destroyed by an elephant!

11.3.2 Freezing or splitting panes

When entering data, it is useful to keep the headings of columns always visible as you scroll down the screen. This can be achieved by freezing the panes.

11.3.3 Drop-down lists

There are several ways to avoid typing a sequence more than once. If the *species* column has a repeating list of text strings, these could be typed just once and then the spreadsheet's *Fill* option from the *Edit* menu could be used to fill the remaining cells.

We illustrate the situation shown in Fig. 11.1, where the species names do not form a repeating sequence. It is however more typical because it follows the species randomization order in the field. The same four species plus the control are each repeated four times within the column. When the same text string is entered many times, typing errors inevitably occur. This is shown in Fig. 11.1 where the species name, *A.polycantha* has been mistyped as *A.Polyantha* in block 4.

Figure 11.6 shows the creation of a drop-down list. The five species names for block 1 are entered into cells D2:D6. These five names can then be used to create the drop-down list.

Once the drop-down list has been created, selecting a cell in that column will bring up an arrow to the right of the cell. Clicking on the arrow will display the drop-down list (see Fig. 11.7). An appropriate selection can then be made.

New data entered in these cells must either be a string from the given list, or the cell can be left blank. Entry into the cells can be forced by unchecking the *Ignore blank* field in the *Data Validation* dialog box shown in Fig. 11.6.

The rest of the species can now be selected from the drop-down list. For example, Fig. 11.7 shows the selection of *A.indica* for plot 201.

Fig. 11.6. Creating a drop-down list for species.

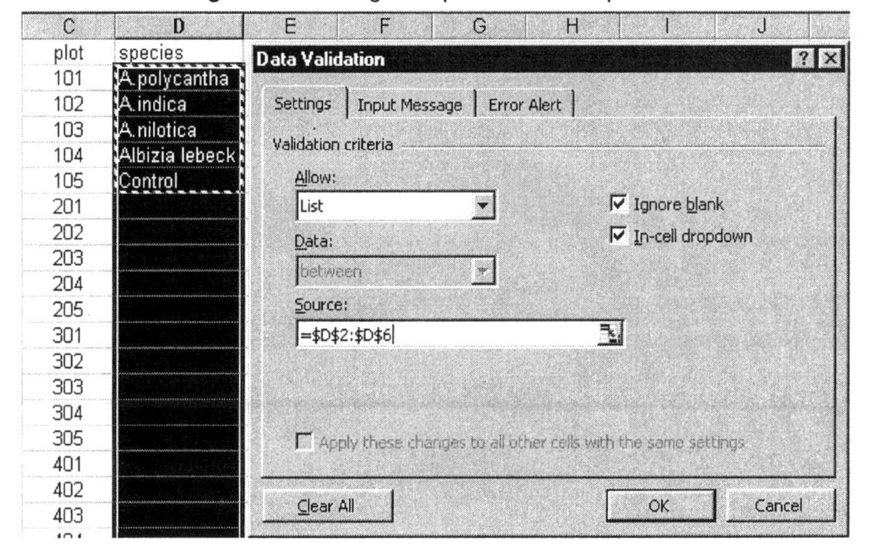

Fig. 11.7. Selecting a species from a drop-down list.

C	D
plot	species
101	A.polycantha
102	A.indica
103	A.nilotica
104	Albizia lebeck
105	Control
201	
202	A.polycantha
203	A.indica
204	A.nilotica Albizia lebeck
205	Control

Using drop-down lists for data entry helps to avoid spelling errors.

11.3.4 Data validation

Validation checks can and should be set on ranges of cells within the spreadsheet. A range could be an entire column or row, several columns or rows, or just a single cell. The validation rules apply when new data are entered.

One validation tool available in EXCEL is the facility to set up range checks for numerical data. For example, the measurements recorded for the variable *rcd* are expected to be in the range from 10 to 26. A range check is simple to set up from the *Data Validation* menu, as shown in Fig. 11.8. At the same time, an *Input Message* and an *Error Alert* can be set up; their effects are shown in Fig. 11.9. The *Input Message* is displayed when each cell in the column is activated (Fig. 11.9a). This reminds the person entering the data of the range of values allowed. The *Error Alert* message is displayed when a value outside this range is typed (Fig. 11.9b). Note that only the data cells are highlighted, not the variable name at the top.

11.3.5 Adding comments to cells

EXCEL has a facility for adding comments to a cell. These differ from values within the cell. Comments should be used for any unusual observations or questions concerning a particular data value. In our example, when entering the data for *rcd*, a decision had to be made for *plot 101*, where two values were entered on the data-recording sheet. We chose to calculate the mean and add a comment to the cell, as shown in Fig. 11.10.

If there had been several plots with two values recorded, two columns of *rcd* data could have been entered and a third column could have been used to calculate the mean.

Fig. 11.8. Setting up range checks.

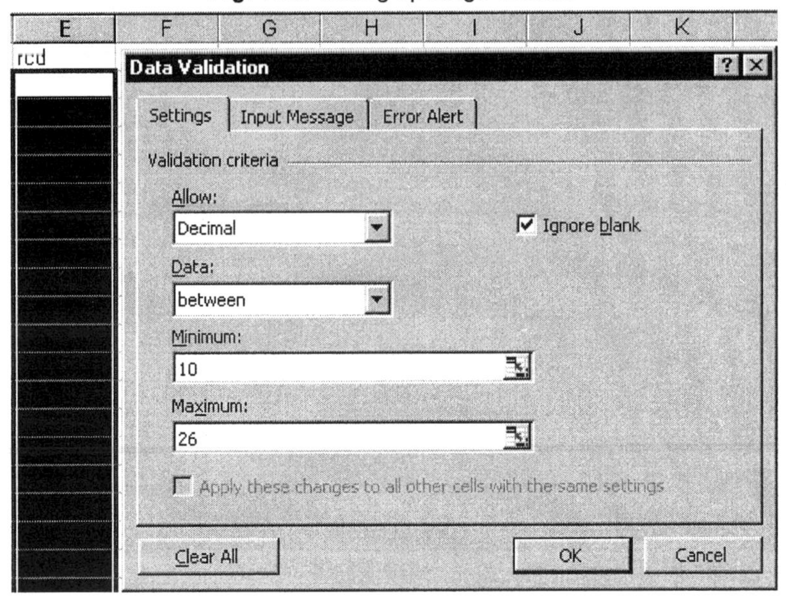

Fig. 11.9. Validation at data entry.

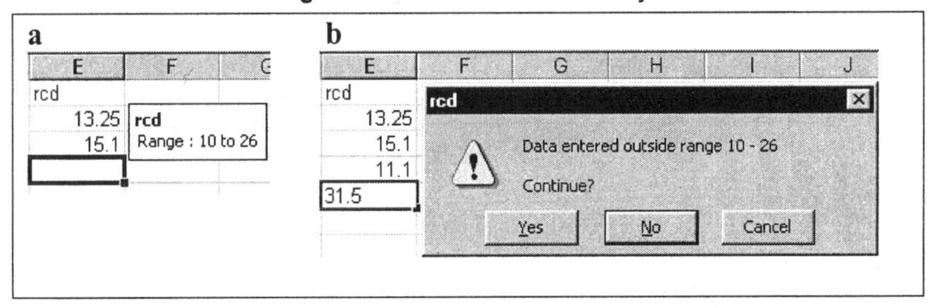

Comments are also useful in explaining why certain values are missing. For example in Fig. 11.10, the value 'Dead' had been entered in cell D10. With the exception of the column header, all cells in a column should have the same data type. The data type for column D is numeric. The string 'Dead' in cell D10 is therefore inappropriate for the cell, but can be put as a comment to indicate why the value is missing. Fig. 11.10 shows comments that have been added to cells. A comment is set on a single cell but can also be copied to a range of other cells.

Fig. 11.10. Examples of a comment in a cell.

E	F	G
rcd		
13	Mean of 12.7, 13.3	
15.1		
11.1	350	20

control		
Albizia lebeck	12	
A. polycantha		Dead
A. nilotica	10.1	343
A. nilotica	10	330

11.3.6 Formatting columns

EXCEL suppresses trailing zeros by default. For example, in Fig. 11.10 the value 13.0 is displayed as 13. This can easily be changed, as shown in Fig. 11.11, which also shows that there is flexibility in how the data are displayed (font size, cell borders, etc.).

Fig. 11.11. Formatting data.

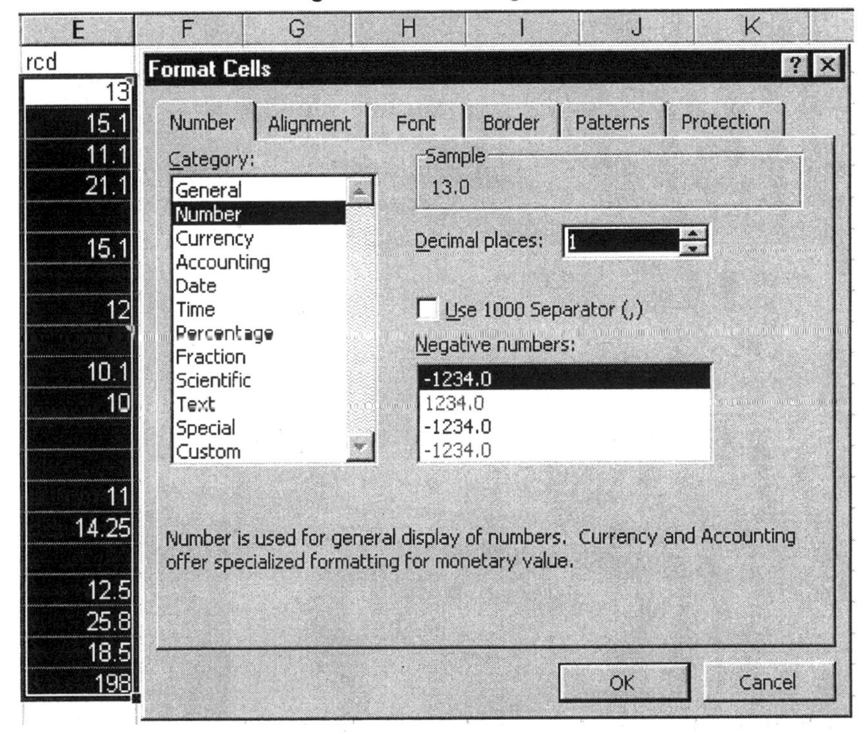

11.3.7 Data auditing

The suggestions given above are all intended to facilitate data entry. However, there is also a facility known as auditing, for checking data that have already been entered. Use this facility whenever you add validation rules following the data entry, or when you make changes to existing rules. Validation rules are very flexible. The type of data such as whole number, decimal, text, etc., can be specified; data, which are not of the correct type will be rejected on data entry. You can then set further restrictions to accept only those numbers within a given range, or text strings of a particular length, for example.

We illustrate auditing on the data in Fig. 11.12, where validation rules have been added for both *species* and *rcd*. Figure 11.12 shows the Auditing Toolbar and some errors circled for these variables.

Fig. 11.12. Auditing of existing data.

C	D	E	F	G	H
species	rcd	height	branch	Crown_0	Crown_90
A. polycantha	12.7,13.3	438	23	673	730
A. indica	15.1	415	17	374	354
A. nilotica	11.1				
Albizia lebeck	21.1				
Control					
A. indica	15.1	470	19	420	395
control					
Albizia lebeck	12	300	12	394	322
A. polycantha	DEAD				
A. nilotica	10.1	343	22	420	401
A. nilotica	10	330	23	443	402
Albizia lebeck					
Control					
A. indica	11	410	21	415	440
A. polycantha	14.25	635	23	852	880
Control					
A. nilotica	12.5	373	23	602	500
A. polyantha	25.8	630	25	920	750
A. indica	18.5	404	22	420	370
Albizia lebeck	198	465	10	352	340

11.4 The Metadata

In the previous section we looked at some methods of validating data, both during data entry and afterwards. However, the data we started with are still not complete. We suggest adding rows and columns to the spreadsheet before entering the main body of data. These extra rows and columns will store documentation that provides background information about the data, i.e. the metadata.

Figure 11.13 shows a blank spreadsheet divided into input areas. The numbers of rows and columns in these areas are not fixed but can increase or decrease depending on the data to be entered.

Figure 11.14 shows the data from Fig. 11.3 with the metadata added into these extra rows and columns. It is useful to adhere to a strict code as to what should be entered into each section of the spreadsheet. Keep in mind the following three aims:

- Encourage completeness
- Avoid unnecessary duplication
- Minimize errors

Fig. 11.13. Input areas in an EXCEL spreadsheet.

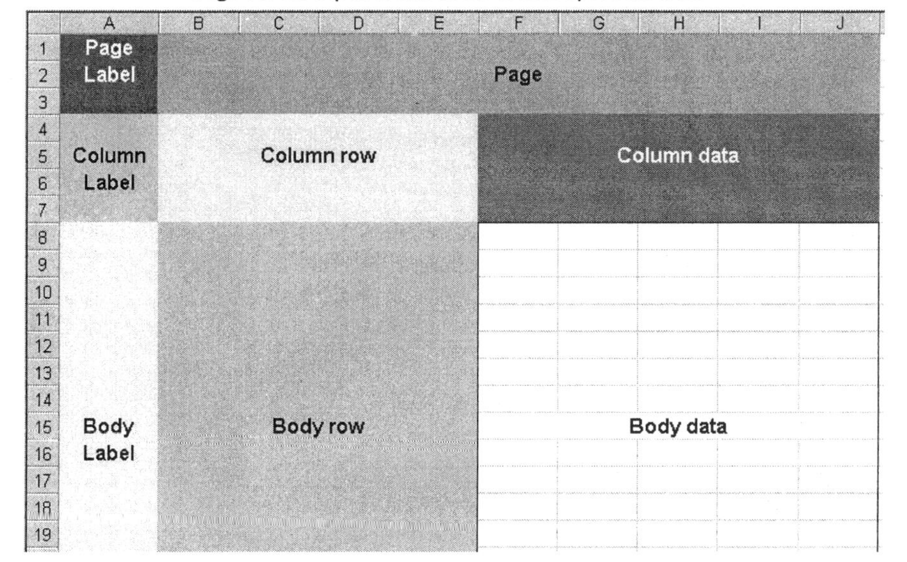

11.4.1 Body data

The *Body data* area contains all values that have been observed or measured. With reference to the data in Fig. 11.3, this is the range of cells starting from cell E2 and including all cells below it and to its right. In Fig. 11.14 it is the range with cell F14 in the top left-hand corner. Values in columns A to D of Fig. 11.3 are not measurement data but correspond to information relating to the design of the experiment. These, with the exception of the first row, fall into the *Body row* area of Fig. 11.13. The *Body label* area can be used to add more information about the body data.

11.4.2 Column information

In the *Column label*, *Column row* and *Column data* areas we add information to describe the measurements. Here a measurement corresponds to a unique identifiable column of data.

The column headers from the first row of Fig. 11.3 go into the column data areas of Fig. 11.13 (i.e. F4:J4, which become F13:J13 in Fig. 11.14). We still need more information. For instance, the data file does not clarify what specific measurements are being made, so it is necessary to include a description of what *rcd*, *height*, *branch*, etc. actually represent. It is also necessary to specify the units of measurement being used for each variable. The descriptions and units are shown in Fig. 11.14 in the range of cells F10:J12.

11.4.3 Page information

All that remains to be entered now is documentation related to the whole dataset. Effective documentation requires that, given a dataset, one should be able to retrieve the original protocol plus all other related data. This documentation appears in the *Page* area of the spreadsheet with the *Page label* area providing labels for each piece of information. The minimum documentation is the study code and its title. It is also useful to include the location where the data were collected and the name of the person responsible for the study. The study objectives, the design and other protocol details should be entered if available.

Fig. 11.14. Metadata.

	A	B	C	D	E	F	G	H	I	J
1	Study code	TS695								
2	Study Title	Evaluation of rotational woodlands and fodder for soil fertility								
3	Site	Shinyanga, Tanzania								
4	Scientist	Robert Otsyna								
5	Project	4.1								
6	Project title	Systems evaluation and dissemination, developing choices for farmers								
7	Objectives	Assessment of tree growth								
8	Design	Random complete block								
9	Date	36493								
10	Trait-title					Root collar diameter	Tree height	No. of branches	Crown cover diameter	
11	Units					(cm)	(cm)	()	(cm)	
12	Orientation								0 degrees	90 degrees
13	Trait-name	block	plotwb	plot	species	rcd	height	branch	Crown0	Crown90
14		1	1	101	A.Polycantha	13	438	23	673	730
15		1	2	102	A.Indica	15.1	415	17	374	354
16		1	3	103	A.Nilotica	11.1	350	20	268	375
17		1	4	104	Albizia lebeck	21.1	553	17	700	620
18		1	5	105	Control					
19		2	1	201	A.Indica	15.1	470	19	420	395
20		2	2	202	Control					
21		2	3	203	Albizia lebeck	12	300	12	394	322

11.4.4 Entering the date

All observed raw data should be associated with a date of observation. Before entering the date of measurement ensure that the computer date setting matches the expected date input format. Remember that 9/2/99 could refer either to 9 February 1999 or 2 September 1999 depending on the date settings on your computer. To ensure you have the correct date, choose a display setting that includes the name of the month, e.g. February 9, 1999.

If all observations are taken on the same date, the date should be entered in the *Page* section with an appropriate label in the *Page label* area. This is shown in Fig. 11.14. If this is not the case, the dates should be entered into the *Body row* area with an appropriate label in the *Column row*. Where there are a limited number of dates, for example where the values were measured on one of just three or four dates, a drop-down list should be created with these dates and values chosen from the list on data entry.

11.4.5 Using multiple sheets

Entering the data plus the metadata as described above gives us Fig. 11.14.

An alternative is to put the *Page information* on a separate sheet in the EXCEL workbook. This is often convenient when there is a lot of information at the dataset level. This can be taken further, so the *Column data* section, that describes the measurements that were taken, and the units, etc., could also be on a separate sheet.

In such cases, where the metadata are on separate sheets you may either keep the data sheets simple, as in Fig. 11.3, or include a small *Page* section in each data sheet to describe the type of measurements that are entered in that sheet.

11.5 Checking the Data After Entry

We have demonstrated in Section 11.3 how auditing can be done to check existing data against the validation rules. Here we give some further steps that may be undertaken to validate the data after they have been entered.

Fig. 11.15. Scatter plot of root collar diameter.

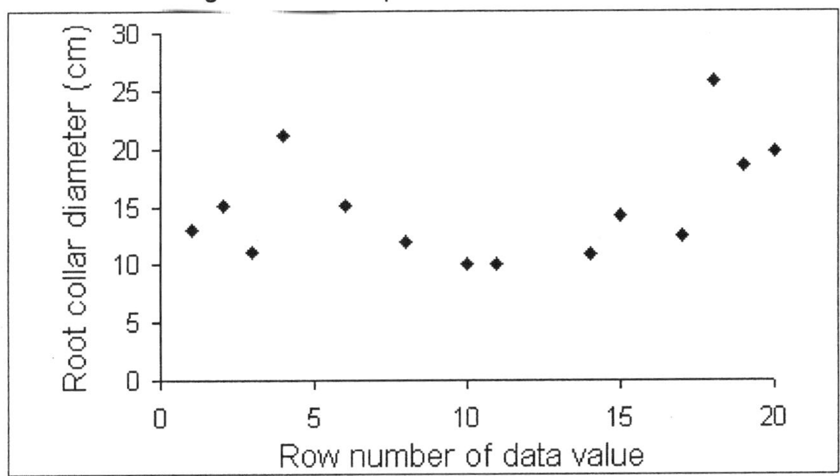

11.5.1 Use of plots to highlight outliers

Scatterplots are useful tools for helping to spot suspicious inputs (i.e. outliers). Many would be trapped at the data entry stage if validation rules had been set up. However, there may still be some values that differ substantially from the rest. Fig. 11.15 shows a scatter plot of root collar diameter (referred to earlier as *rcd*) where the x-axis corresponds to the order in which the values appear in the data file. This plot shows that all data records lie within a range from about 10 cm to about 25 cm.

Fig. 11.16. Tree height plotted against root collar diameter.

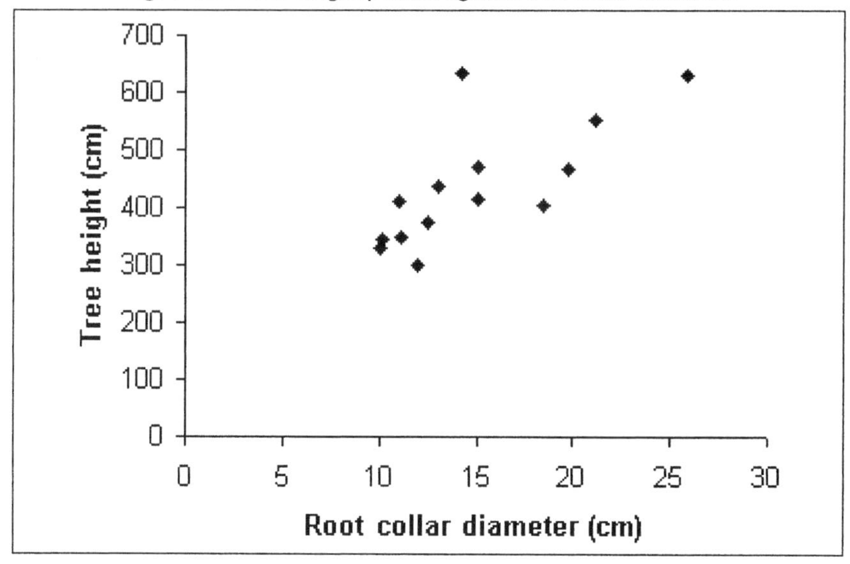

Such a plot is quick and easy to produce and can be done for all observed variables. It is also useful to produce scatterplots of pairs of variables if our prior knowledge expects the two variables being plotted to show a definitive pattern. Figure 11.16 shows tree height plotted against root collar diameter. There is at least one suspicious observation.

Fig. 11.17. Use of AutoFilter to spot unexpected entries.

a

	A	B
1	block	plotwb
2		(All)
3		(Top 10...)
4		(Custom...)
5		1
6		2
7		3
		4
		5
8	2	2
9	2	3
10	2	4
11	2	5
12	3	1

b

	A	B
1	block	plotwb
2		(All)
3		(Top 10...)
4		(Custom...)
5		1
6		2
7		3
		4
		5
8		15
9	2	3
10	2	4
11	2	15
12	3	1

11.5.2 Using AutoFilter as a data checking tool

In our example dataset, we know that there are four blocks with five plots in each block. In Fig. 11.17a, the AutoFilter option has been used for *plotwb* to display a list of values in that column. It shows that only values 1 to 5 have been entered in the column. If, for example, the value 15 had been entered instead of 5, the list would also contain the value 15 (as in Fig. 11.17b) and so the error could easily be spotted.

11.5.3 Tabulation

Frequency tables can be created easily using EXCEL's Pivot table facility.

Fig. 11.18. Use of a Pivot table to spot data entry errors.

	A	B	C	D	E	F	G
1	**Pivot table for Figure 3 dataset**						
2		Data					
3	block	Count of block	Mean rcd	Mean height	Mean branch	Mean Crown0	Mean Crown90
4	1	5	15.08	439.00	19.25	503.75	519.75
5	2	5	12.40	371.00	17.67	411.33	372.67
6	3	5	11.75	458.33	22.33	570.00	574.00
7	4	5	19.15	468.00	20.00	573.50	490.00
8	Grand Total	20	14.95	436.86	19.79	518.07	491.36
9							
10							
11	**Pivot table for Figure 1 dataset**						
12		Data					
13	block	Count of block	Mean rcd	Mean height	Mean branch	Mean Crown0	Mean Crown90
14	1	5	15.77	439.00	19.25	503.75	519.75
15	2	6	11.80	360.75	19.00	419.25	380.00
16	3	4	12.63	522.50	22.00	633.50	660.00
17	4	5	63.70	468.00	20.00	573.50	490.00
18	Grand Total	20	28.81	436.86	19.79	518.07	491.36
19							
20	PivotTable						
21	PivotTable ▾						
22							

Table mean / Metadata

In Fig. 11.18, the counts of the number of observations within each of the four blocks plus the means of the variables are shown for the data in Fig. 11.3 and Fig. 11.1. Using Pivot tables is another way of spotting data entry errors. For example, from the table shown in Fig. 11.18 it is possible to drill down to look at subsets of the data that correspond to particular cells of the table.

Earlier we claimed that the data entered in the layout given in Fig. 11.2 was not to be recommended. However this may be a convenient way to examine the data after they have been entered. From the standard list format shown in Fig. 11.3, we can use a Pivot table to display the information as shown in Fig. 11.19. This is almost the same as what was shown in Fig. 11.2, but here we are separating the data entry from the analysis.

11.6 Conclusions

Our main conclusion is that a spreadsheet package, such as EXCEL, can be used for effective data entry, particularly for datasets with a simple structure.

Fig. 11.19. Pivot table giving an alternative layout for the data.

	A	B	C	D	E	F
1	Mean of Crown_0	block				
2	species	1	2	3	4	Grand Total
3	A.Indica	374	420	415	420	407
4	A.Nilotica	268	420	443	602	433
5	A.Polycantha	673		852	920	815
6	Albizia lebeck	700	394		352	482
7	Control					
8	Grand Total	504	411	570	574	518
9						
10	Mean of Crown_90	block				
11	species	1	2	3	4	Grand Total
12	A.Indica	354	395	440	370	390
13	A.Nilotica	375	401	402	500	420
14	A.Polycantha	730		880	750	787
15	Albizia lebeck	620	322		340	427
16	Control					
17	Grand Total	520	373	574	490	491

Even for simple data entry, it is important to separate the task of organizing the spreadsheet from that of actually entering the data.

EXCEL provides a range of aids to effective data entry, some of which have been described in Section 11.3. If you wish to know the EXCEL commands for the tasks described, use the EXCEL Guide on which this chapter is based. It can be found on the web at www.ssc.rdg.ac.uk

It is simple and useful to include the metadata within the data entry process. Compare Fig. 11.14 with Fig. 11.1 to review the potential of EXCEL for complete data entry for simple datasets.

It is also possible to use EXCEL for the entry of more complex datasets, as we show in the next chapter. This takes more planning and we recommend that, for such tasks, consideration also be given to the use of a database package such as ACCESS, or a specialized data entry system such as EPI INFO, which is distributed by the Centers for Disease Control and Prevention (CDC), Atlanta, USA. See Chapter 23 for more information.

In supporting the use of a spreadsheet package for data entry we have been driven, to some extent, by its popular use. It is clear that many people will continue to use a spreadsheet for their data entry and this chapter suggests ways of making the entry effective. There are however limitations. For example, there are no easy facilities for skipping fields, conditional on the entry of initial codes and are no automatic facilities for double data entry. The graphics in EXCEL as illustrated in Section 11.5 are meant primarily for presentation and there are no boxplots or other exploratory techniques that could assist in data scrutiny.

Spreadsheets are intended as jack-of-all-trades software. They are certainly not the master of data entry. Hence, if the data entry component is large or complex, a spreadsheet should not be the only contender for the work.

Chapter 12

The Role of a Database Package

12.1 Introduction

In Chapter 10, we emphasized the importance of having a good strategy for data management in research projects. We also stated that when spreadsheets are used, they should be used with the same discipline that is imposed automatically when using a database package. In Chapter 11, we explained what we mean by 'using a spreadsheet with discipline'.

This chapter is to help researchers and research managers to decide whether they need a database package to manage their data. Most projects will already make some use of spreadsheets, and so the question is 'If we are reasonably confident working with a spreadsheet, why do we need to learn about a database package as well?' We use Microsoft ACCESS as an example, but the concepts are general and apply equally to any relational database package.

There are many textbooks on database packages, but they concentrate primarily on how to use the software. We concentrate on whether the software is needed and if so, what skills do different members of the project team require to use the software efficiently.

12.2 Data Management in EXCEL

In this section, we review some aspects of data management within EXCEL, with the assumption that the reader has some familiarity with a spreadsheet package. We use a survey for illustration to complement our use of an experiment in the previous chapter.

12.2.1 Survey data in EXCEL

The data used in Fig. 12.1 are taken from an activity diary study carried out in Malawi. Individuals within households kept a record of activities carried out at four different times of the day. Households are grouped into mbumbas or clusters. A cluster is a set of households for a mother, her adult daughters, their husbands and children. There are therefore three levels of data, namely *Mbumba*, *Household* and *Person*. In an EXCEL workbook, it is convenient to store each level of data in a separate sheet. Each sheet is given an appropriate name.

Fig. 12.1. Extract from EXCEL showing many worksheets in one file.

	A	B	C	D	E	F	G	H
1	ID	Hsehold	Mbumba	PersonNo	IndividualName	Age	Relationship	Gender
31	2311	3	2	11	Roderick	5	3	1
32	2312	3	2	12	Regina	2	3	2
33	2413	4	2	13	Mr Mukhumba	30	1	1
34	2414	4	2	14	Olaliya	25	2	2
35	2415	4	2	15	Donata	8	3	2
36	2416	4	2	16	Gladys	6	3	2
37	2417	4	2	17	Charles	3	3	1
38	2518	5	2	18	Hilda	20		2
39	2619	6	2	19	Uncle	65	7	1
40	3101	1	3	1	Mai	70	4	2
41	3102	1	3	2	Elizabeth	45	1	2
42	3103	1	3	3	Enoch January Manyela	23	3	1
43	3104	1	3	4	Binette January Manyela	21	3	2

Mbumba / Household / Person / Codes

In this survey, the *Mbumba* level includes the name of the mbumba, and its location, etc. At the *Household* level, the family name is stored. The *Person* level includes the name, age and gender of the individual. The unique identifier for the person is a combination of mbumba number, household within mbumba and person within mbumba. Thus, person 2518 is the 18th person in mbumba number 2 and is in the fifth household in mbumba 2. Figure 12.1 shows the details of the *Person* level sheet. The mbumba and household numbers are also stored at this level and these act as a reference to the household and mbumba level sheets.

Much of the data concerned activities which were recorded at four different times during the day. They are stored on a fourth sheet as shown in Fig. 12.2, though a better way to store these data is demonstrated later. The recording of these activities has introduced a fourth level to the data, namely a *Time-of-day* level.

Fig. 12.2. Extract from the Activities worksheet in the EXCEL file.

	A	B	C	D	E	F	G	H	I	J	K	L	M
1	PERSON	DATE	TIME OF DAY	ACTIVITY 1	ACTIVITY 2	ACTIVITY 3	ACTIVITY 4	ACTIVITY 5	ACTIVITY 6	ACTIVITY 7	ACTIVITY 8	ACTIVITY 9	ACTIVITY 10
2	2101	01/03/98	1	14									
3	2101	01/03/98	2	30	29								
4	2101	01/03/98	3	17									
5	2101	01/03/98	4	16	29	30							
6	2205	01/03/98	1	29	28								
7	2205	01/03/98	2	29									
8	2205	01/03/98	3	17	11								
9	2205	01/03/98	4	30	29								
10	2206	01/03/98	1	16	14								
11	2206	01/03/98	2	27									
12	2206	01/03/98	3	16	13								
13	2206	01/03/98	4	16	17	28	29						
14	2518	01/03/98	1	45									

Person Activities / Activity

Codes have been assigned to the activities, and a coding table is stored in a fifth sheet in the same file. A maximum of ten activities at any one time of the day is assumed.

12.2.2 Data entry forms

A useful feature is the ability to use a form for data entry. Choosing *Form* from EXCEL's *Data* menu produces the form shown in Fig. 12.3.

12.2.3 Linking data from different sheets

We have mentioned that each person is assigned a unique identifier. This identifier is used in the *Person Activities* sheet and acts as a link to the *Person* level data. Using this link we are able to view data from the *Person* level alongside data in the *Person Activities* sheet. For example, Fig. 12.4 shows the *Person Activities* sheet with additional columns for *Age* and *Gender*. We have used EXCEL's VLOOKUP function to display data stored in the *Person* level sheet. The key point here is that these data are only stored once; in the *Person* level sheet, but using VLOOKUP we are able to view them in other sheets. Storing a data value just once helps to minimize errors. This has been achieved by dividing data into levels and storing each data item at the appropriate level.

Fig. 12.3. Data entry forms in EXCEL.

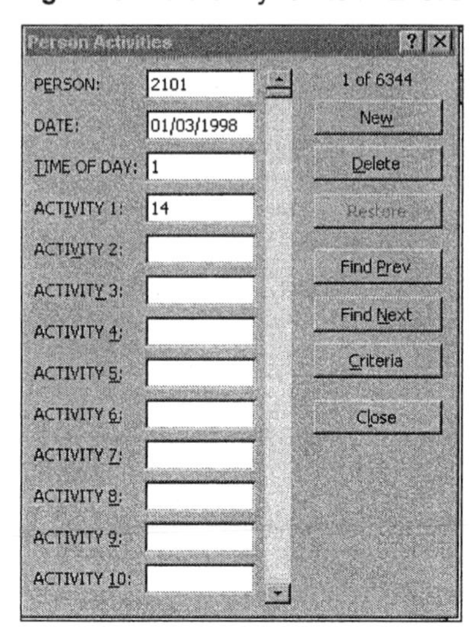
Person Activities

PERSON: 2101
DATE: 01/03/1998
TIME OF DAY: 1
ACTIVITY 1: 14
ACTIVITY 2:
ACTIVITY 3:
ACTIVITY 4:
ACTIVITY 5:
ACTIVITY 6:
ACTIVITY 7:
ACTIVITY 8:
ACTIVITY 9:
ACTIVITY 10:

1 of 6344
New
Delete
Restore
Find Prev
Find Next
Criteria
Close

Fig. 12.4. Using VLOOKUP to combine data from separate worksheets.

C2 = =VLOOKUP(A2,Person!$A$2:$H$93,6,FALSE)

	A	B	C	D	E	F	G	H	I	J	K	L	M	N	O
1	PERSON	DATE	***Age***	***Gender***	TIME OF DAY	ACTIVITY 1	ACTIVITY 2	ACTIVITY 3	ACTIVITY 4	ACTIVITY 5	ACTIVITY 6	ACTIVITY 7	ACTIVITY 8	ACTIVITY 9	ACTIVITY 10
2	2101	01/03/98	55	2	1	14									
3	2101	01/03/98	55	2	2	30	29								
4	2101	01/03/98	55	2	3	17									
5	2101	01/03/98	55	2	4	16	29	30							
6	2205	01/03/98	40	1	1	29	28								
7	2205	01/03/98	40	1	2	29									
8	2205	01/03/98	40	1	3	17	11								
9	2205	01/03/98	40	1	4	30	29								
10	2206	01/03/98	31	2	1	16	14								
11	2206	01/03/98	31	2	2	27									
12	2206	01/03/98	31	2	3	16	13								
13	2206	01/03/98	31	2	4	16	17	28	29						

Person Activities / Activity / Mbumba / F

12.2.4 Activity level data

In this survey, respondents were asked to list the activities they carried out at particular times of the day as shown in Fig. 12.4. This is an example of a multiple response question that is common in surveys. A respondent could list one or more activities and the number of activities is different for each person.

One way of entering and storing the activity data is shown in Figs 12.2 and 12.4, but this results in a non-rectangular data block, because not all respondents have 10 activities at each time of day.

An alternative way of entering these data is to consider an *Activity* level rather than a *Time-of-day* level, as shown in Fig. 12.5, where each row of data now refers to an activity rather than a time of day. This layout uses more rows of data but has the advantage of a simple rectangular structure with no arbitrary limit on the number of activities. We shall see later that this structure is the natural one to use if the data are to be stored in a database package.

Fig. 12.5. A single activity per row.

	A	B	C	D
1	PersonID	Date	TOD	Activity
2	2101	01/03/98	1	14
3	2101	01/03/98	2	30
4	2101	01/03/98	2	29
5	2101	01/03/98	3	17
6	2101	01/03/98	4	16
7	2101	01/03/98	4	29
8	2101	01/03/98	4	30
9	2105	01/03/98	1	21
10	2105	01/03/98	1	27
11	2105	01/03/98	2	17
12	2105	01/03/98	2	29
13	2105	01/03/98	3	26
14	2105	01/03/98	3	17
15	2105	01/03/98	3	28
16	2105	01/03/98	4	28
17	2105	01/03/98	4	29
18	2205	01/03/98	1	29
19	2205	01/03/98	1	28
20	2205	01/03/98	2	29
21	2205	01/03/98	3	17

12.2.5 Pivot tables

Once the data are entered, they need to be analysed. Simple analyses usually consist of summary tables and graphs; both are standard features of spreadsheet packages. Figure 12.6 shows a summary table that uses EXCEL's powerful *Pivot table* feature. These are in effect cross-tabulations with the advantage of being interactive; you can easily swap rows and columns for instance. This Pivot table was created using the activity data, where a subset of the activities has been chosen and are shown as row headings. Individuals have been grouped into boys, girls, men and women depending on their age and gender, and these groupings appear as column headings in the table. The cells of the table show the number of records falling into each category. Such tables can also give percentages and other summary values. If the original data are changed the Pivot table can be refreshed to reflect these changes.

Fig. 12.6. Pivot table in EXCEL.

	A	B	C	D	E	F
1	Count of PersonID	Group				
2	Activity Text	Boys	Girls	Men	Women	Grand Total
3	Drawing water	6	121	106	510	743
4	Gathering firewood and chopping wood	22	19	31	102	174
5	Ironing, visiting a taylor, paying debts	1	2	13	2	18
6	Preparing food, making porridge	19	140	86	851	1096
7	Shopping at the market or grocery	3	30	51	107	191
8	Grand Total	51	312	287	1572	2222

12.2.6 Review of EXCEL

When a spreadsheet is used with discipline, it is adequate for entering and managing data that have a simple structure. We define a simple structure as one which does not have too many levels. When the structure is more complex than this, spreadsheet data entry and management becomes more difficult. In our example which had four levels we saw that the multiple response question on the activities carried out at a particular time of day, was easily handled by entering the activity data into a separate sheet. When surveys have more than one multiple response question, the data entry requires yet more tables.

A second problem can be seen in Fig. 12.3 where we used a data entry form. When there is a lot of data it is sensible to make the data entry process as simple as possible, i.e. to make the form on the screen resemble the questionnaire form, and this can not be done effectively using EXCEL alone. Microsoft ACCESS allows the creation of forms from within EXCEL via the Microsoft ACCESS Links Add-In for EXCEL. When you use this feature, EXCEL will create an ACCESS database file with your current worksheet as a linked table; changes made to the data within ACCESS will be reflected in the EXCEL file. With this feature, you have more flexibility on the design of the form and can exploit all the form design features of ACCESS. It should be noted, however, that validation rules set up in EXCEL are not carried through to ACCESS; you would need to set checks in the ACCESS form itself.

Another possible limitation, when dealing with complicated data structures, is that we often want to summarize the data in many different ways. In EXCEL, this is usually equivalent to writing several different simple reports, with each one going on to a new sheet. Once we have several report sheets it is important to document the workbook well, so that we can refer back to this in future.

EXCEL and other spreadsheets have major strengths. These include the fact that what you are doing is always visible. Spreadsheets are also powerful and very flexible. Set against this is the fact that it is difficult to work with discipline if datasets are large, complex or both in their structure. Then a structured approach is needed for entry and management to fully exploit the data. A database package provides this structure.

12.3 Components of a Database Package

In this section we build on the ideas from Section 12.2, and look at designing a database, data input, and data use in the context of a database package. The terminology inevitably includes standard database jargon. Readers will now be able to understand consultants as well as the literature that extols the virtues of databases!

12.3.1 Designing the database

In a database package, data are stored in *tables*. The example in Section 12.2 had four tables respectively for the *Mbumba*, *Household*, *Person* and *Activity* levels. In a database package, the tables must be created before the data can be entered. As a minimum, you must specify the number of fields or columns needed for the data, give a name to each field and define the data type, e.g. text or numeric. This already imposes much of the discipline that we have already encouraged for the use of a spreadsheet. The table design screen in Fig. 12.7 is where the field names and data types are set up

Fig. 12.7. Table design in ACCESS.

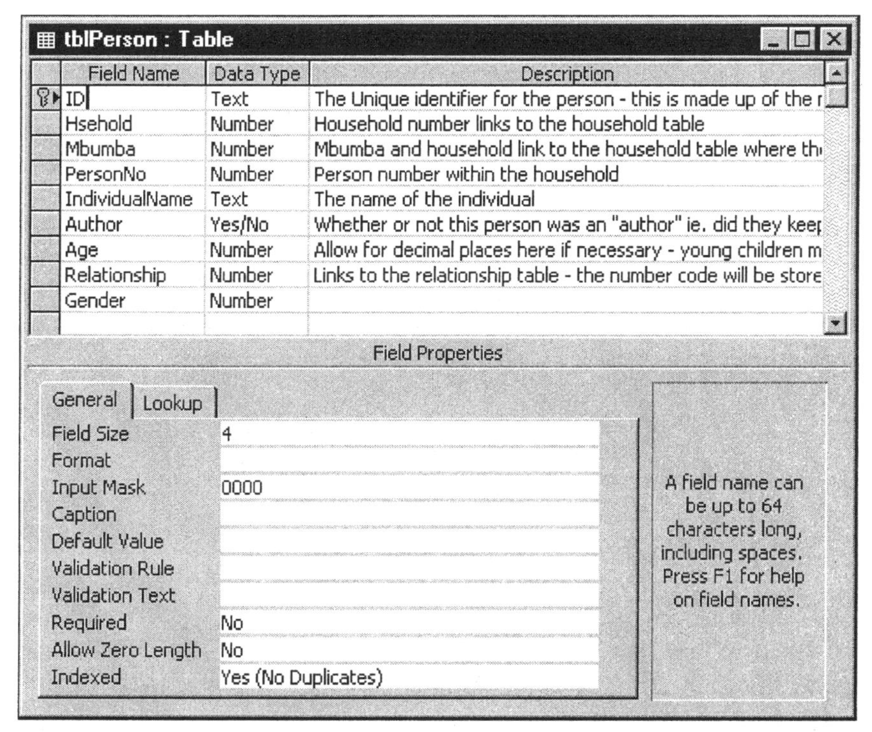

This screen shows information on the *Person* level table. Each field must be named and the data types defined. Once the table is created, we can enter data

through the datasheet or spreadsheet view (Fig. 12.8). The datasheet resembles the EXCEL worksheet. It is tailored for the data you want to enter; each column refers to a field in the table and will only accept data of the type specified in the table design. There is no limit to the number of rows you can enter other than the physical limit on the size of your disk. One difference between the datasheet in ACCESS and the worksheet in EXCEL is that there is no automatic numbering of the rows in ACCESS. However, information at the bottom of the window indicates the total number of records and which record or row you are on.

Fig. 12.8. 'Datasheet' or spreadsheet view of Person level data.

tblPerson : Table

ID	Hsehold	Mbumba	PersonNo	IndividualName	Author	Age	Relationship	Gender
2101	1	2	1	Mai Mazinga	☐	55	1	Female
2102	1	2	2	Mercy	☑	18	3	Female
2103	1	2	3	Tokozani	☐	15	3	Male
2104	1	2	4	Charity	☐	1	6	Female
2105	1	2	5	Unknown	☐			
2205	2	2	5	Mr Nangwale	☑	40	1	Male
2206	2	2	6	Martha	☐	31	2	Female
2207	2	2	7	Enifa	☐	11	3	Female
2308	3	2	8	Frank Filipo	☐	30	10	Male
2309	3	2	9	Femia	☐	27	2	Female

Record: 1 of 94

As with a spreadsheet, it is important that a database package is used with discipline. Minimal discipline – defining the number of fields and their data type – is enforced, but you should normally go beyond this minimum. As an example, we explain why it is important that all tables have what is called a *primary key*.

All data, whether stored in a database, a spreadsheet, or elsewhere, should have a unique identifier for each record. This may be a single field or a combination of fields. In EXCEL and other spreadsheets, there is no way to enforce uniqueness for this identifier and therefore duplicates can occur. In ACCESS and other database packages, however, you can and should set a primary key for each table. This is either a single field or a combination of fields, which acts as a unique identifier. It is always unique; duplicates in the primary key are not allowed. At the *Person* level, the unique identifier is the ID. Referring again to Fig. 12.7 this field has a symbol, which looks like a key, to the left side of the fieldname, to indicate that this is the primary key field for this table. In many cases, the choice of the primary key field is obvious.

Now consider a situation where the primary key field is not so clear-cut. Data at the *Activity* level include *PersonID*, *Date*, *TOD*, and *Activity*. An extract of these data is shown in Fig. 12.9.

Clearly, none of these fields is unique by itself. Thus, we must look at combinations of fields and when we do, we find that the only unique combination is that of all four fields. This combination can be defined as our primary key, but multi-field primary keys of more than two fields become difficult to handle and can easily lead to mistakes when setting up relationships.

Fig. 12.9. 'Datasheet' view of Activity level data.

PersonActivities : Table

PersonID	Date	TOD	Activity
2101	01/03/98	1	14
2101	01/03/98	2	30
2101	01/03/98	2	29
2101	01/03/98	3	17
2101	01/03/98	4	16
2101	01/03/98	4	29
2101	01/03/98	4	30
2102	01/03/98	1	14
2102	01/03/98	1	27
2102	01/03/98	1	21
2102	01/03/98	1	13
2102	01/03/98	2	17
2102	01/03/98	2	28

Record: 14133 of 94537

An alternative is to use an *autonumber* field as the primary key. This will assign a unique number to each record. However, we still want to ensure that the combination of the four original fields is unique. To do this we define an *index*.

An index can be created for any field and any combination of fields and speeds up the process of sorting and selecting. Once an index has been created, it can be made unique, making it impossible to enter duplicates into that field or combination of fields.

Figure 12.10 shows the table design screen for the Activity level data and includes the *autonumber* field that we have added as the primary key. The *Index* window shows that there is an index called *Identifier* which is a combination of the original four fields. The *Unique* property has been set to *Yes* for this index.

Fig. 12.10. Table design with Index window.

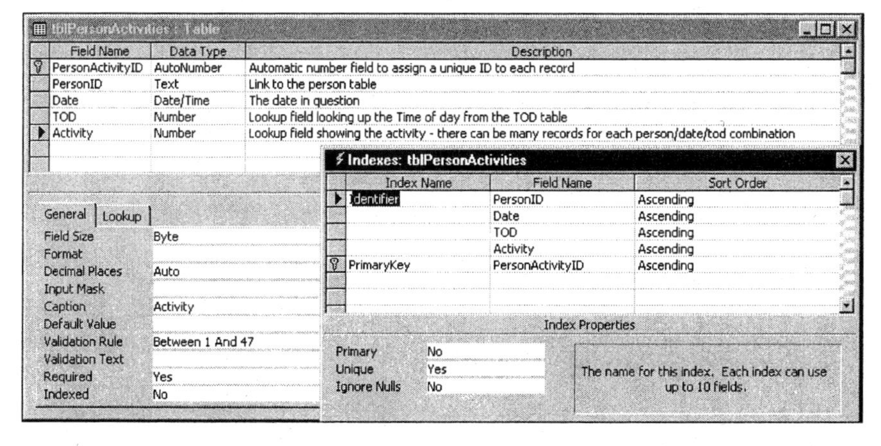

An important extra that comes with relational database packages such as ACCESS, is the ability to create relationships or links between tables of data. These links can be built into the design of the database. Figure 12.11 shows the same structure of data that we developed in EXCEL but this time in ACCESS. The four levels are translated into four interrelated tables. The relationships are all 'one-to-many' in that a single record in one table is related potentially to many records in another table. For example, each household has many individuals.

Fig. 12.11. Database structure in ACCESS.

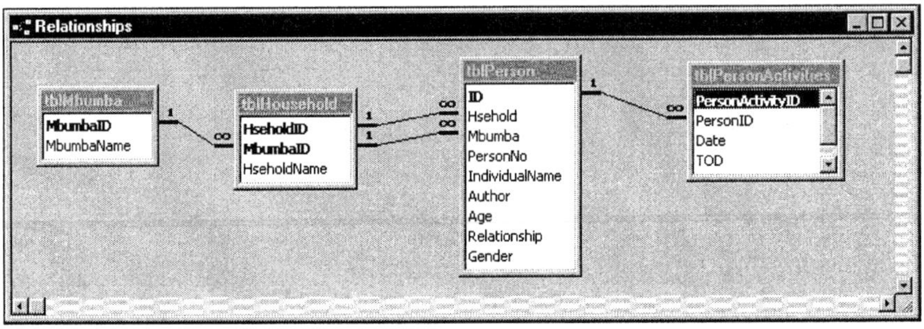

Relationships

tblMbumba: MbumbaID, MbumbaName

tblHousehold: HseholdID, MbumbaID, HseholdName

tblPerson: ID, Hsehold, Mbumba, PersonNo, IndividualName, Author, Age, Relationship, Gender

tblPersonActivities: PersonActivityID, PersonID, Date, TOD

ACCESS includes a set of rules known as *Referential Integrity*. When enforced this helps to validate relationships by preventing you from entering a record in a table on the 'many' side of a relationship where there is no corresponding record in the table on the 'one' side. For example, you would not be able to enter details of an individual until there was a household defined for that individual.

Fig. 12.12. Extract from the table of Activities.

tblActivity : Table

ActCode	ActText
1	Cultivating/checking the field
2	Planting/transplanting crops
3	Tilling/ridging
4	Sowing seeds/preparing seeds/materials for planting/maintaining nursery
5	Weeding
6	Banking
7	Fertilising a field/dimba or applying manure/watering crops
8	Applying pesticides/removing infected plants
9	Harvesting or hauling crops from the fields, e.g. maize, cassava, potatoes, etc.
10	Buying agricultural inputs/collecting starter packs and other benefits
11	Feeding animals/Dipping animals/Gathering animal feed/ Cleaning a khola/Leading livestock to a khola
12	Gathering firewood and chopping wood
13	Drawing water
14	Sweeping, washing, weeding, cleaning, chasing mosquitoes, killing insects, etc.
15	Harvesting green crops for relish
16	Preparing food, making porridge
17	Eating
18	Going to a milling machine
19	Pounding, winnowing, unsheathing, shelling, drying maize
20	Making fire
21	Bathing self or child, Shaving

Multiple tables are valuable and not too difficult to use. Consider for instance the *Activities* in our example set of data. The activities are coded from 1 to 47 and the code is stored in the database. It would be relatively easy to add a two-column table containing these codes and their associated descriptions. Figure 12.12 shows some of the data from the *Activities* table, while Fig. 12.13 shows how this table and similar tables for the *Time-of-day* and *Family relationship* can be added to the database structure.

Unlike a spreadsheet, where seven tables with data would be confusing, this is quite a simple structure for a database. A database would typically have between 5 and 50 tables.

Fig. 12.13. The complete database structure.

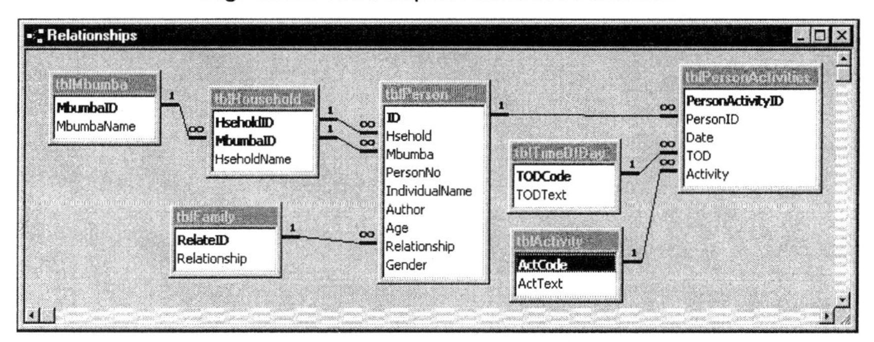

12.3.2 Input into the tables

We mentioned earlier that data can be entered directly into the table via the datasheet. We saw an example datasheet in Fig. 12.8. When there is just a small amount of data, this is easy and adequate. Figure 12.14 shows all five records from the *Mbumba* table in spreadsheet view.

Fig. 12.14. Mbumba level data.

tblMbumba : Table

MbumbaID	MbumbaName
1	MUTHOWA
2	MAZINGA
3	MARICHI
4	CHIMVULA
5	SIMEON

Record: 1 of 5

For larger volumes of data, it is more common to set up special data entry forms. These need more practice than in EXCEL but simple forms are very easy to design. The form in Fig. 12.15 is for entering data on individuals. This form was generated automatically from the corresponding table using one of the *autoform* wizards in ACCESS. It is similar in structure to the form in Fig. 12.3 that was given in EXCEL.

Fig. 12.15. A simple Person level data entry form.

tblPerson

ID: 1101
Hsehold: 1
Mbumba: 1
PersonNo: 1
IndividualName: Mr Muthowa
Author:
Age: 70
Relationship: 1
Gender: M

Record: 1 of 94

Figure 12.16 shows the same form after a few simple design changes. It is easy to start with an automatically generated form and change the layout to match the questionnaire. The ease of producing forms of this type in ACCESS is one of the reasons for its popularity.

Fig. 12.16. Variation in a Person level data entry form.

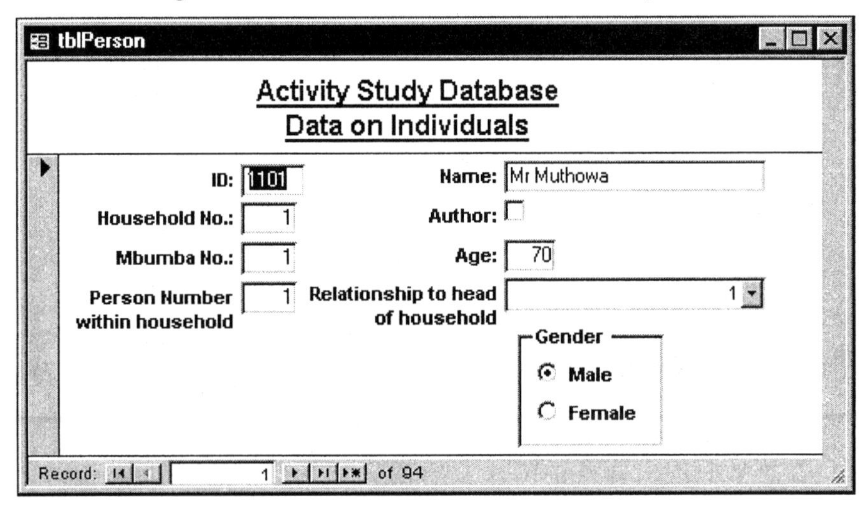

tblPerson

Activity Study Database
Data on Individuals

ID: 1101
Household No.: 1
Mbumba No.: 1
Person Number within household: 1
Name: Mr Muthowa
Author:
Age: 70
Relationship to head of household: 1
Gender: Male / Female

Record: 1 of 94

However, a survey form often includes data from more than one table. In our case, the *Person* form included space to record the activity level data. Ideally, we would therefore want to enter data from a single questionnaire into two or even three tables at the same time. This further step does require some degree of expertise but is still relatively easy in a database package such as ACCESS. This is important as it makes the data entry much easier and hence more reliable.

Figure 12.17 shows a form that was used in this study. The top part of the form is for entering data on the individuals. The bottom half of the form is for entering activity data. This is actually a *sub-form* and data entered here are stored in the *Activity* table.

Because of links between the main form and the sub-form you only see the activity data for the individual displayed in the main form. Generally, there is a one-to-many relationship between data in the main form and data in the sub-form. In Fig. 12.17, this particular individual has several activities for the morning of 1 June 1998. Thus, the multiple response question on the different activities in each time period translates into a separate record for each response.

In our use of EXCEL we emphasized the importance of distinguishing between the person who designs the system for data entry and the staff who do the actual entry. This is now a much clearer distinction with a database package. If there is a complex survey or database, it becomes a skilled task to design an effective data entry system.

Fig. 12.17. Person level form with an Activity level sub form.

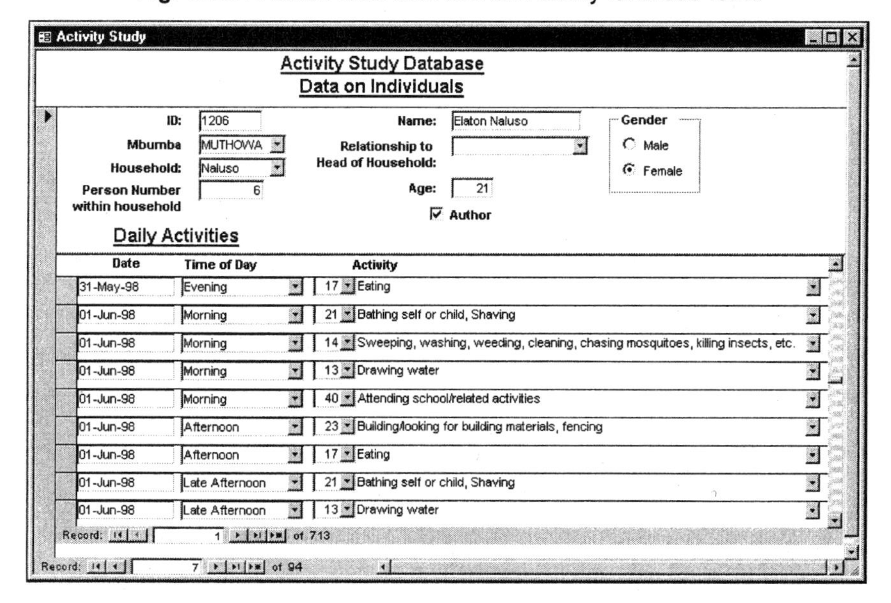

12.3.3 Validation and verification

Validation means checking the data as they are entered, and verification is checking the data once they have been entered. The auditing process described in Section 11.3.7 can be thought of as verification according to this definition.

In the entry of survey data it is important that the data are verified. This may be by providing checks as the data are entered, or by a double entry system where two data entry operators enter the same data into separate files, the files are compared, and differences checked against the original paper version of the data.

Double data entry is an automatic feature in some software that is designed for survey data entry. One example of such software is EPI INFO which combines many of the database concepts already described with easy facilities for double data entry. This is free software which can be downloaded from the Internet.

EPI INFO offers a rigorous method of data verification. After records have been entered and saved in a file, there is an option to re-enter and verify records in the existing data file. Data are entered exactly as for new records. When a field entry matches the data in the file the cursor moves on exactly as for new entries. When an entry does not match, a message appears and the operator is given a chance to re-enter the value or to compare the original entry with the new one and make a choice.

Data from EPI INFO can be imported into ACCESS. It is therefore possible to use software such as EPI INFO for the main data entry phase and then transfer the data to ACCESS for storage and management.

In ACCESS, validation rules can be set on individual fields. Figure 12.10 shows a validation rule of *Between 1 And 47* for the *Activity* field. It is also possible to set validation rules on the table. This might be used for instance, where the value in one field cannot exceed the value in another field. For example, if we were storing the number of people in a household and the number of children, there cannot be more children than people and so we can set a validation rule of *[People]>[Children]* for the table.

Database packages such as ACCESS were primarily designed for business users where the process of entering data and using the data is an on-going cycle. The case for double entry is less clear in these circumstances, and double data entry checking procedures are not provided by ACCESS or other similar database packages. In surveys and scientific work, where there is a recognized data entry phase, verification is necessary and double data entry often desirable. It is possible, via the use of a series of queries, to compare the contents of two databases used for double entry, but the process requires a systematic approach and a good level of organizational skills.

12.3.4 Using the data

In Fig. 12.6, we showed how a Pivot table was used in EXCEL to summarize and present the data. In ACCESS, we use *queries* and *reports* to do the same.

A simple query provides a way to view or summarize subsets of data from a given table in the database. An example is shown in Fig. 12.18, which is similar to the Pivot table produced by EXCEL in Fig. 12.6.

In a database the tables are linked, and therefore it follows that queries can involve data from multiple tables. Figure 12.19 shows the results of a query that includes data at both the *Person* level and the *Activity* level. The query counts the number of activities for each individual.

Fig. 12.18. Crosstab query.

Activities : Crosstab Query

	Activity Description	Boys	Girls	Men	Women	Total
▶	Gathering firewood and chopping wood	106	251	245	1843	2445
	Drawing water	240	1118	1423	4612	7393
	Preparing food, making porridge	118	1055	1013	7715	9901
	Ironing, visiting a taylor, paying debts	9	8	93	25	135
	Shopping at the market or grocery	64	187	598	810	1659

Record: 1 of 5

The results from a query can be used in a report, used as the basis for further queries, viewed with a form, exported to another package or stored in a new table.

Fig. 12.19. Query counting activities for selected individuals.

Query1 : Select Query

	IndividualName	Age	Sex	CountOfActivity
▶	Elaton Naluso	21	Male	713
	Enoch January Manyela	23	Male	2350
	Joseph Ephraim	35	Male	3040
	Lyton Yasini	18	Male	964
	Matheus Anderson	30	Male	1236
	Mercy	18	Female	1150
	Mr Chigonamadzi	60	Male	2983
	Mr Mazinga	37	Male	1087
	Mr Naluso	37	Male	886
	Mr Nangwale	40	Male	1192
	Roderick Mkwezelamba	20	Male	928
	Simeon Magomero	31	Male	2846

Record: 1 of 12

Another way of using the data in ACCESS is to create reports. A report provides a snapshot of the data at a given time. Reports can be designed to show the same sort of data that you would see in a query but they extend the idea of a query by allowing a display of the data or summary to suit your needs. The extract below in Fig. 12.20 is taken from a report that lists the activities for each individual and for each time period.

Unlike EXCEL, when you save queries and reports you do not generally save the results. Instead, you save the instructions that produce the results. Whenever a query or report is run, the data are taken from the underlying table(s). Thus, the results always reflect recent changes in the data. This is a little like refreshing a Pivot table in EXCEL so it reflects any changes in the data. The results of a report can be viewed on screen, sent to a printer or saved in a *snapshot* file. Some versions of ACCESS include a *Report Snapshot Viewer*, which is used to view these snapshot files.

Fig. 12.20. Report listing activities for each time period.

Activity Study

Name: Mai Mazinga

Age: 55

12 January 1998

Morning

1 Attending a funeral rite

2 Harvesting green crops for relish

3 Preparing food, making porridge

Afternoon

1 Preparing food, making porridge

2 Eating

Late Afternoon

1 Gathering firewood and chopping wood

2 Preparing food, making porridge

3 Eating

Evening

1 Playing, resting, sleeping, basking in the sun, playing games

13 January 1998

14 January 1998

Morning

1 Eating

2 Cultivating/checking the field

3 Hearing cases

4 Preparing food, making porridge

Afternoon

1 Preparing food, making porridge

2 Gathering firewood and chopping wood

3 Playing, resting, sleeping, basking in the sun, playing games

Late Afternoon

1 Preparing food, making porridge

2 Eating

3 Chatting, visiting friends, families, places, going out for social activities

Evening

1 Playing, resting, sleeping, basking in the sun, playing games

Because ACCESS stores the instructions to run the queries and reports, it is possible to do a pilot survey, or collect a few records initially and develop all the queries and reports, based on just these few records. The data are used to check that you are producing the correct style of table or summary. When you have entered all the real data, you can then run the queries and/or reports to produce the results.

12.3.5 Objects in ACCESS

ACCESS refers to tables and forms as *objects*. An ACCESS database can include up to six different types of object. We have so far mentioned four of these namely: tables, forms, queries and reports. The remaining two, *macros* and *modules* can be used to automate tasks and pull the other objects together into a user-friendly database application. Use of these objects is not essential for good data management practice. All the objects within a database are accessible from the main database window, an example of which is shown in Fig. 12.21.

The objects are grouped by type and by clicking on the appropriate tab it is easy to move say from the list of tables to the list of forms. This is an example of a data management system.

12.3.6 Exporting from databases

One aspect that often discourages users from adopting a database package is the difficulty they perceive in extracting data in a format ready for analysis. However, by its very nature ACCESS is more flexible in this respect than EXCEL.

Using queries, it is easy to extract subsets of the data based on given criteria, view data from linked tables, summarize data, and perform simple calculations and summaries. Data produced from queries, can at the click of a button, be exported into EXCEL.

Fig. 12.21. Database window from ACCESS.

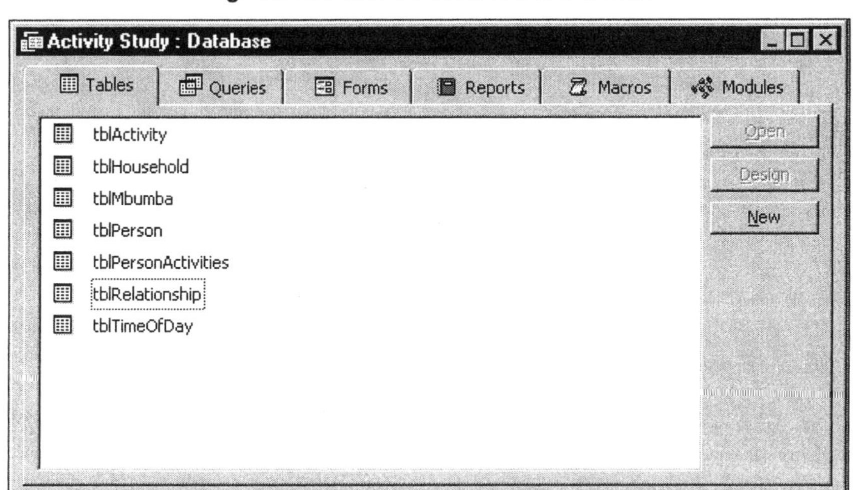

Many statistics packages now use ODBC (Open DataBase Connectivity) to read data directly from database packages. Transferring data between packages is no longer the problem it once was.

Data can therefore be stored and managed in ACCESS, and subsets extracted to EXCEL or another package for analysis as and when needed. Ease of data transfer enables you to use the best features from each package.

12.3.7 Review of ACCESS

Database packages are designed to handle large and complex volumes of data. We believe that avoiding them in favour of spreadsheets is risky when exploiting research data fully. Database packages impose much of the disciplined approach that we have been encouraging. Using a database does not guarantee that you will have complete, error-free data, but used efficiently, it is possible to move nearer to that goal. Simply by designing the tables for your data you are forced to think about both the data and its structure.

We saw in Section 12.3.5 how the different objects in ACCESS are kept separate and easily accessible within the database. This differs substantially from EXCEL where data and results, Pivot tables, calculations, charts, and so on, are all stored in the same way as sheets within the workbook. Unless you are well organized and document all your work it is not always easy to find the sheet you are looking for. In ACCESS, the data and results are separated. In general, the results are not stored in the database but are generated each time the query or report is run.

A database may have to be used as a final step to leave a usable archive after the project has finished. In such cases, it is more efficient to use a database from the start, so the project team can benefit from the system for data management.

However, some expertise is needed to construct an effective database. Sometimes we find that inexperienced users do not add the relationships of the type shown in Figs 12.11 and 12.13. A database without relationships is just like a spreadsheet, except it is harder to see all the data.

12.4 The Data Flow

In this section, we follow the flow of data during the lifetime of a research project and the role of the database package in this process.

We consider four aspects: data entry, data management, data analysis and data archiving. For large volumes of data or data collected at more than two levels, using a database package for the entry and management is recommended. One of the roles of data management is to provide good quality data for analysis. The use of a database package does not necessarily guarantee this but when used effectively with validation checks, primary key fields, referential integrity on relationships and so on, we are at least moving in the right direction.

ACCESS is not generally sufficient for data analysis. Cross-tabulations are possible using queries, but EXCEL's Pivot table feature is far more flexible. Charts in ACCESS are extremely limited. This is the point where subsets of data should be exported to other packages. It is important to realize that once data are exported you have duplication; if you notice an error in the exported data the correction must be made in the database and the data exported again. If this is not done, data integrity can be compromised. The database should contain the final copy of the data.

Data archiving can be viewed as simply a copy of the database containing all the project data, but it can be much more. Ideally, it should also include copies of graphs, results of analyses and copies of programs run on the data. An archive CD should include all the data files and output files, whether they are in EXCEL, SPSS, ACCESS or another program. This all needs to be documented and one way to do this is to use a database. Earlier, we saw how easy it is to add tables to the database; why not add a table to store information about the analyses that have been carried out. A record could include the name of the data file, the name of the results file, the software used, the type of analysis, the date the analysis was carried out, the person running the analysis, and so on. In addition to text and numeric data, ACCESS can store images. It is therefore feasible to scan photos and maps and store them in the database as images.

12.5 Adopting a Database Approach

In this section, we consider team members who have some skills in EXCEL and are considering incorporating a database package into their work. With a spreadsheet, such as EXCEL, it is often adequate for individual staff members to start without a formal training course and simply add to their knowledge of the package as the need arises.

Spreadsheets are normally used on an individual basis, with data to be shared often being copied to each person. Databases can be used in the same way, but it is more effective to share the data from within a single database. This is the natural way to operate when computers are networked, but applies even if the database is on a single stand-alone machine.

Thus, the establishment of one or more databases will normally involve decisions of responsibility for the entry, validation and use of the data. This extra formality is usually also important to ensure good quality data.

When a database package is introduced, alternative ways range from employing an outside consultant to proceeding in a similar step-by-step approach as often used for EXCEL.

12.5.1 Employ an outside consultant

Here an outside consultant or database professional is employed to construct each database for the project. You inform the consultant on the data elements that need to be stored and specify how you would like to enter, view and extract the data. He/she then creates the database structure together with a set of queries, forms and reports. The consultant could also produce an interface to your database so that reports can be run and data extracted at the click of a button. This effectively turns your database into an application. At this level, all the project team needs to know is how to run this application. This option requires very little time and effort from the project team members.

However, this route would not usually be recommended. If none of the team members understand sufficient database principles, it is often difficult to specify exactly what is needed. Changes and additions are often required during the life of the project and it is both time-consuming and expensive to return each time to an outside consultant.

Another potential drawback is that, although it is easy to find database consultants, most are experienced in business applications. You may therefore be posing new challenges for them, both in the data entry requirements and in the necessary queries and reports.

12.5.2 Working in partnership with an outside consultant

A basic database knowledge is needed by project team members in order to work constructively with a consultant. For staff that are already familiar with

WINDOWS and EXCEL this might typically be a short course, with some of the time being spent on the construction of queries and reports.

The difficult part of the work is setting up the initial database, with the relationships and the data entry forms. A consultant could be used for this work. The system delivered should contain some queries and reports, and verification of the data should also be taken into consideration.

Once the structural work is done, it becomes relatively easy for the project staff to add extra queries or reports as needed. They could even make minor modifications to the structure. However, there is a difference between these two types of task and care needs to be taken. An error in a query will only affect the person who wishes to run the query, but an error when changing the worksheet structure could render the database unusable.

12.5.3 Construct the database in-house

The third option in creating a database is to construct the entire database in-house. This is the obvious approach if one of the project team is a database expert, but otherwise we counsel caution. It is just as easy to construct a poor database today, as it was to write a poor computer program in earlier days. The diagram of relationships could end up resembling a plate of spaghetti, making it difficult to generate reports or to modify the structure in any way.

12.5.4 Recommendations

In project teams that do not include a database expert, we recommend a partnership approach. The major change in database software in recent years has been the ease with which users who have relatively little experience can modify a system once it is in place.

Whereas with EXCEL there might be equal training, if any, for all team members, with ACCESS it is normally more appropriate to select a subset of the team for training in the basics of database management. They, possibly in conjunction with a consultant, could then deliver a one-day course to the rest of the team on the principles of the current database system for their data, once a test version is available.

Data entry staff would need special training. Their task should be easier because of the facilities available in a database system to facilitate efficient data entry. If the data entry is not simple, then the project team should request that further improvements be made.

Chapter 13

Developing a Data Management Strategy

13.1 Introduction

The issues raised in the last three chapters show the elements of good data management for natural resources research. In this chapter, we pull these ideas together and describe the components needed for an effective data management strategy.

So far, we have considered data management at the project level. The project may be small, such as an individual doctoral or master's thesis with limited activities, and mainly the responsibility of a single person. Larger projects may involve multiple activities with more than one researcher. In these cases the project is the natural level at which to develop and apply a data management plan, and this should be done as part of a data policy that is agreed upon at the department or institute level. The requirements for building a data management strategy are listed in Section 13.2 and the possible components are described in Section 13.3. This provides the framework for a project's data management plan, as considered in Section 13.4.

13.2 Requirements for Building a Data Management Strategy

In developing a strategy, it is useful to consider three groups of staff, namely:

- *Organizers*: the people who handle raw data on a routine (e.g. daily) basis. They set up filing systems, enter and check data and maintain data banks.

- *Analysers*: the people who analyse and interpret data. They are responsible for reducing raw data to useful information that can then be reported.
- *Managers*: the people responsible for providing an enabling environment for the first two groups. They must also ensure that all commitments to stakeholders are met.

Next, we consider two case studies to illustrate the importance of producing a data policy at a level higher than the project level. In the first example a research project report contained a summary table of results (Table 13.1). In the study, 64 farmers had been asked to rank four developments in descending order of importance; and 43 of them had ranked irrigation as the most important. We wished to ask some further questions of the data. For instance, we were interested to know about the 21 farmers who did not put irrigation first, in comparison with the other 43.

Table 13.1. The number of farmers giving particular ranks to different developments.

Rank	Irrigation	Fertilizer	Seed	Extension
1	43	10	6	5
2	10	24	16	14
3	8	23	20	23
4	3	7	22	32
N	64	64	64	64

When we looked for the original data, we found that the scientist concerned had left the institute where the research had taken place, and so we contacted her at her new workplace. She explained that none of the raw data had been taken away with her but had all been left in the institute on the computer that she had been using. It then transpired that, since the project had finished, the computer's hard disk had been reformatted, prior to the computer being allocated to someone else.

The second example concerns the coordinator of a regional crop network who used his final year to prepare a review of the results of trials that had been undertaken in multiple sites in about ten countries over several years. All the data had already been analysed for each individual trial, hence the task was simply to conduct and report on the combined analyses. However, this did not get very far. The assistants were used to the routine of data entry and analysis for individual trials, but not at handling the data management tasks that were now needed.

These two examples illustrate some of the components that need to be considered in developing a data management strategy. The first case study, where the data were destroyed after the scientist had left, indicates why a data policy usually needs to be formulated at a level in the organization that is more permanent than a typical project. The key problem in the second example was the lack of data management skills among the project staff, combined with ill-defined reporting and support mechanisms for project teams within the institute.

The need for training indicates that implementing a sound strategy will not be cost-free for the institute. As we mentioned in Chapter 10, there is a parallel between the management of financial data and the need for effective management of research data. It is generally accepted that accountants need years of training; likewise at least some data management training is needed for research staff.

The four requirements are: commitment, skills, time and money.

Commitment must come from all members of the project team, particularly the senior managers. These are the people with control over the financial budget and staff workloads. Each team member must play a part in data management and have a clearly defined role and responsibilities. The task should not be given to just one individual. Regular audits should be carried out on the data management strategy and data management should be part of staff performance reviews.

Fig. 13.1. Project life cycle: from problems, through data transformations to knowledge.

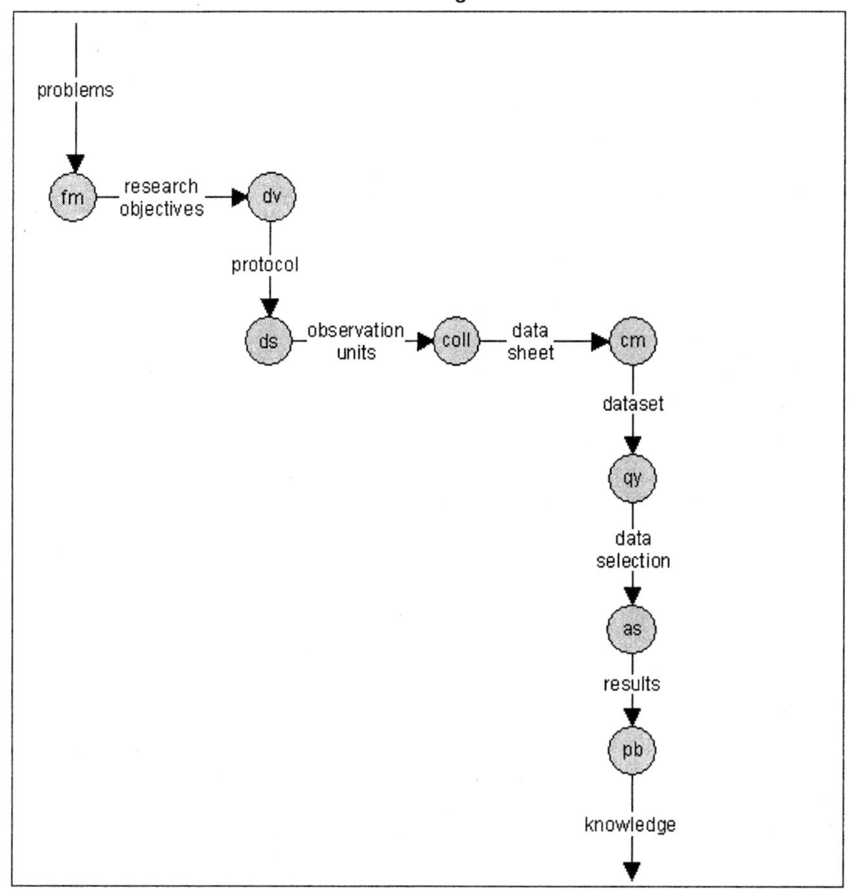

All staff must be equipped with the skills needed for their data management responsibilities. They must also be open to new techniques and be prepared to upgrade their strategy. It is particularly important that junior staff have the necessary skills to undertake the data management tasks expected of them. For example, data entry is often considered a trivial task that can be carried out by unskilled employees. However, a data entry clerk who understands the meaning of the data is more likely to spot errors before they get into the computer.

Time must be allocated for the data management tasks. Each team member must be aware of the data management activities for which they are responsible, and be given sufficient time to do them. In the past, data management was often not given the high profile it needs. Both funders and researchers considered the results of the analyses to be the key output from the project, and hence that output was allocated most of the resources in terms of both time and money. Now the data itself is increasingly recognized as an important project output. Hence, the resources should usually be available so that a well-documented functional data archive can be produced.

Where time must be given to tasks in a project, the funding must also be available to support it. All project proposals should demonstrate the existence of a data management plan. This plan should correspond to a section of the budget and indicate clearly how the corresponding funds will be used.

To explain the key components of a data management strategy, we will use a schematic diagram of a project life cycle (Fig. 13.1). In this cycle, we first *formulate* (fm) the research objectives to address particular problems. Given these objectives, we *develop* (dv) the research protocol, from which we *design* (ds) the study and in particular, identify the observation units (e.g. the focus groups, households or experimental plots). We then *collect* (coll) the data. They are *compiled* (cm) into well-structured datasets. We can then *query* (qy) these datasets and select subsets, which we *analyse* (as). Results of our analyses are then reported or *published* (pb), leading to knowledge.

We can take this conceptualization further, as shown in Fig. 13.2 where we have added some extra inputs. *Errors* are a negative input. They can come anywhere, but are illustrated with errors at the data collection stage. Aim to minimize their occurrence by training and motivating staff. Also, try to minimize the steps between their occurrence and their detection so as to increase the chance of being able to correct them. This underlies the importance of feedback loops in data management. The loop in Fig. 13.2 indicates that we might detect errors when compiling the results.

Skills are a positive input to the process. The example shown is of skills that influence the analysis. Data management skills are often associated with the stages of compiling (cm) and querying (qy) the datasets.

The immediate outputs from one step are rarely sufficient to perform the next step. We therefore need to be methodical about the way we organize all the products from each step, so they are available for future reference. For example, to analyse the selected data effectively you need information generated earlier, such as the research objectives. In Fig. 13.2, this is indicated by the *historical references* loop. When the project information is not well managed, it becomes difficult to complete the analysis when the data originators are either not available or cannot remember all the details that are critical for correct analysis and interpretation.

Fig. 13.2. Project life cycle: additional inputs and feedback loops.

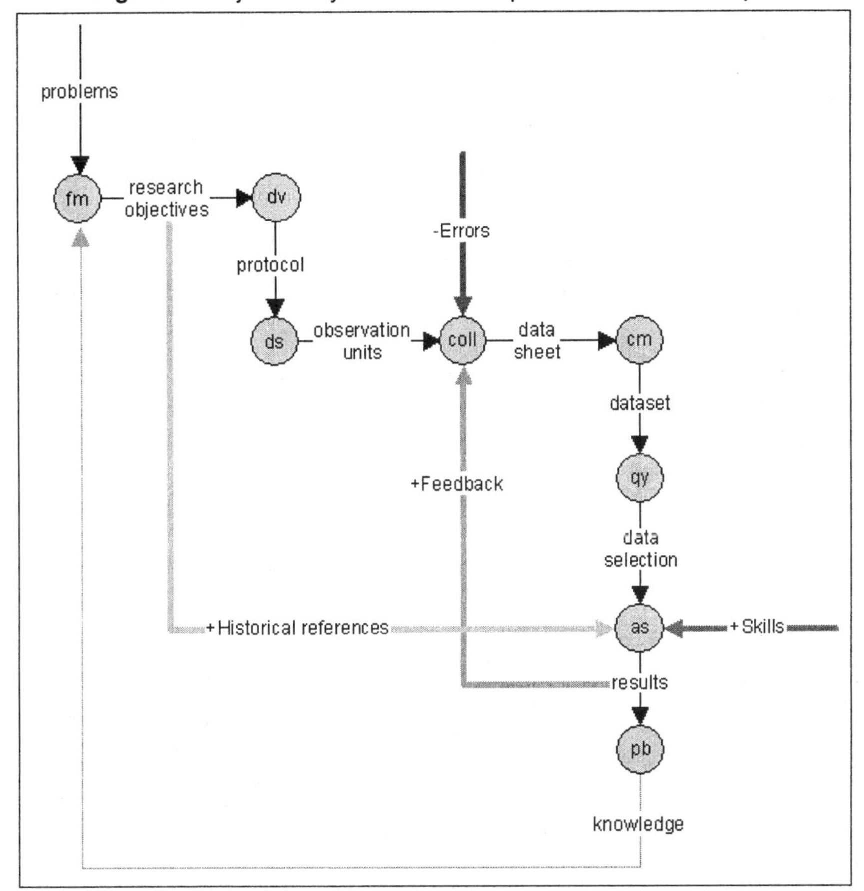

At the analysis stage, we use not only data from the current project, but also data and knowledge produced from earlier projects. This is sometimes known as secondary data analysis and can help establish a starting point for the analysis of the current project data. Of course, this is only possible if the datasets from these earlier projects have been well-managed and documented and are in a usable form.

With the outline as above, we identify the four key components of a data management strategy as follows.

- Transformations and their products: the data resulting from each of the transformations shown in Figs 13.1 and 13.2.
- Metadata: the documentation giving a description of all the information about the data. We described the metadata in Section 10.2.

- Data management plan: a reference for the project, against which the steps in the above components can be evaluated to monitor performance. We describe this component in Section 13.4.
- Data policy: the set of agreed principles that guide the structure of the plans and the metadata for each project. We describe this further in the following section.

13.3 Data Policy

A data policy describes the directions that guide the activities of the department, division or institute to ensure that projects have a consistent approach to data management. They can be considered as the overall ideals for the institution, while the data management plan for each project details how these ideals are to be implemented for that particular case. A typical policy would include the following directives:

- Minimize the times between data collection, entry and analysis.
- Standardize quality control procedures for all datasets.
- Maintain adequate backup protocols for all datasets.
- Establish and be able to distribute high-quality, long-term datasets.
- Ensure data and other product materials are archived in a timely fashion, and review the archive periodically to ensure continued readability.
- Facilitate data access and usability through improved metadata.

In Fig. 13.3, we give an example of a data policy. Among the points in this data policy, note that it proposes a decentralized system, with all the metadata stored centrally, but with pointers to the actual data. For example, a project may use climatic data as one input. The central system would store the metadata, which describes exactly which data were used, and indicates the location in the national meteorological centre where the data are held. The same might apply to projects that involve data from multiple institutes. Each institute would keep a copy of the metadata, but the actual data for each activity would be kept in the institute responsible.

One contentious issue that the data policy has to address clearly and unambiguously is that of data ownership and access. There are several competing traditions and expectations which need resolving, and require clarification at the start of any project. For example, many institutes treat research data as belonging to the institute rather than any individual, as it is the institute which normally has the legal responsibility for the resources used to generate it. However, there is a long-standing tradition in academia that raw research data are the property of the scientist who generated them. A conflict between these views can occur when the scientist moves to a new institution, or when the institute engages with new partners to work on the project.

Many NRM projects are multidisciplinary and multi-institutional. The terms and conditions of sharing data between scientists and institutes have to be clearly spelled out. The conflict here is often between the need for sharing the data, in order to meet project objectives, and the need of the scientists who generated the data to get recognition (through publication) for their contribution. It is becoming more common to find arrangements that put time limits on the exclusive use of a dataset by the researcher who generated it.

Fig. 13.3. Sample data policy.

Introduction

The data policy of <Name of institute> is intended to augment long-term success of research, ensuring that valuable data collected by researchers are properly and effectively utilized and managed, thereby making the best use of allocated project resources.

Principles

Data management issues will be addressed at the project level. Project level data management is a continuous process spanning the life of the project and beyond and is essential for the dissemination and utilization of project results. This process should be captured within a data management plan, which must be formulated and provided as part of each research project proposal.

<Name of institute> recognizes that project data management is a specialized activity requiring skills and resources that may not be available or accessible. However, developing a data management plan is a value adding activity that will:

1. Reinforce the principles of data management, thereby improving the longevity and effective use of information resources.
2. Provide a framework against which advice can be given.
3. Identify and document problem areas where there is a need for improved data management capacity.

<Name of institute> will use primarily a decentralized data management and distribution system. The centralized component will be a comprehensive inventory of metadata with pointers to the data location and key contact persons.

Implementation

Principal investigators should formulate a data management plan as part of the research proposal for a project. The data management plan should outline how the project will address:

1. Information and data flow.
2. Data documentation (e.g. metadata).
3. Data quality.
4. Technical issues (databases etc).
5. Dissemination of final project data.
6. Longevity and final archiving.
7. Performance indicators.

Increasingly, those who fund research are attaching contractual conditions on data sharing and requiring that data go into a public archive, even in some cases as soon as they are collected, not just at the end of a project. Data management policies and plans need to accommodate this.

In most cases, it should be a principle of data management that data collected using public resources is for the public benefit, and this will be maximized if anyone who can add value to a dataset has access to it.

The data policy is also likely to describe the roles and responsibilities for individuals, either within or outside the project team. Examples are as follows:

Data owner:

The <Name of appropriate institute> is the owner of all the research data and holds copyright to its policies, manuals and compilations of its information.

Data custodian:

A manager of the institute who has been delegated responsibility for a portion of the information resource on behalf of the institute, to ensure its integrity and accuracy. Responsibilities include:

- Identify items of corporate data and distinguish primary source data.
- Identify and document who is allowed access to the data and what level of access they should have.
- Authorize downloads and uploads of corporate data.
- Identify and document the process for authorizing and granting access to individuals.
- Implement processes that maintain the integrity, accuracy, precision, timeliness, consistency, standardization and value of data.
- Arrange appropriate training for staff and others to ensure data are captured and used accurately and completely.
- Understand and promote the value of the data for institute-wide purposes.
- Ensure compliance with the principles of the Privacy or Data Protection Act (or both) and other relevant statutes.

Data user:

An individual who has permission from the data custodian to access and use the data. It is the responsibility of all levels of management to ensure that all data users within their area of accountability are aware of their responsibilities as defined by this policy. A data user:

- Is responsible and accountable for all data access made through their user account and the subsequent use and distribution of the data.
- May not use the data for their own personal gain or profit, or for that of others.
- May not access data to satisfy their personal curiosity.
- May not disclose data to unauthorized persons without the consent of the data custodian.
- May not disclose their password to anyone.
- Must abide by the requirements of the Privacy or Data Protection Act (or both) and other relevant statutes.

Security administrator:

A person responsible for the administration of the data. Responsibilities include:

- Provide access to users as specified by the data custodian.
- Ensure that the proper logical safeguards exist to protect the data, and that appropriate disaster recovery procedures are in place.
- Provide adequate procedural controls to protect the data from unauthorized access.
- Designate a custodian for all corporate data.

Information systems group:

The information systems group in the institute has a data management role. Responsibilities include:

- Promote the management of research data as a corporate resource.
- Understand and promote the value of data for institute-wide purposes and facilitate data sharing and integration.
- Document and promote the logic and structure of institute data.
- Manage the use of common standard codes and data definitions throughout <Name of institute>.

13.4 Data Management Plan

This is a plan that describes how data will be recorded, processed and managed during the life of the project and archived at the end of the project. It will normally include the following components:

- Clearly defined roles for all project staff, including who keeps what, who is to organize the data entry, data checking, etc.
- A regular backup procedure including details of when backups are made, where they are stored, etc.
- Details of quality control checks to be done on the data and the mechanisms for recording and correcting errors.
- Putting data management on the agenda of project meetings to keep all team members abreast of the current situation.
- Details of how an archive is to be produced at the end of the project.
- Details of how the archive is to be maintained.

13.5 Further Points

Once individuals, project teams and organizations have an effective data management strategy they should find it saves rather than costs money. Without it, staff at all levels, can spend large amounts of time manipulating and rescuing data for what are only short-term benefits.

However, some researchers and organizations might consider that building a strategy, from their current situation, will be a daunting and time-consuming process. We offer the following thoughts on how they might proceed.

Document the current procedure

What you are doing now gives the starting point for your strategy. Determine what works well and what is not working.

Seek consensus

A strategy is unlikely to succeed unless all team members, or at least a majority, are in favour. Seeking a consensus makes all team members feel they have some control over the decisions made, rather than having them imposed upon them. The procedures are then more likely to be followed.

Establish a data management forum

This gives data management a higher profile. If discussed regularly the topic will stay on the minds of team members. Use meetings to update staff on progress and to set deadlines for particular tasks.

Standardize

Use similar data management plans for all projects. This saves time and also makes staff more familiar with the procedures.

Obtain funding

Include a detailed data management plan in any project proposal and include it in the budget.

Be realistic

An effective data management strategy cannot be developed overnight. Many organizations have datasets in a disorganized form from projects that are long since finished. Documenting these and producing order from chaos is a daunting task. Start with new projects to gain experience, then proceed step-by-step on the older sets. Start by producing an inventory of the old datasets, and then decide which sets need to be made available. Metadata can then be produced for these sets. Keep reviewing the work to check it is worth the time.

Chapter 14

Use of Statistical Software

14.1 Introduction

In preparation for Part 4 of this book, we now look at the statistical software that can support data analysis.

Many users are familiar with a spreadsheet, such as EXCEL, and have become comfortable using it for their work. A spreadsheet is often adequate and suitable for the simple descriptive analyses described in the early sections of Part 4.

Once a dataset has been entered, a spreadsheet can be used to rearrange it so that the data are in a suitable layout for analysis. A descriptive analysis would consist mainly of graphical and tabular displays of the data, including summaries, as dictated by the objectives of the study. We give an example in Section 15.6.

Many studies need more than a descriptive analysis and dedicated statistical software should be considered for this. Most statistics packages have a user-friendly interface and some are no harder to learn than EXCEL. These packages can also read data from spreadsheets and can write results back. Therefore, a statistics package can be an additional tool, rather than a substitute for a spreadsheet. However, the opposite does not apply; a spreadsheet is no substitute for a statistics package. Once you are comfortable with a statistics package, you can also use it for descriptive analyses. So you can, if you wish, use a single package for all your analysis.

Good statistics packages provide a unified approach to the statistical modelling ideas that are described in other chapters of this book, particularly Chapters 19 and 20. They also cope easily with practical complications, such as missing values, which are the rule rather than the exception when real data are collected. When fitting models, good packages also provide diagnostics so users can check whether the data are consistent with the assumptions of the model. For example, in a regression study, one common assumption is that after fitting the model, the residuals are equally variable.

Like EXCEL, statistics packages are user-extendable, so new methods can be added as they become available. Most of them can also be used either with WINDOWS-style menus or by writing commands. Hence, routine analyses can easily be automated, to avoid repeated pointing and clicking.

In some institutions there is an expectation that good statistical (and other) software should be available without cost, and requests for software are sometimes met with the response that the funds are not available. Yet software costs need only be a very small part of any project's total costs, and paying for appropriate software is a matter of priorities rather than overall lack of funds.

Organizations should develop a statistical software strategy and we give an example in Section 14.2. In Section 14.3, we suggest a mix of statistical software that may be suitable for the analysis of NRM data.

14.2 A Strategy

In Chapter 13, we recommended that parts of the data management strategy should be decided at the institute, rather than the project, level. Similarly, some aspects of a statistical software strategy should be organized at a level that is higher than an individual project. This will encourage coherence in the software that is used, both at an individual and an institutional level. It is not intended to stop individual users from using particular packages; quite the opposite.

We take a university environment as an example. There the mix of statistical software packages might be dictated by the following needs:

- To support the teaching and learning of statistics in a modern and effective manner.
- To be the tools for the analysis of data collected in student projects and in staff research, primarily from surveys and experiments.
- To equip students with the necessary skills for the job market, through training in data analysis and the use of statistical software.
- To be the essential software tools for the university statisticians and others to support research and to further their own professional development.

For statistics training and support, it is likely that some software licences will be needed for the organization as a whole. The following criteria are relevant to the choice of software.

A common starting point

Most students and researchers are already familiar with a spreadsheet. For the reasons described above, plus the preparation for the job market, a spreadsheet should probably be a part of the strategy.

Multiple packages

Modern statistical packages tend to be similar in the basic facilities they offer, but can differ in their more advanced features. Given the ease and similarity of use of most of these packages, there is no need to search for a single 'winner' but instead to have several, choosing at any one time the most appropriate for the required task. Indeed the capability of students and researchers to use more than one package should help them with their current work within the university and in the future when they may have to adapt to further packages.

Cost and type of licence(s)

If commercial software is provided with donor support, then priority should be given to those packages that are bought, rather than rented on an annual basis. The latter are harder to sustain once the external support has ceased.

Public domain or open-source software is particularly attractive, because upgrades can often be downloaded free of charge. However, software acquired free of charge may have poor or out-of-date documentation or help, and will almost certainly offer a more limited range of facilities. It is the total cost that counts and software is not free if more time is then needed to prepare user guides, or if full analyses cannot be done.

Software for individual machines

At least some of the statistical software should be available to be downloaded to individual computers belonging to staff or students. Distance learning may become more important in the future and this aspect will also permit students or researchers to use the software away from their place of work.

Special software

In addition to the organization level software, it is assumed that some additional packages will be needed for particular departments and individuals. The policy should encourage and support this.

Value of a software strategy

Particularly within developing countries, the availability of a software strategy provides a framework within which software for individual projects can be proposed to donors. A new project may improve or change the software that is used by the organization, or by a particular group, but it would do this in recognition of the software that is currently available.

Without a strategy, there is often a hotchpotch of statistical software for individual groups that is unsustainable once a particular project has finished. The organization of short training courses or statistical support will also be more difficult, if there is no software that is available throughout the organization.

14.3 Which Statistics Packages Could Be Used?

NRM projects will define their strategy for analysis within the institutional environment where the project is conducted. Sometimes project funds are used to buy or licence software. Ideally, it should only be necessary for a project to purchase the specialist software that it needs. If a project also has to buy general software, it should contribute to the institute or department's strategy, rather than defining its own strategy simply because the project is well-resourced. This supports the project team's role in strengthening the department, rather than reinforcing divisions between projects.

Most projects will require access to one or more statistics packages because they need more than descriptive methods for the analysis. We list some of the contenders below. We need the software to process both the activities that are survey-like, and those that are more experimental.

Unlike most other application areas, some of the popular statistics packages have been around a long time. The two industry standard packages are SAS and SPSS, both of which started in the 1960s. With the advent of personal computers in the 1980s many new statistics packages appeared. But while it seems easy to write a simple statistics package, it is difficult to write a good general-purpose package that gets the right answer for all sorts of data. Consequently the older packages remain, joined by a few important newcomers.

SAS and SPSS are both modular, so you need to decide which components you need. These packages are normally licensed, so you pay an initial fee in the first year, but then to maintain the licence you must pay an additional, usually smaller, fee each year. While paying rental is sometimes difficult, you do receive all upgrades as part of the cost.

Some other packages are licensed or purchased with a one-time payment. STATA is a package that includes good facilities for processing survey data, and in Version 8 its graphical features were redesigned. GENSTAT is particularly strong for the analysis of experimental data. It used to be known as a package that encouraged good statistics but was difficult to use. Version 7, released in 2003, is much easier for non-statisticians. We use GENSTAT for the examples of the statistical modelling approach illustrated in Chapters 19 and 20.

Among the important relative newcomers are S-PLUS, which is a commercial product, and R, an open-source package. These are both based on the S language, and are object-orientated packages. We use S-PLUS for many of the figures in this book. The older packages have languages that are more procedural. They were invented in the pre-object era.

There are more packages vying for attention. MINITAB is popular in colleges and universities. STATISTICA and SYSTAT are two other powerful packages.

All these packages try to keep reasonably up-to-date on modern developments. They differ in price, in ease of use, and in the clarity with which they display results. They also differ in the extent to which they have implemented more advanced methods or specialist techniques.

There is healthy competition between the commercial statistics packages, and steady changes. Hence, our comments are relatively general, and may still become out of date, even by the time this book is published. The ease and effectiveness of the analysis can be considerably affected by the particular packages that you use. However, the strengths and weaknesses of the alternative packages, together with the different facilities and expertise available in different institutes, means that each package mentioned above could be a suitable contender to be part of your strategy. You may well need more than one statistics package, though we would expect that teams would rarely need more than two.

To help you to define a strategy, we give some questions that you might pose to each contending statistics package. Each feature listed below is desirable, but as far as we know, there is no single package that could confidently answer 'yes' to all the points. Of course, some of these points may not apply to your project, and you may be able to add more.

- In Part 4, we emphasize the importance of taking the structure of the data into account as part of the analysis. By structure, we partly mean the different groups that you may wish to recognize in fitting the models. These might be districts or age-categories. These grouping variables are called factors; they are also called class-variables in some packages. Does your package make it easy to consider this element of the structure? Take regression modelling as an example, which we look at in Chapter 19. Can you easily fit and choose between regression models that include factors?
- When collecting qualitative or other survey data, you may have to analyse some multiple response data (Section 17.3.5). Does the software have any facilities to make it easy to process this type of data?
- In analysing a survey, you may wish to present a table of percentages, and it would be useful to include a measure of precision to compare different groups, as we discuss in Chapter 16. Commonly, a survey will use a multistage sampling scheme (Section 6.2.3), rather than simple random sampling. Is it easy to give this measure of precision, called a standard error, that takes the structure of your survey data into account?
- In an experiment you may have collected data that have a hierarchical structure, e.g. as in a split-plot design (Section 5.4.3). Is it easy to produce the correct analysis and are the results clearly displayed? Can they include the right measures of precision for comparing any two treatments?
- When there are complications, does the software leave you in control, or give the impression that there is a single right approach? We take a subject called repeated measurements as an example. This is where you return repeatedly to measure the same unit, perhaps a tree at different stages of growth, or a village as the season progresses. In Part 4, we will emphasize the importance of analysing the data according to the objectives of the study. With repeated measures data, we find that the automatic methods of analysis are often over-complicated, and can be difficult to relate to your precise

objectives. Hence, we would prefer statistical software that encourages a range of different approaches.

The last point highlights a potential risk in the use of statistical software. Sometimes we find the analysis is clearly driven by the package that was used, rather than by the aims of the project. In one extreme case, a different package was used in the second year of a three-year project in which a similar experiment was analysed each year. The style of the report changed completely, indicating that the presentation was determined by the statistics package. The tool had become the master! This should not happen.

In addition to the general statistical software there may be a need for one or more special packages. At a simple level, you may find a special graphics package suits your needs more than a spreadsheet, or any of the statistics packages. Or you might get an additional statistics package primarily for its graphics capabilities. Software for mapping data, or a full geographical information system (GIS) may be another useful component. Like statistics packages, GIS software is becoming cheaper and easier to use and is now more accessible to non-specialists.

Part 4: Analysis

Chapter 15

Analysis – Aims and Approaches

15.1 Taking Control

Some researchers find that the analysis stage is the most exciting part of the research project. For these researchers all the planning and fieldwork has been completed to reach this point, and it is here that the results are discovered and new truths revealed. Others, who are not so comfortable with numerical methods, or who may lack an understanding in statistical methods, view the analysis with dread. They may feel that they have identified the answers by observing the results in the field, without having to prove it with statistical analysis. For them, the statistical analysis may just seem to be a ritual, needed to satisfy the demand for *p*-values by journal editors.

We feel the most appropriate position is between these two. The analysis step should certainly be exciting as results become clear, and it should not be 'routine' (in the sense of blindly using the same few techniques every time). Nor should it be a process of generating *p*-values (or any other particular statistic) simply to meet a possibly misconceived demand. However, a good researcher will have been actively involved in earlier steps of data collection and management, and will therefore most likely have a good idea of some of the key results. Using this, and other information about the data origins and underlying theory, will improve the analysis. This is one reason why researchers should be closely involved with analysing their own data, and not just hand data over to a statistician.

Throughout this book we aim to put statistical ideas across in a practical, non-mathematical way. Some people may think that by 'taking the maths out of stats' will make it easier. And though it does, the statistical analysis can still be tricky, particularly if the dataset is complex. Careful thought, imagination and curiosity are all required at the exploratory stage. Later, an understanding of how to interpret results is needed.

The analysis has two overarching objectives. The first is to compile the evidence the researcher will use to support the position or theory being promoted. The research should have been initiated with some outcomes in mind; in natural resources research this often concerns the effect of changing resource management. The evidence must stand up to the most critical examination. It is the analyst's job to start the criticism: to think of alternative explanations and interpretations, and to examine each carefully. The analyst therefore has to be objective and self-critical, despite having probably already generated an idea of the key results while collecting the data.

The second aim of the analysis is to identify and interpret any unexpected patterns or relationships in the data. Unanticipated or unknown phenomena may be apparent in the data, and could well be important in the overall research; these unforeseen observations are sometimes the most important outcomes from a research activity. The analyst therefore has to work in a way that will allow such things to be seen, and be open-minded enough to spot them, interpret them and perhaps abandon long-held ideas.

Statistical theory is sometimes presented in terms of decisions, with the analysis leading directly to decision-making. The decision may be acceptance/rejection of a hypothesis or may concern a management action (e.g. to clear a forest or change a policy). NRM research should lead to some real changes in the management of resources, though a simple decision framework, such as standard hypothesis testing, is often too limiting for the statistical analysis to be useful. Research hypotheses are not usually just accepted or rejected in the face of empirical evidence, they are more likely to be partly accepted but with revisions and modifications of the conditions under which they are reasonable. Decisions about resource management should be based on empirical evidence and on understanding of underlying theory, though they will also be based on many other considerations. These can rarely be incorporated directly into a quantitative decision-analytic framework. For these reasons we do not find such a framework particularly helpful when analysing data.

15.2 Qualitative or Quantitative Analysis

Statistics is mainly concerned with numerical analysis of quantitative data. Sections 17.1 and 17.2 discuss some aspects of the analysis of qualitative data, and the ways in which such information can be made quantitative through coding. Other approaches to the analysis of qualitative information are sometimes used in the social sciences. If a researcher has collected narratives or made observations on several people, communities or situations, they start the analysis by looking for common patterns. They identify deviations from the common pattern and describe them, perhaps as differences between groups. They may seek explanations of the patterns, and additional evidence to support the explanations.

While the analysis does not use numbers, the approach has clear parallels with the analysis of quantitative data. For example, we describe common patterns as means, the deviation by variances and look for explanations for the variation in the structure of the data (groups) and covariates. This similarity in approach is important; it means some of the ideas behind good statistical practice are equally important when doing a qualitative analysis. We illustrate this with an example in which the observations are simple graphs. The ideas are the same with narrative data, or other types of qualitative information.

Part of a study in Malawi involved an examination of trends in fertilizer use in the country. In 23 villages in three regions (north, centre and south), households were classified into one of three agricultural sustainability groups (1, 2 or 3) on the basis of farming practices. Each of these groups was then asked to describe the trend in fertilizer use from 1970 to 2000, and recorded it as a simple trend line, with time on the horizontal axis. These graphs could be coded so a quantitative analysis is possible. One way to do this would be to record a +1, 0 or -1 depending on whether the gradient of the line is positive, zero or negative for the start, middle and end of the time period. Then these numbers could be analysed using statistical methods described later. However, we can also just look at the trend lines and draw some conclusions.

Fig. 15.1. Reported trends in fertilizer use from 1970 to 2000, reported by groups from 23 villages in 3 regions of Malawi. Groups practised farming of low (1), medium (2) or high (3) sustainability. The horizontal axis of each graph is time, the vertical axis goes from low to high fertilizer use.

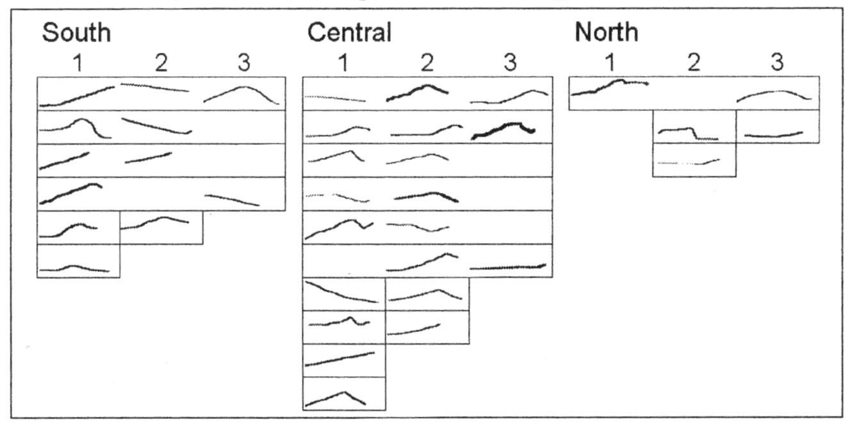

Figure 15.1 shows all the trend lines. Some features of the whole set are apparent. A common pattern is for an increasing use of fertilizer at the start of the period, but then a levelling off or decrease towards the end. However there is much variation, with some groups reporting little change throughout, and others, a steady decrease.

The next step is to use the structure in the data to explain some of this variation. There are several things we know about each trend line: which village it came from, which region that village is in, and which sustainability group produced it. Figure 15.1 is arranged to show this structure. Each box with a solid outline shows the results for one village. Trend lines in the same box are from different sustainability groups in the same village. Now some more features of the data are revealed. For example, there is a rather small sample from the north. Hence, with the level of variation present, we may not be able to reach any clear conclusions about the north. Villages in central Malawi are mostly consistent in describing a recent decrease in fertilizer use. In the south there is more variation in recent experience.

There are two ways to compare sustainability groups. We can compare groups from the same village or those from different villages. Comparing groups in the same village should give a more precise picture of the differences, as some of the factors causing variation (maybe distance to a fertilizer retailer, or effectiveness of the local extension officer) may have similar effects on all groups. Using these comparisons in central Malawi shows that groups 1 and 2 reported much the same trends. There was a larger difference between groups 1 and 2 in the south. Some villages only had one sustainability group, which adds additional information, so we also need to make comparisons between villages.

The analysis so far highlights the considerable variation in trends that is not explained by the region or by consistent differences between sustainability groups. The next step would be to seek explanations based on other things we know about the villages or groups: how large they are, for example, or whether they grow maize for sale.

From now we concentrate on quantitative, numerical approaches. For qualitative data, the methods we discuss can be used after the information is coded.

15.3 Steps

Every analysis is unique but there are some common steps to keep in mind to maximize the chance of doing a good analysis.

1. Making objectives specific. It is impossible to do a good analysis without knowing what you want to find out. The first step is therefore to specify precise analysis objectives. These objectives will drive the rest of the analysis, so cannot be left in general terms such as 'To find out what influences tree growth' or 'To understand adoption of integrated soil fertility management'. Such general objectives need breaking down into clear components, so the analysis will have several specific objectives. An example is in Section 15.6.2. The analysis objectives should closely follow the study objectives, which were established when it was first being designed. We expect researchers to give some thought to the analysis objectives at the planning stage, but the objectives will probably need to be made more precise, and perhaps modified in the light of field experience once the study is underway.

2. Preparing data. However well-managed the data (Part 3), there will always be a step of data preparation. Most analyses work on a 'matrix' of data (as can be pictured in a spreadsheet, for example). The data preparation step involves choosing the rows and columns of this matrix and formatting it for use with the statistical software. It may also involve calculating variables from the raw data to be used in the data analysis.

3. Descriptive and exploratory analysis. Descriptive analysis generates the main summaries, tables and graphs that 'tell the story' to meet the objectives. Statistical tests and formal inference procedures are not needed, and much of this step can be done by someone who is not familiar with them, as long as they are comfortable with using numerical data to make a point. Exploratory analysis also reveals further and unexpected patterns and relationships. The techniques are limited only by the user's imagination, as long as the results can be interpreted.

4. Confirmatory analysis. The confirmatory analysis provides the confirmation that the patterns are 'real', given the level of uncertainty in them. The uncertainty arises because of measurement error and the random and unexplained variation, which is the 'noise' present in any natural system. Statistical models allow important effects to be quantified along with their uncertainty. They also allow complex patterns of multiple causes and effects to be unravelled.

5. Interpretation and use. Without interpretation the analysis is pointless. In step 1, general objectives were broken down into a series of specifics. This last step will include a synthesis, which reassembles the components into a comprehensive picture. The synthesis is not just internal, but involves merging information from the current study with other knowledge to produce overall conclusions.

Rarely are these five steps followed in a linear way. Iteration, returning to earlier steps and repeating, will almost certainly be needed. For example, the data preparation step may lead to the realization that an objective cannot be met with the data available, and it will need revising. The confirmatory analysis might suggest that some summary statistics used in the previous step are inappropriate, and need revising. Many of these decisions will depend not just on the numbers, but also on their context. This means that analysis will be interactive (each step depending on the results of earlier steps) and involve the researcher, not just a statistician.

15.4 Resources for Analysis

In Parts 2 and 3 of this book we have stressed the importance of project teams allocating sufficient time and other resources for the planning and data management phases. The same applies to the analysis.

There are two extremes to avoid. The first is the research student who does a small amount of fieldwork and then spends the remaining three years on a very detailed and unwarranted analysis. The opposite, which is more common, is for a research team to spend most of the resources on data collection, and then find that there is almost no time or skilled staff available for the analysis.

The ideal is for the data collection and analysis to be as closely integrated in time as possible, but consistent with the resources for a full analysis and reporting by the end of the research. This means that much of the choice of methods for the analysis should be done early in the project, so that the initial processing can be done while other data are still being collected.

Allocating the resources for the analysis includes the following components:

- Organizing the data management to avoid using time sorting out the data, that should be spent on the analysis.
- Clarifying the statistical software issues (see Chapter 14) before starting the analysis.
- Allocating resources with the right level of expertise. Sometimes technicians are expected to do the analysis because they have been on a statistical software course. The responsibilities for the analysis must rest with the researchers, possibly in collaboration with statisticians or statistical auxiliaries.
- Being realistic about the time needed for the analysis, interpretation and reporting. The time depends on the complexity and volume of the data, and the number and complexity of objectives. It also depends on whether the methods envisaged for the analysis are routine for the research staff or are new to them.
- Accepting practical constraints, such as when analyses involve team members who are based in different locations or countries. However, there are strong reasons (error checking, and keeping the context clear) for doing the analysis physically close to the source of the data.

Resource allocation for analysis is sometimes difficult, because project managers cannot answer all the points above, and also expect interesting features and new objectives to emerge during the analysis that are difficult to plan for. A good manager, will recognize the need for insurance, such as keeping some resources (time and money) for contingencies.

15.5 What Do You Need to Know?

Looking in statistics books, or at statistical software, can reveal a bewildering selection of statistical tools, many of them presented in a mathematical way. While all statistical analysis can be described using mathematics, this is often not the most helpful presentation for those not used to working with formulae. The concepts behind a set of basic tools can be easily understood and their use facilitated by good modern statistical software. We attempt here to describe some of these concepts, without using mathematical formulae. This basic toolbox is analogous to the 'do-it-yourself' (DIY) toolbox kept at home for simple jobs; anyone can learn to use them, and without them you are fairly handicapped. The content of this basic toolbox will to some extent, depend on the application area: someone regularly working with biodiversity data may need different techniques than a rangeland manager or community forester. However, there are several common approaches, ideas and concepts that all resource management investigators need.

Since software takes care of most of the calculations, the challenge to using these basic tools shifts from what was emphasized in older books and training courses. They were principally concerned with the mechanics of 'doing the sums'. The new challenges are: firstly, to know what you want to do; secondly, to get the software to do what you want, rather than what it is easy to do; and thirdly, to understand the results, particularly when the critical numbers may be hidden amongst a large volume of irrelevant output.

Some trainers in statistics do not agree with our approach. They feel that researchers still need to concentrate on the basic theory so they have a solid foundation on which to build further methods of analysis. However, many researchers are put off by the theory and therefore understand little of the statistics, but still go ahead and use methods for analysis available in software. In addition, many such trainers do not cover sufficient of the basic toolkit in their courses to enable users to handle real data.

The home toolbox is not sufficient for every job. There will be tasks that require something more advanced: a craftsman's tool, and perhaps a craftsman to wield it. This might be anticipated from the start; for example, if a hierarchical or multi-scale design is used, a corresponding analysis will probably be needed. It might be a surprise revealed in early stages of the analysis, or it might only become apparent after extensive simple analyses have been made. Scientists first need to recognize when they have such a problem. While there is no limit to the range of problems that might require special tools, there are some situations that commonly occur and scientists need to be aware of the nature of these problems and the possibilities for solving them. The strategies available are either to get help from a statistician or to learn the special techniques, which might be appropriate if the scientist is regularly going to be faced with similar problems. In any case, the more scientists understand about the analysis problem and potential solutions, the more likely they are to be able to find appropriate help.

The rest of this part of the book discusses some of these tools. In Chapter 16 the essential ideas from the DIY toolbox are presented. Their application to the analysis of surveys and experiments is covered in Chapters 17 and 18. Chapter 19 then shows how many of the ideas in the DIY toolbox can be put into a common framework of statistical modelling, which allows a wider range of problems to be tackled. The modelling framework also provides the basis for much of the craftsman's toolbox, described in Chapter 20. We start with an example to illustrate the steps given in Section 15.3.

15.6 Getting Started

In the following chapters we emphasize many aspects of 'good practice' in statistical analysis; ignoring these may result in misleading or incorrect results. In many studies the analysis can start quickly and simply, using basic methods and common sense. We often hear that work on a research activity has come to a halt because researchers 'cannot analyse the data'. In many of these cases we find that a systematic approach using merely simple summaries and graphs can provide much of the key information. Lack of familiarity with more advanced methods, or access to a statistician or good statistical software should not be an excuse for delay in undertaking this simple analysis. We use a survey to illustrate the process. All the tables, summaries and graphs shown were generated using standard spreadsheet software.

15.6.1 Example

The researcher described the problem and background as follows:

My objective is to study the factors that influence the planting of timber trees on farms.

I divided the villages into four groups according to elevation and their proximity to the main road. Thus, I had:

Group 1: Cabacunggan, Hinaplanan, Kalawitan, Gumaod

Group 2: Patrocenio, Anei, Poblacion

Group 3: Lanesi, Luna, Tamboboan

Group 4: Rizal, Madaguing

From a census list, I randomly selected between 30 and 40 farmers in each group. Each farmer was interviewed, and their farm parcels were surveyed. Overall 139 farmers were interviewed and 217 parcels surveyed. There are two data sheets, one with observations on household and the other with observations on individual parcels. I still do not know if the analysis should be done at the household level, or at the parcel level, or both.

In this example the data were already coded (Table 15.1). Thus the answers to questions such as 'Who manages the plot?' have been classified into discrete values. Further reflection might suggest modifications, for example, why is the lowest level of adoption 0-5 trees, rather than 0 trees? Unless the questionnaire is pre-coded, coding is an important step in the analysis (see Section 17.2.7).

15.6.2 Setting detailed objectives

The study was designed to meet some general objectives. To make progress with the analysis it is essential to give more details, by breaking them down into specific questions or hypotheses. In this example the researcher gives a detailed objective as:

Land more suitable for tree crops, rather than intensive annual crops, is steep, far from the house and without soil and water conservation. Is that observed in current tree planting? If so, does the extent depend on the type of ownership?

Table 15.1. Variables used in the example.

Variable name	Description	Category codes
Adoption	Number of trees planted per hectare	0: 0-5 1: 6-20 2: 21-100 3: >100
Manage	Who manages the plot	Owner Tenant Partially rented out Rented out
Disthouse	Distance of plot from the house (m)	0: 0-100 1: 100-500 2: 500-1000 3: 1000-2000 4: >2000
Slope	Slope (%)	a: <3 b: 3-18 c: 18-30 d: 30-50 e: >50
Swc	Existence of soil and water conservation	y: yes n: no

Tables 15.2a and 15.2b. Outline tabulations for the study.

(a)

	Percentage adoption of tree crops			
	Owner		Other	
	steep	less steep	steep	less steep
near				
far				

(b)

Percentage adoption of tree crops				
Manage	Slope	Distance		Overall
		near	far	
Owner	steep			
	less steep			(100%)
Tenant	steep			
	less steep			(100%)
Partially rented	steep			
	less steep			(100%)
Rented	steep			
	less steep			(100%)
Overall	steep			
	less steep			(100%)

It is often useful to outline the tables or graphs that would give the answers. This can be done without numbers, even prior to collecting the data. Tables 15.2 a and b show two examples.

These outline tables may well have to be modified when you see the data, but preparing them helps to focus attention at the start. If the tables were to be produced, then further coding of the data would be needed. Table 15.2b indicates that we would give percentages separately for each type of management.

15.6.3 Select the data to use for the analysis

In this case there are two units of analysis: parcels and farmers. Therefore we have to choose the level or unit of analysis and the variables to be used. Questions that relate to the physical characteristics of the land, which varies by parcel, should probably be studied at the parcel level. Questions relating to social and economic measures, which vary across farmers, should be studied at the farmer level. In this case the main questions relate to the parcels, so that will be the level.

Sometimes questions can involve both levels and then we have to be more careful, because there are usually alternative questions that have subtly different meanings. For example 'Do tenants plant more trees than owners?' and 'Does land managed by tenants have more planted trees than land managed by owners?' are superficially similar. However the first is a household level question, while the second would be studied at the parcel level.

The answers to the questions may already be in the data file, or they may be new. New variables can be constructed by recoding, by combining or by summarizing. In this example, recoding could be to define slopes less than 30% as less steep, and the others as steep.

Combining variables is useful when many measurements are of similar things. For example, we may construct a 'land class' variable with a few different levels from steepness, existence of soil and water conservation measures *(Swc)* and the time the plot has been cultivated.

Summaries are needed if variables from a lower level unit are used for an analysis at the higher level. Here, an analysis at the household level would use summary values for that household of variables measured at the parcel level.

In some analyses it is useful to think of one variable as a 'response' and the remaining variables as 'explanatory'. The idea is that the explanatory variables should explain some of the variation in the values of the response. Here the response variable is *Adoption* (coded number of trees per hectare). Potential explanatory variables are *Manage*, *Distance*, *Slope*, and *Swc*.

15.6.4 The tables and graphs

In this section we produce some tables and graphs that correspond to the questions. In general we are looking for the following:

- The major patterns that answer the questions.

- Any odd or surprising observations that might be errors to correct.
- Indications that the results will be clearer if the variables are modified (e.g. recoded).
- Indications that the questions need modifying.
- Patterns suggesting new questions that might be more informative.

Thus, in this part of the analysis, we iterate between objective setting, selecting data and exploratory analysis.

As the key question is of adoption, we start with a table of counts to get an idea of the data available. Table 15.3 indicates a problem.

Table 15.3. Counts of *Adoption* by management categories.

	Adoption:	0: 0-5	1: 6-20	2: 21-100	3: >100	Total
Manage:	Owner	58	32	40	18	148
	Partially rented	2	-	1	1	4
	Rented	3	-	3	-	6
	Tenant	46	6	6	1	59
Total		109	38	50	20	217

Of the four levels of *Manage*, two (*Partially rented* and *Rented*) have too few observations from which to make much sense. *Tenant* has 46 out of 59 parcels with no adoption of trees. This is clearly a different pattern to that of the *Owner*. Ownership status clearly has a huge impact on adoption. The number of adopting tenants is small so it is unlikely that we will find variables that explain this. Hence the part of our objective relating to ownership cannot be resolved with these data, and in the subsequent analyses we will look at the main objective for the owners only.

For the remaining three variables we start by looking at them individually. This can hide some complex patterns, but we try to understand the major relationships first (Table 15.4).

Table 15.4. Counts of *Adoption* by soil and water conservation measures (*Swc*).

	Adoption:	0: 0-5	1: 6-20	2: 21-100	3: >100	Total
Swc:	no	55	25	31	14	125
	yes	3	7	9	4	23
Total		58	32	40	18	148

These counts are useful, but the pattern is clearer if percentages are displayed, as in Table 15.5.

Table 15.5. Counts of *Adoption* by *Swc*, shown as percentages.

	Adoption:	0: 0-5	1: 6-20	2: 21-100	3: >100	Total
Swc:	no	44	20	25	11	100
	yes	13	30	39	17	100
Total		39	22	27	12	100

There are higher levels of adoption on parcels with soil and water conservation than on parcels without, contradicting the hypothesis. The pattern shown in Table 15.5 may be spurious, because it is related to the other variables.

Alternatively, the hypothesis could be revised, perhaps as follows: '*Swc* and timber trees will tend to be found on the same parcels, since both are indicators of adoption of sustainable practice.'

Table 15.6. Count of *Adoption* by slope.

	Adoption:	0: 0-5	1: 6-20	2: 21-100	3: >100	Total
Slope:	a: < 3%	8	3	8	2	21
	b: 3-18%	29	14	22	8	73
	c: 18-30%	14	12	7	7	40
	d: 30-50%	4	3	3	1	11
	e: >50%	1	-	-	-	1
	no	1	-	-	-	1
	blank	1	-	-	-	1
Total		58	32	40	18	148

In Table 15.6 one blank and one unknown code ('no') have to be ignored, as does the single observation in class *e*. There are few observations in class *d* so this can be combined with *c*. We represent the data as percentages in Table 15.7.

Table 15.7. Count of *Adoption* by slope, shown as percentages.

	Adoption:	0: 0-5	1: 6-20	2: 21-100	3: >100	Total
Slope:	a: < 3%	38	14	10	10	100
	b: 3-18%	40	19	30	11	100
	c&d: 18-50%	35	29	20	16	100
Total		39	22	27	12	100

The patterns are not clear. There is no obvious change in intensity of adoption with slope class. However if we simplify the adoption measure by combining levels, then a possible pattern does emerge (Table 15.8).

Table 15.8. Count of *Adoption* (as percentages) by combined slope classes.

	Adoption:	0&1: 0-20	2&3: > 20	Total
Slope:	a: < 3%	52	48	100
	b: 3-18%	59	41	100
	c&d: 18-50%	65	35	100
Total		60	40	100

There seems to be higher adoption on less steep parcels, again contrary to the original hypothesis. Following a similar process, Fig. 15.2 shows the effect of *Distance* on intensity of adoption. A graph is useful here because distance from the house is a quantitative variable.

Fig. 15.2. Per cent *Adoption* plotted against distance from house (coded).

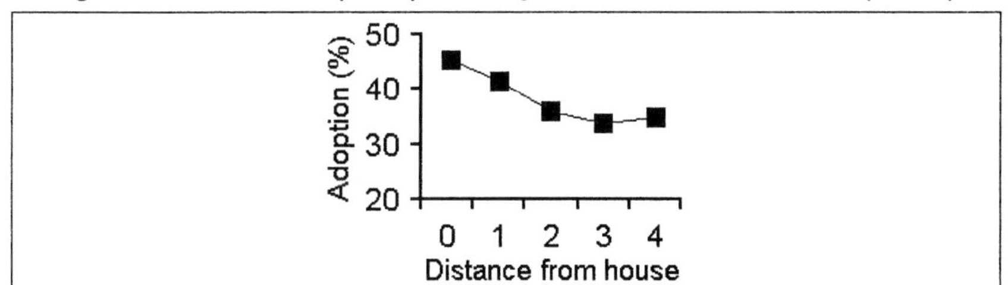

Intensity of adoption decreases (on the whole) with distance from the house.

More complex interactions can now be explored. In Table 15.9 we look at just one, that of *Slope* and *Swc*.

Table 15.9. Per cent *Adoption* by *Slope* and by *Swc*.

	Adoption:	0&1: 0-20		2&3: > 20	
	Swc:	no	yes	no	yes
Slope:	a: < 3%	52	-	48	-
	b: 3-18%	61	50	39	50
	c&d: 18-50%	73	36	28	64

There are no parcels in *Slope* class *a* with *Swc*. This is not surprising as it is not needed on (nearly) flat ground. The pattern of higher adoption intensity on parcels with *Swc* appears to be stronger on the steep fields.

15.6.5 What's next?

This analysis is by no means complete. The next steps will usually include:

- Looking at the several variables simultaneously.
- Confirming that the patterns are real, given the level of uncertainty.
- Interpreting and finding explanations for the results.
- Preparing tables and graphs in styles that are suitable for presentation.
- Repeating the process for other questions, both those known at the start of the investigation and those that emerge during analysis.

These are topics we discuss in the following chapters. However, these initial steps do permit initial conclusions to be drawn and a start can be made on writing a draft report, or preparing an initial presentation. Inability to complete the full sequence of steps in the analysis should not prevent a start being made!

Chapter 16

The DIY Toolbox – General Ideas

16.1 Opening the Toolbox

This chapter outlines the essential elements of a successful statistical analysis. The ideas underpin Chapters 17 to 20 as they apply to data collected in any planned study. Starting with the preliminary stages of exploratory work we continue through to formal analysis methods. The concepts underlying estimation and hypothesis testing are particularly important; we introduce them here in a simple setting but they extend to all areas of data analysis. They set the scene for discussions on specific issues, such as how to carry out a modelling exercise, or how to tackle the analysis of experimental data, which are discussed in later chapters.

16.2 Summaries to Meet Objectives

The first step is to generate summaries that meet your objectives. These objectives are often encompassed within one of the five categories described below. A survey example is outlined in Section 15.6, and an example for an experiment in Section 18.2.

16.2.1 Summarizing average and variation

Numerous measures of 'average' (or central tendency) of a set of numbers have been described, but the arithmetic mean is by far the most commonly used, with the median probably second. Making sure that the average helps meet the objectives is not always so straightforward. Take the objective of producing a summary that describes the 'typical' area under improved fallow for the data in

Table 16.1. The overall average of the area under improved fallow across all 60 farms is 300 m^2. However, that figure includes 15 farms that do not have any improved fallow (an area of zero). Implicit in the objectives could be to report the 'typical' area for those farms that actually have some improved fallow (areas of non-zero), which is 400 m^2. Both these summaries are valid and useful, but for different purposes.

Table 16.1. Results from a survey measuring the area under 'improved fallow' on a sample of 10 farms in each of 6 villages, divided between two districts.

		Area (m^2) under improved fallow on 10 farms										Aver-age	Per-cent	Average area of non-
District	Village	1	2	3	4	5	6	7	8	9	10	area	zero	zero
north	1	0	0	0	0	250	730	1050	1070	1200	2600	690	40	1150
	2	0	0	0	150	370	370	480	800	820	1070	406	30	580
	3	0	0	0	0	0	0	140	210	370	740	146	60	365
	All north											414	43	731
central	1	0	70	90	120	140	190	280	310	320	400	192	10	213
	2	85	90	130	140	140	170	210	250	270	410	190	0	190
	3	0	100	120	150	150	170	220	240	290	320	176	10	196
	All central											186	7	199
Overall												300	25	400

Most sets of data show variation of different types, and a summary of variation must reflect the real intention when calculating and using it. Referring to Table 16.1, we could summarize the variation by giving the overall standard deviation of the 60 areas, namely 427 m^2. However the data in Table 16.1 show variation between farms within a village, variation between villages and variation between districts. Ignoring this structure and simply putting them all together is unlikely to meet any real objectives. Many books will recommend alternative, 'robust' summaries, rather than means and standard deviations: for example, use of the interquartile range to measure variation. These measures may well be useful, but they do not solve the problem of representing variation that is relevant to the objectives.

Zero values such as in Table 16.1 are often considered to be a problem in summarizing data. Usually the 'problem' is that the objectives are not specified sufficiently. Often there are actually two objectives, one relating to the number of users of improved fallow, and the other relating to the area of those who do use improved fallows. Notice that 25% overall (15 out of 60) of the sample do not use improved fallows, but the percentage varies between zero in one village (central 2) to 60% in another (north 3) and much of this variation occurs between districts. Which of these summaries helps to meet objectives and understand relevant patterns in the data? Make sure the summaries of 'average' and 'variation' that you use actually relate to your objectives, if necessary separating variation of different types (such as between farms and between villages).

16.2.2 Comparison of groups

Most studies are comparative; this is the basis of all experiments. Many observational studies also require comparison of different parts of a dataset to meet objectives. In Table 16.1 we might need to compare farmers in the two districts. This particular comparison demonstrates the importance of choosing a relevant summary; the difference between the two districts depends on whether the zeros are included or not. Note also that summaries that may be hard to interpret overall, can be helpful when comparing groups. For example the overall standard deviation of 427 m^2 for the 60 farms in Table 16.1 is hard to make sense of, but the corresponding standard deviations for the north (576 m^2) and central (106 m^2) districts do show that farms in the north tend to be more variable.

Groups may be formed by a single factor or classifying variable, such as district. If there is more than one classifying variable then other ways of defining groups emerge. Suppose each farm was also classified by 'gender of head of household'. We could then produce summaries to compare genders. We could also define groups by crossing the two factors of gender and district, to end up with four groups (m-north, f-north, m-central, f-central in Table 16.2). The idea of looking at crossed factors and summaries for both the margins and the body of the table is important, and arises in many statistical analyses.

Table 16.2. Cross-tabulation of data by gender and district.

		District		
		north	central	
Gender	m	m-north	m-central	*Gender margin*
	f	f-north	f-central	
		District margin		*Overall*

In Table 16.1 there were two classifying factors present: district (with values north and south) and village (with values 1, 2 and 3). We could consider crossing these two factors, to end up with six groups (the six villages). However, the two factors of district and village should be considered as nested, rather than crossed. There is no direct comparison that can be made between villages in different districts, they just happen to have received the same label when sampling and processing the data. We can calculate and interpret summaries for each district, and for each village within districts, as shown in Table 16.1.

The factors that might be used to form and compare groups describe part of the structure of the data. The discussion here and in the previous section highlights the importance of recognizing the structure in the data. Analyses that ignore the structure rarely meet their objectives.

16.2.3 Changes in a continuous variable

The objectives may require an understanding of how a response changes as a continuous quantity varies. For example, how does 'area planted to improved fallow' vary with farm size, or length of time since forest conversion, or distance from the main road? The problem could be turned into a 'comparison of groups' problem by categorizing the continuous quantities: distance to the road might be coded as 'near' (<0.5 km), 'mid' (0.5 to 2 km) and 'far' (>2 km). Then the group comparison ideas of the previous section are relevant. However, this may be unsatisfactory for several reasons. We lose information by turning a distance into simply 'near', 'mid' or 'far'. We also might be complicating a simple situation: for example, a linear relationship between the variables can be summarized by the slope, but forming it into groups would result in multiple comparisons to make.

Changes in a response are often revealed with graphs, particularly scatter-plots with the response on the vertical axis. Suitable numerical summaries, such as the slope of a straight line, can then be chosen. It may be hard to summarize the patterns in a graph when several quantities vary simultaneously (analogous to the crossed and nested factors forming groups in Section 16.2.2), though some tools are available (Section 20.2). Numerical summaries in this situation depend on the ideas of statistical modelling.

16.2.4 Association between units

The previous sections describe the summaries we might use when we know something of the structure or expected relationships in the data and are looking for patterns in them. A different problem arises when the objective is to identify any possible structure. This might be the situation in a study of plant communities, in which the species composition of many communities is measured, and we want to find communities that are similar. If we can reduce the measurements to a single variable (for example the percentage of ground covered by grass species) then looking for groups of communities could be easy. We can simply list the communities in order of increasing grass ground cover and look for clusters and breaks in the spread of data values. With two variables (e.g. percent ground cover by grass species, and number of legume species present) a scatterplot can be drawn, which again will show up any clusters of similar plant communities.

Once there are more than two variables of importance, the approach has to be more subtle. One method is to find a few new variables that describe most of the structure. Another is to define distances between plant communities and from these inter-community distances, produce clusters of similar observations. These multivariate approaches are discussed further in Section 20.3. Both are designed for detecting structure. They are not useful when you know *a priori* what the structure is; you will simply end up reproducing it inaccurately.

16.2.5 Association between variables

Another type of objective requires finding associations between the variables rather than between the units or objects of the study. For example, a study on poverty in communities in Kenya found ways of measuring the level of social, financial, physical, and natural capital in each community. The interrelationships between these are important, but it does not really help to think of one as a 'response' and others as 'explanatory'. It is hard to say whether variation in social capital leads to variation in financial capital or the other way round. Correlations can be used to describe relationships between such variables, at least when the relationship is linear. A scatterplot can be used to check linearity. With more than two variables in the system we may see A being correlated with B, and B with C. Hence there will also be a correlation between A and C. Can this be separated from the direct relationship of A with C? This is the role of partial correlations, introduced in Section 20.3.3.

16.3 Response Variables and Appropriate Scales

Implicit in the previous section is the idea that there are 'responses' of interest: a variable (or variables) for which we want to summarize and understand patterns and changes. In many studies the response variable is a natural consequence of the objectives and the measurements taken. For example, crop yield is a response in many agricultural investigations. In other cases the response may be constructed from several quantities that have been measured on each unit, for example, financial returns per hectare calculated for each farm from input and output quantities and prices.

A direct measurement of the response needed to meet the objectives may not have been possible, either for practical or conceptual reasons. Then a 'proxy' or indicator is used. The idea is to use something that reflects the response of interest, and which is likely to parallel the way the response would change under different conditions. Tree diameter is often used as a proxy for tree biomass because, unlike biomass, it is quick and easy to measure non-destructively, and calibration studies have shown simple and consistent relationships between biomass and diameter in many situations. Other proxies may be more complex constructions, such as a measure of wealth based on possession of a range of assets. Sections 17.3.8 and 17.3.9 consider proxies and indicators further.

The nature of the data and objectives can sometimes make the choice of response daunting. In a study of tree diversity in Ugandan farming landscapes, the species of every tree on 250 farm plots was recorded. With so many observations per plot, the choice of response was bewildering. However, a closer look at the objectives showed that three key response variables could be defined:

- Tree density (number of trees/plot area) to answer the question, 'Where are the trees?'
- Number of distinct species to answer the question, 'Where are the species?'
- Shannon diversity index to answer the question, 'Where is the diversity?'

Each response variable could be summarized to show changes associated with the grouping factors *distance to forest* and *landscape position* (valley, hill-side or hill-top). When important changes in these were noted, additional response variables were defined, to understand what was contributing to the change.

The primary reason for choosing a response variable must be the objectives of the study. However, there are cases in which choice of scale can simplify the analysis. For example, the four group means in Table 16.3(a) show differences as both A and B change. The A effect is much higher when B=H than when B=L. The data in Table 16.3(b) do not show this interaction: the difference between A=L and A=H is 4.0, for both levels of B. The data in Table 16.3(b) are the same as in 16.3(a), but on a different scale: they are the square roots. Choosing an appropriate scale, provided it makes sense, may often simplify the description of patterns in a response. Some transformed scales may result in numbers that are hard to explain (what does the square root of a kilogram measure?). On the other hand we have no difficulty with quantities routinely measured on a transformed scale, such as pH to measure acidity (the logarithm of a concentration) and should at least think about alternatives.

Table 16.3. (a) Example data. (b) Same data, transformed to square root scale.

(a)

		Factor A	
		L	H
Factor B	L	2.0	29.2
	H	8.4	47.6

(b)

		Factor A	
		L	H
Factor B	L	1.4	5.4
	H	2.9	6.9

16.4 Doing the Exploratory Analysis

How should the exploratory analysis be done once your objectives have been clarified, response variables selected and data prepared? A good exploratory analysis should of course be carried out with care and good sense, but at the same time you can be imaginative in finding effective ways to summarize and present the information. The methods and procedures will inevitably be unique to each problem, so generic instructions cannot easily be given. However, there are some elements of good practice, in addition to those already discussed in this chapter; these are listed here.

16.4.1 Critical examination of each step and each conclusion

As an analyst, you should be the first to criticize all that you do with your data. Does it really make sense to average across those subsets? Could that difference just be due to a sampling artefact? Are there other explanations for the pattern?

16.4.2 Look for explanations – drilling down

The most insightful analyses are those that provide some explanation for patterns observed, and do not just report the patterns. For example, finding more bird species in forest than in agricultural plots is perhaps interesting, but finding that this is explained by greater diversity of vegetation heights is more useful. This might be done by drawing a scatterplot of species numbers by habitat height diversity, distinguishing between forest and agricultural plots. The result in

Fig. 16.1a would certainly be evidence, not just for a relationship between species number and habitat diversity, but that it 'explains' the difference between forest and agricultural land. Figure 16.1b would lead to a different conclusion; there is a relationship but it does not explain the forest-agriculture differences. The most convincing evidence would come from a result such as that in Fig. 16.1c. Here there is an overlap in the height diversity of agricultural and forest plots, and those plots of different types but the same diversity have similar species numbers. Another possibility is shown in Fig. 16.1d. In this case, the evidence for the relationship is weak; the species numbers could be explained by any number of differences between forest and agricultural plots.

Fig. 16.1. Scatter clouds of bird species numbers (vertical axis) and habitat height diversity (horizontal axis) for agricultural (grey) and forest (white) plots.

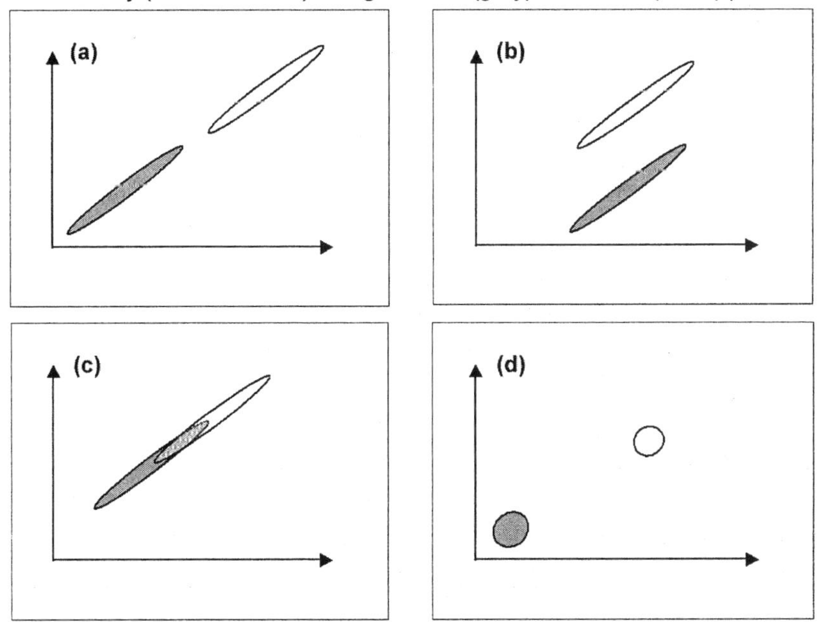

This type of analysis requires 'drilling down'. In other words, instead of looking at overall patterns we take into account other properties of the observations to help explain the patterns. This idea should be familiar since many analysts already use different symbols to indicate different groups on a scatterplot; probably the simplest example of drilling down!

16.4.3 Looking for consistency

Many people still describe statistical analyses as 'looking for significant differences'. A more useful idea is looking for 'significant sameness', or consistency of patterns across different groups of observations or parts of the study. The relationships and patterns that we have most confidence in using are those that are seen repeatedly in different contexts. This means the analysis should be looking for consistency of pattern and relationships across different, but sensible, groupings of the data.

16.4.4 Missing variables and spurious correlation

A study of 60 communities in Indonesia measured the extent to which community members cooperated on resource management problems, on a scale from 0 to 100. Communities were classified as ethnically homogeneous or heterogeneous, giving the results in Table 16.4.

Table 16.4. Ethnic classification of Indonesian community members.

Ethnicity	Mean cooperation (sample size)
Homogeneous	50 (20)
Heterogeneous	30 (40)

The results clearly suggest homogeneous communities are more cooperative. Now a second factor is included, namely whether the communities are recent immigrants or not, with the results in Table 16.5.

Table 16.5. Ethnic classification of Indonesian community members.

	Recent immigration?			
Ethnicity	No		Yes	
Homogeneous	60	(15)	20	(5)
Heterogeneous	60	(10)	20	(30)

Now what should the conclusion be? The rows of the table are identical for homogeneous and heterogeneous communities, so there is no evidence of a difference in cooperation between these: the differences in cooperation can be 'explained' by recency of immigration. Notice how misleading the conclusions from Table 16.4 might be if Table 16.5 had never been compiled. Similarly, misleading results can be obtained when looking at relationships between continuous variates, rather than factors as in this case; a phenomenon sometimes referred to as 'spurious correlation'.

Yet the results of Table 16.4 are valid in one sense. The relative proportions of recent immigration in homogeneous and heterogeneous communities in the sample (Table 16.5) may reflect that of the population. This would be the case if communities were a simple random sample within these two ethnicity strata. Then 'typical', or average cooperation in homogeneous communities in the population as a whole is higher than in heterogeneous communities. If the objectives related to that, then Table 16.4 would be sufficient. However, the extra insight provided by Table 16.5 will usually be important.

16.4.5 Achieving a balance

The previous two points suggest that a good analyst will 'drill down' and look at the observations contributing to results in ever increasing detail, and will look at more and more variables simultaneously to avoid the 'spurious correlation' problem. But there are dangers. Looking in ever more detail will mean that effective sample sizes get very small, until we start trying to understand patterns from single observations. Including more and more variables into an analysis can lead to the same problem. If two more two-level factors were to be included in the analysis of the data in Table 16.5 there would be 16 cells overall, with some of them possibly containing 0 or just 1 observation. We cannot expect to learn much about general patterns from single observations! A balance has to be reached, between looking in detail and having sufficient observations to be able to draw meaningful conclusions.

16.4.6 Attention to outliers and oddities

Single or groups of observations that seem odd, or out of line with the rest, need attention. They may represent mistakes, perhaps in data entry, and obviously then should be corrected or omitted. They may represent observations from a different population or stratum, which should not be considered within the main study. For example, two plots described as 'agricultural' in the bird diversity study showed low height diversity but a high number of bird species. Investigating further revealed that the plots contained some open water (cattle watering ponds), and should therefore be treated separately. In hindsight we can see the study design was flawed. If plots with open water are important, then they should have been included as a separate stratum to ensure a sufficient sample of them was included. If they are not important they should have been excluded by using a tighter definition of an 'agricultural plot'.

16.4.7 Variation and confounding

Many studies involve looking at relationships between two variables. These can only be detected if there is some variation in the 'explanatory' variable. During the exploratory analysis, make sure you are aware of how much variation there is. For example, if all the agricultural plots are maize fields and show no variation in height then don't expect to see a relationship between bird species numbers and height diversity in agricultural plots. Related to this is the problem of confounding. Figure 16.1d shows a common problem. Forest and agricultural plots show different bird species numbers. Is this related to height? Within each group there is no relationship with height (perhaps because there is not much variation in height). Across the two groups there are differences in height, and it is tempting to conclude that a relationship exists. However the same pattern would be seen by plotting against any other variable that changes between forest and agricultural plots: number of plant species, ground cover, human disturbance are all sensible. It is impossible, with this data, to distinguish between effects of any of these. They are described as 'confounded', or confused. Your exploratory analysis should reveal patterns of confounding between important variables.

16.4.8 Iteration

The analysis is unlikely to proceed as a simple linear process of preparing data, producing the summaries, formalizing the analysis, then writing up results. Instead it will take a series of iterative steps, with insights from one stage leading to backtracking and revising an earlier stage. To do this effectively, you will have to keep a clear focus on the objectives, and careful track of what you have done and why. Remember you will have to report your conclusions at some stage. You will then have both to provide the evidence (the results of the analysis), and justify the steps used to produce it. Things which seemed obvious and memorable at the time, such as choosing to drop some observations, may not seem so months later, when you have to defend them.

One of the messages of this book is that there are no recipes for effective statistical analysis. You cannot follow a series of predefined steps and be sure you have finished because you have completed the last one. Instead there are general principles of being interested in the data, asking sensible questions of it and trying to understand what it is telling you. Keep going with the analysis until you have answered your questions (or know that they cannot be answered with that data). Critically examine each step, and make use of the 'so what' test. This is one of the most useful of statistical tests and is simple to implement. When you have applied any of the statistical analyses, ask yourself, 'SO WHAT?' In what way does that part of the analysis help you towards your objectives?

16.4.9 Sensible use of a computer

The whole process will be easier to do efficiently if the data are well organized on a computer and you have access to suitable software with the knowledge of how to use it. This is discussed in detail in Chapters 10 to 14.

16.5 Understanding Uncertainty

In this section we review the basic concepts of estimation and hypothesis, or significance testing. Our aim is to discuss the key ideas of statistical inference in a way that is easy to understand. These topics are not unique to the analysis of natural resource management data. However, they do commonly lead to misunderstandings and inappropriate use of analysis tools, and this lack of understanding contributes to the scepticism felt by some scientists of the role of statistics in their work.

16.5.1 Applying estimation ideas

Estimating characteristics of a population from a sample, is a fundamental purpose of statistical work, whether the activity is a survey, an experiment or an observational study. We use 'population' here in the statistical sense, meaning the whole ensemble of things we are interested in (farmers, forest plots, communities, watersheds, etc.).

Point estimation arises when a quantity, calculated from the sample, is used to estimate a population value. Usually we estimate the population mean (μ) by the sample mean, $\bar{x}$, given by

$\bar{x} = \sum x / n$ (where n = sample size and $\sum x$ = the sum of the data points).

Similarly, we estimate the population standard deviation (σ) using the sample standard deviation, s, where

$$s^2 = \sum(x - \bar{x})^2 / (n - 1)$$

For example, consider estimating the average maize production per farmer among (the population of) smallholder farmers in a selected agro-ecological region. To do this, suppose a sample of 25 farmers is randomly selected and their maize yields are recorded. The average of the resulting 25 yields is calculated, giving say $\bar{x}$ = 278 kg/ha. This value is then taken as the estimate of the average maize production per farmer in the selected region. It estimates what one would expect for an individual farm. Similarly the sample standard deviation is an estimate of the amount of variability in the yields from farm to farm.

Other estimates of the population mean are possible. For example, in many surveys the observations are not sampled with equal probability. In this case we might use a weighted mean, $\hat{x} = \sum wx / \sum w$ instead of $\bar{x}$, with weights (w) that compensate for the unequal probabilities.

Population proportions (π) can also be estimated. For example we may wish to estimate π, the proportion of families who own their own land, or the proportion of respondents who indicate support for community co-management of neighbouring forest areas during a semi-structured interview. Then we might use $p = m/n$, as the estimate, where m is the number of persons making a positive response out of the total number (n) who were interviewed. For example, if $m = 30$, out of $n = 150$ interviews, then we estimate the proportion as $p = 0.2$, or 20%.

As a point estimate this is the same as a measurement (x), where $x = 1$ if community co-management was supported, and $x = 0$ otherwise. The estimate (p) is then the same as $\bar{x}$ given earlier, despite the 'data' originally being 'non-numeric'. Much qualitative material can be coded in this way. Many of the ideas of precision and uncertainty discussed here are therefore just as relevant to participatory and informal survey methods as they are to more formal forms of research.

If we have a contingent question, a follow-up only for those 30 who 'qualify' by supporting co-management, we might find, for instance, 12 out of the 30 who are prepared to play an active role in co-managing forest reserves. Arithmetically, the ratio, $r = 12/30 = 0.4$ has the form of a proportion, but it is actually the ratio of two quantities which might both change if another sample of size 150 is taken in the same way from the same population. If the follow-up question is important, then we must ensure that the original sample is large enough that there are an adequate number of respondents who qualify (here there were 30) for the precision required of the study.

Our main objective is not always to estimate the mean. For example, in recommending a new variety of maize to farmers, we may wish to ensure that the new variety gives a better yield for at least 90% of the farmers, compared to the existing variety. One way to proceed is first to calculate the difference in yields for each farmer. If from experience, or from our sample, we can accept a normal model, i.e. that the population of yield differences has approximately a normal distribution, then the required percentage point is found (from standard statistical tables) to be μ - 1.28σ, where μ is the mean difference and σ is the standard deviation of the differences.

In this case we can still use our estimates of μ and σ to estimate the required percentage point, or any other property. In general, we call the (unknown) quantities μ and σ, the parameters of the model. If we assumed a different probability model for the distribution of the yields, then we would have to estimate different parameters. The formulae would change, but the ideas remain the same.

If in the example above, we were not prepared to assume any distribution, then we could still proceed by merely ordering the differences in yields for each farmer and finding the value exceeded by 90% of the farmers. This is a non-parametric solution to the problem; we return to this approach in Section 20.4. Generally this approach requires more observations than a 'parametric' or 'model-based' approach, such as that in the preceding paragraphs.

For reference later, we explain here the commonly-used term 'degrees of freedom'. Degrees of freedom can be interpreted as 'pieces of information'. For example, with the sample of 25 farmers used earlier in this section, we have a total of 25 pieces of information. In any study it is usually important to have sufficient pieces of information remaining to estimate the (residual) spread of the population. In our simple example, the spread is estimated by s, and in the formula we divided by $(n-1)$. This is because the spread is being measured about the sample mean, $\bar{x}$. The sample mean is one of our 25 pieces of information, so we have $n-1$ or 24 degrees of freedom remaining to estimate the variability.

16.5.2 Standard errors

When some quantity (e.g. a mean, a difference, or a proportion) is estimated, it is important to give a measure of precision of the estimate. An estimate with very low precision may not be very useful, and could be misleading if its low precision is not appreciated. High precision for an estimate is reassuring, though it might indicate that more data than necessary were collected.

The measure of precision of an estimate is called the *standard error* (s.e.) of the estimate. The smaller the standard error, the greater is the precision of the estimate. Thus a small standard error indicates that the sample estimate is reasonably close to the population quantity it is estimating.

As an example, suppose we select a random sample of 12 farmers (n = 12) and measure their maize yields per hectare, we might find $\bar{x}$ = 1.5 t/ha and

$s = 0.6$ t/ha. Then our estimate of μ is given by $\bar{x} = 1.5$ t/ha and its standard error is given by the formula

$$\text{s.e.} = s/\sqrt{n}$$

In this case the s.e. is $0.6/\sqrt{12} = 0.17$ t/ha.

From the above formula it is clear that we get precise estimates either because the data have small variability in the population (i.e. s is small) or because we take a large sample, (i.e. n is large). For example, if we had taken a larger sample of 108 farmers instead, which had given rise to the same mean and standard deviation, then the standard error of the mean would have been 0.058. Equally, if yields had been less variable at $s = 0.2$ t/ha then with 12 farmers, we would also have had an s.e. of 0.058.

Depending on the investigation, we are often interested not so much in means, but in differences between means (e.g. differences in mean yield). In simple situations, such as an experiment where there is equal replication of the treatments and n replicates per treatment, the standard error of the difference (s.e.d.) between two means is

$$\text{s.e.d.} = s\sqrt{(2/n)}$$

i.e. about one-and-a-half times the standard error of each individual mean.

The formulae for the standard error of a proportion and a ratio that were considered earlier are more complicated, but the point about precision being related to sample size and variability of the data is general. When the design of the study or the model is complex, standard errors cannot be easily computed by 'hand' and suitable software is used to obtain standard errors for estimates of interest.

In this section we have repeatedly mentioned that the data are a random sample from the population. The reason that randomness is important is that it is part of the logic of the standard error formulae. This is because as our sample was collected at random, it is one of many that might have been chosen. Typically, each sample would have given a different mean, or in general a different estimate. The standard error measures the spread of values we would expect for the estimate for the different random samples.

The idea of the standard error as a measure of precision can help researchers plan a data collection exercise, as is described in Section 16.9.5. In any study, σ is the unexplained, or residual, variation in the data; an effective study is one that tries to explain as much of the variation as possible. Continuing the example above, we might find that the farmers use different production systems, thus giving us two sources of variation. There is variation between production systems and there is variation between the farmers using the same production system.

Suppose the overall variation of the yields, ignoring the different production systems, is estimated as $s = 0.6$ t/ha while the within production-system variability is $s = 0.2$ t/ha. If we were planning a new investigation to estimate average maize production we could either, ignore that there are different production systems, and take a simple random sample from the whole population, or we could take this into account and conduct a stratified study. The standard error formula shows us that in this instance we would need nine times as many farmers in the simple random sample compared to the stratified study to get roughly the same precision.

Chapter 21 describes how the standard error is used in the reporting of the results. The next section of this chapter, which is on confidence intervals, shows how the standard error is used to describe precision. The width of a confidence interval is often a simple multiple of the standard error.

16.5.3 Confidence intervals

The confidence interval provides a range that is highly likely (often 95% or 99%) to contain the true population quantity or parameter that is being estimated. The narrower the interval the more informative is the result. It is usually calculated using the estimate and its standard error.

When sampling from a normal population, statistics textbooks explain that a confidence interval for the mean μ can be written as

$$\bar{x} \pm t_{n-1} \times \text{s.e.}(\bar{x})$$

where t_{n-1} is the appropriate percentage point of the t-distribution with $(n-1)$ degrees of freedom.

A 95% confidence interval is commonly used, for which t-values are 2.2, 2.1 and 2.0 for 10, 20 and 30 degrees of freedom respectively. So we can usually write that the 95% confidence interval for the mean is roughly

$$\bar{x} \pm 2 \times \text{s.e.}(\bar{x})$$

The example in Section 16.5.2 involving 12 farmers gave $\bar{x}$ = 1.5 t/ha with s.e. = 0.17 so 2 × s.e. = 0.34. The 95% confidence interval for μ is therefore between 1.16 and 1.84 t/ha; so we can say that this range is likely to contain the population mean maize yield. (The exact 95% interval is 1.12 to 1.88 t/ha, and is found by multiplying by the t-value of 2.19 instead of by the approximate value of 2. The exact t-value and interval can be found using a statistical software package.)

More generally, for almost any estimate, whether a mean or some other characteristic, and from almost any population distribution, we can state that the 95% confidence interval is approximately

$$\text{estimate} \pm 2 \times \text{s.e.(estimate)}$$

Hence it is useful that statistical software routinely provides the standard error of estimates. With the example of Section 16.5.1 of p = 30/150 = 0.2, or 20% of the 150 farmers, the standard error is about 0.03, resulting in a confidence interval of about 0.14 to 0.26.

Care must be taken in knowing what a confidence interval really shows. A 95% confidence interval does *not* contain 95% of the data in the sample that generated it. Very approximately, the latter is the interval $\bar{x} \pm 2s$, and is sometimes called a prediction or tolerance interval. In our example of farmers above, with $\bar{x}$ = 1.5 t/ha and s = 0.6 t/ha, this interval is 0.3 to 2.7 t/ha, and we would say that most of the farmers have yields in this range.

Users often confuse the confidence interval for the mean, with an interval that contains most of the data, because the objectives of the study often relate to parameters other than the mean.

In our example above, the 95% confidence interval for the mean is 1.12 to 1.88 t/ha with the sample of 12 farmers (Fig. 16.2). With more data, this interval would be narrower, as is seen by comparison with the confidence interval for a sample with 108 farmers, where the appropriate calculation gives a 95% confidence interval for the mean of about 1.4 to 1.6 t/ha. The prediction interval, within which 95% of farmers lie, does not get narrower as the sample size increases; it is just estimated with increased precision.

When estimating proportions, this interpretation of the confidence interval is more obvious. With the example of the 150 farmers, 20% of them said 'Yes', giving a confidence interval for the true proportion of 14% to 26% or 0.14 to 0.26. But each individual farmer said either 'Yes' (x=1) or 'No' (x=0), as illustrated in Fig. 16.2.

Fig. 16.2. Data points and confidence intervals for means (upper plot: maize yields for a sample of 12 farmers) and proportions (lower plot: yes/no responses for a sample of 150 farmers).

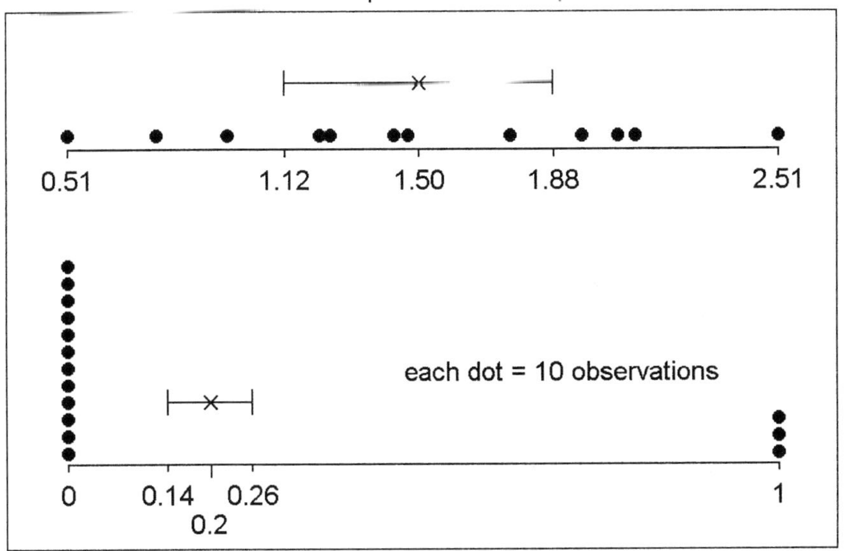

When the assumptions about the data may not be quite right, scientists may feel they ought to abandon the ordinary confidence interval and use some different procedure altogether. It is usually more constructive to proceed instead by using the usual method, but noting that the true coverage of the '95%' confidence interval may not be exactly 95%. For most purposes, the 95% figure is used to provide a conventional measure of uncertainty about an estimate, rather than the basis for decision-making. Knowingly the approximate magnitude of the uncertainty is usually just as useful as knowing its exact value.

16.6 Hypothesis Testing and Significance

For good reasons, many users find hypothesis testing challenging as there is a range of quite complex ideas. We begin with a simple illustration.

16.6.1 Simple illustration of hypothesis testing

A researcher facilitates an on-farm trial to study the effect of using tephrosia as a green manure for fertility restoration. She claims the use of the manure will increase pigeon pea yields, i.e. pod weight. In the trial, pigeon peas are grown with and without the tephrosia in two plots on each of eight smallholders' fields and the values recorded are the differences in yields.

We test the correctness of the researcher's thesis. In this case the 'null hypothesis' is usually that the true mean increase is zero, $\mu = 0$. The 'true' mean increase is the increase for the population of smallholders; our eight are assumed to be a random sample from this population. Another way of interpreting the 'true' mean here is that it is the increase you would expect on any smallholder's plot if you ignored plot-to-plot variation. The alternative hypothesis is usually that the true mean increase is other than zero.

The null hypothesis is often given, quite explicitly, with the alternative hypothesis being vague. This is for two reasons:

1. Standard statistical tests calculate the probability of getting a sample as extreme as the one observed, assuming the null hypothesis is true; this calculation has to be done using explicit values for the parameter(s) of the null hypothesis distribution.
2. Hypothesis testing adopts the legal presumption of 'innocent until proven guilty'. The null hypothesis that $\mu = 0$ remains valid, unless the data values are inconsistent with it.

In much of applied science, the null hypothesis makes little sense. For example, it is really not tenable that growing pigeon pea with tephrosia makes no difference in yield whatsoever. The researcher and farmer are usually more interested in a different question, namely, is the difference in yields caused by the tephrosia large enough to be interesting. We return to this point later.

Textbooks often distinguish between one-sided and two-sided tests. In this example we might consider the test of the null hypothesis that $\mu = 0$, against the one-sided alternative that $\mu > 0$, on the assumption that there is no logical reason that the manure will decrease yields. Usually a one-sided test merely halves the significance level, so what was significant at 4% with a two-sided test, becomes significant at 2% with a one-sided alternative. As will be seen below, we are not keen for readers to become too attached to a particular significance level; so halving a value is not important enough for users to spend much time on this idea.

Example 1

Suppose in the illustration above, the differences in pod weight (kg) between 'treated' and 'untreated' plots were as follows:

3.0 3.6 5.4 −0.4 −0.8 4.2 4.8 3.2

The computer output from an analysis of these data would look something like:

```
Test of mu = 0 vs mu not = 0

Variable     N    Mean     StDev    SE Mean
podweight    8    2.875    2.290    0.810

Variable     95.0% CI              T       P
podweight    ( 0.959, 4.791)      3.55    0.009
```

The t-test used to investigate the hypotheses follows the general formula:

(estimate – hypothesized value) / s.e.(estimate)

Here we are interested in the mean difference in pod weight and whether this is zero, so our test statistic is:

$$t = \left(\overline{x} - 0\right) / \text{s.e.}\left(\overline{x}\right)$$

i.e. (2.87 – 0)/0.81 = 3.55 (the value given as `T` in the computer output).

By comparison with the t_7 distribution, a value as extreme as 3.55 has a probability of 0.009, i.e. less than 1 in 100, of occurring. So if the null hypothesis is true, then there is a chance of just less than 1 in 100 of getting the sample we found. Either something unlikely has occurred or the null hypothesis is false. This event is sufficiently unlikely that we declare the result to be statistically significant and reject the null hypothesis.

In Section 16.5.3 on confidence intervals we used a t-value of 2 to give approximate 95% confidence intervals. Similarly we find here that values larger than 2 are extreme, (at about the 5% level of significance) and hence cast doubt on the hypothesized value.

16.6.2 Understanding significance

The classical argument is that we should approach this type of decision-based testing in an objective way, by pre-setting the significance level, or p-value at which we would choose to reject the null hypothesis. If we were working to a significance level of 5%, or $p = 0.05$, we would reject it at the 5% level and report that $p < 0.05$. Rather than following such a stringent approach, we recommend that decisions be made on the grounds that a p-value is low.

Example 2

We have the same hypothesis as in Example 1, but suppose we collected a slightly more variable sample. The data values might be:

3.0	3.6	6.8	−1.6	−2.0	5.8	7.1	0.3

Computer analysis of these data gives the results shown in the box overleaf.

The standard error of the mean is now larger than in Example 1, and we get a t-statistic of 2.23 with a probability of 0.061. If we used the 5% level as a strict cut-off point, then we would not reject the null hypothesis. This does not mean we accept the null hypothesis as 'true' and users who write as if it does are showing a serious weakness of interpretative skills. The probability of getting such a sample under a hypothesis of no effect is still low, so there is some suggestion of a treatment effect, but not low enough to meet our criterion at the 5% level.

```
Test of mu = 0 vs mu not = 0

Variable      N   Mean   StDev   SE Mean
podweights    8   2.87   3.64    1.29

Variable      95.0% CI           T      P
podweights    ( -0.17, 5.92)    2.23   0.061
```

There is insufficient weight of evidence to draw a conclusion about a difference between the treatments. If a sample of 16 observations had been collected, with the same mean and standard deviation as above, the standard error of the mean would have been lower (at 0.91) and consequently the t-statistic would have been higher (at 3.15). This would have been significant with a p-value of 0.007.

If hypothesis testing is undertaken because a real decision is being made to accept or reject a new variety, for example, not rejecting the null hypothesis may be tantamount to accepting the pre-existing variety. This is not the same thing as accepting that the null hypothesis is correct.

Generally, scientific research does not involve such cut-and-dry decision alternatives. The main purpose of significance testing may just be to establish that an expected effect ('research hypothesis') can be discerned and plausibly shown; not just to be a quirk of sampling. Very tiny effects can be significant if sample sizes are very large; a significant effect also needs to be large enough to be of practical importance before it is 'significant' in the ordinary-language use of the term.

Conversely, a non-significant effect does not necessarily imply that an effect is absent. A non-significant result can also happen if the sample size is too small or if there is excessive variability in the data. In either of these cases, the chance of detecting an effect is low i.e. the effect may be present but the data are unable to provide evidence-based conclusions of its existence. Many researchers find it disconcerting to learn that they can always get a 'non-significant' result by doing a poor study!

Such considerations show it is usually more informative to produce a confidence interval rather than just the decision outcome and p-value of a hypothesis test. In Example 1 above, the 95% confidence interval for the mean is from 0.96 to 4.79 using the method of calculation shown in Section 16.5.3. This indicates that the true mean increase of 0 kg is unlikely, because the 95% confidence for the true mean does not contain the hypothesized value.

Given a calculated t-value or other test statistic, it was traditional to compare this with a 5%, 1%, or 0.1% value in statistical tables. However, since many statistical packages compute exact p-values, results may be accompanied by statements such as ($p = 0.028$) giving a specific numerical figure for the degree of extremeness of the disparity between observed results and null hypothesis expectation. This approach is preferable where possible. It is more informative and easier to interpret, for example allowing a reader to distinguish between different situations in which $p = 0.51$ and $p = 0.051$, both of which would earlier have been reported as 'not significant, with $p > 0.05$'.

16.6.3 General ideas

In a few studies, the objectives of the study correspond to standard hypothesis tests. The examples in the previous section provide one scenario, and the adoption of a new farming practice is another.

Usually, however, the hypothesis testing is just a preliminary part of the analysis. Only rarely can the objectives of a study be met by standard significance tests. The statistically significant result provides objective evidence of something interesting in the data. It serves as a 'passport' to further analysis procedures. Confidence intervals, or perhaps an economic analysis, are then used to describe the nature and practical importance of the results.

When results are not significant, it may indicate that nothing further need be done. Often it enables a simpler model to be used. For example if there is no evidence of a relationship between education level and adoption of a new innovative technology, then the adoption can be studied using all respondents together. If there were a relation, then it might indicate the necessity for a separate analysis (i.e. a separate model) for each education level group.

16.6.4 Recognizing structure

Example 1 above illustrates how a t-test is conducted using differences between plots from eight smallholder farms. The differences were used because a pair of plots was available from each farm. This led to a paired t-test.

Suppose on the other hand we had 16 farms, each with just one plot, and eight were selected to try out the 'treatment', with the remaining farms forming the 'controls'. The analysis then involves the comparison of two independent samples.

It is important to recognize the structure in the data when conducting the analysis. As an example, we show in Table 16.6 what is often lost if truly paired results are treated like independent samples. Here the x- and y-values represent the sediment load in samples collected from a river at two locations, X and Y, on ten separate occasions. The aim was to see whether the two locations differed in the quality of the water.

Table 16.6. Sediment load in paired samples at two locations.

i	1	2	3	4	5	6	7	8	9	10	Mean	s.d.
x_i	174	191	186	199	190	172	182	184	200	177	185.5	9.687
y_i	171	189	183	198	187	172	179	183	199	176	183.7	9.764
d_i	3	2	3	1	3	0	3	1	1	1	1.8	1.135

The difference between the two means is 1.8. In an unpaired analysis, the standard error of this difference is calculated using the standard deviations of the two sets of samples (i.e. 9.687 and 9.764), and found to be 4.3. This leads to a non-significant t-value of 0.41. The correct, paired analysis uses the standard deviation of the differences (i.e. 1.135). The standard error of these differences is $1.135/\sqrt{10}$ or 0.36, leading to a highly significant t-value of 5.0.

The reason that the unpaired analysis gives a non-significant result is that the occasion-to-occasion variation in the samples is included in the calculation of the standard deviations used in the two-sample t-test. Not eliminating this variability means the small but systematic differences between the pairs are not detected. Where true and effective pairing exists, an unpaired analysis is both weak and inappropriate. In general, this paired structure is similar to the idea of blocking in experiments, and stratification in surveys, and needs to be properly accounted for in any subsequent data analysis.

16.7 Analysis of Variance

Practical problems are usually more complicated than the examples shown in the previous two sections. We use analysis of variance (ANOVA) to illustrate how the concepts are applied in larger problems. The same type of generalization is possible for data on proportions or where regression, or other statistical modelling methods would be used. When data are from non-normal distributions, such as survey data on counts, then the ideas of analysis of variance are generalized and are then called the analysis of deviance, which is introduced in Section 20.5. The key concepts remain unchanged.

16.7.1 One-way ANOVA

Table 16.7. Number of fish species in each catch.

Lake:	Tanganyika	Victoria	Malawi	Chilwa
Catches	64	78	75	55
	72	91	93	66
	68	97	78	49
	77	82	71	64
	56	85	63	70
	95	77	76	68
Mean	72	85	76	62

The *t*-test for two independent samples, discussed in Section 16.6.1, generalizes for more than two samples in the form of the one-way analysis of variance. The comparison of a set of independent samples is described as a 'completely randomized design'. We illustrate this with an example using four samples.

In a study of species diversity in four African lakes, data were collected on the number of different species, caught in six catches, from each lake. The values are shown in Table 16.7.

The usual analysis of variance, for instance as generated by a statistical package, will look like that in the following box (**DF** = Degrees of Freedom, **SS** = Sum of Squares, **MS** = Mean Square):

```
One-way ANOVA: catch versus lake

Analysis of Variance for catch

Source     DF      SS       MS       F       P
lake        3    1637      546    5.41   0.007
Error      20    2018      101
Total      23    3655
```

The pooled estimate of variance (s^2) is 101. In the typical computer output above it is shown as the 'Error' mean square (MS), though a more appropriate phrase would be the 'residual' mean square.

The standard error of the difference between any two of the above means is given by the following:

$$\text{s.e.d.} = \sqrt{\left(2s^2/6\right)} = 5.80$$

In the box above, the *F*-value (shown above as **F**) and the probability (shown as **P**) are analogous to the *t*-value and *p*-value in the *t*-test for two samples. Indeed, the comparison of two independent samples is a special case of the one-way ANOVA, and the significance level is the same, irrespective of whether a *t*-test or a one-way ANOVA is used.

With more than two groups a significant *F*-value, as here, indicates there is a difference somewhere amongst the groups considered, but does not say where; it is not an end-result of a scientific investigation. The analysis usually continues with an examination of the treatment means that are shown with the data above. Almost always a sensible analysis will look also at 'contrasts', i.e. comparisons whose form depends on the objectives of the study. For example if lakes in the Tanzanian sector were to be compared with those of Malawi, we could look at the difference in the mean of the first two treatments, compared with the mean of the third and fourth. If this difference were statistically significant, then the magnitude of this difference, with its standard error, would be discussed in the report of the results.

In the analysis of variance a 'non-significant' *F*-value may indicate there is no effect. Care must be taken that the overall *F*-value does not conceal one or more individual significant differences 'diluted' by several not-very-different groups. This is not a serious problem; the solution is to avoid being too simplistic in the interpretation. Again researchers should avoid undue dependence on an arbitrary 'cut-off' *p*-value, such as 5%.

16.7.2 Multiple comparison tests

These tests are often known by their author and include Dunnett's test, Newman-Keuls, and others. They concern the methods of testing differences between means, which require ANOVA-type analyses. Some scientists use them routinely, while others avoid their use.

Our views are perhaps clear from the examples discussed in Section 16.6. Hypothesis testing is usually just a preliminary step, and further analysis, often concerning the treatment means, relates directly to the stated objectives of the study. This will usually involve particular contrasts to investigate differences of interest. We do not recommend multiple comparison methods, because they do not usually relate to the objectives of the research.

The case for the multiple comparison tests rests on the danger of conducting several significance tests on a set of means, for example comparing the largest with the smallest, without adjusting the test for the fact that we have deliberately chosen them as being largest and smallest. The case is clear, but irrelevant in most analyses, because we prefer not to do too many tests. We want instead to investigate the size of differences in relation to their practical importance.

Taking the application of agricultural field trials, the treatment structure will usually be well defined, with factorial structure being the most common. (Reference can be made to Section 5.3.1, on factorial treatment structure.) In such cases, multiple comparison procedures are not relevant. The only type of factor where we might wish to consider multiple comparison methods would perhaps be variety comparison (of maize, say) where we might wish to present the results in descending order of the mean yields. Even here, we are more likely to try to understand the differences in yields as a function of the season length, or country of origin, etc., for the varieties, than to suggest a series of tests.

One instance where multiple comparison methods may be used would be if we wish to group the varieties into sets that have similar responses. Our main concern is that users may be tempted to use a multiple comparison method instead of a more thoughtful analysis, and hence will fail to interpret the data in ways that are needed to meet the objectives of the study. As long as you do not fall into this trap, do both. We predict that when reporting the results in relation to the objectives, you will not use any of the results from the multiple comparison methods. They can then be deleted from the tables in the report!

We return to this problem in Chapter 21, on presentation of results, because some scientists may have withdrawal symptoms if they do not produce tables with a collection of letters beside the corresponding means.

16.8 A General Framework

In order to concentrate on the general concepts, the illustrative examples in Sections 16.5 to 16.7 have all been simple. The concepts include the following:

- The data are (or are assumed to be) a sample from some population, and we wish to make inferences about the population.
- We use our sample to estimate the properties (parameters) of the population that correspond to the objectives of our study.
- The standard error of the estimate is its measure of precision. Sometimes we report the standard error itself and sometimes we report a confidence interval for the unknown parameter.
- We often use hypothesis tests to identify whether differences between parameters can be detected in our study. This testing phase is often the first step in the inference part of the analysis.

We now introduce a common underlying theme to many statistical analyses. That is the concept of the statistical model. Most of the examples in this book can be written in a general way as:

*data = pattern (*or *model) + residual*

This is our assumed model for the population. For example, the problem of sediment in the river can be written as:

sediment = occasion effect + location effect + residual

Our objective was to investigate the difference between the two locations, and the effect was clear. But we also saw in Section 16.6.4 that if we omitted the occasion effect from the model, i.e. if we used the simpler model:

sediment = location effect + residual

then we could not detect the location effect. This showed that we need the occasion component in the model, even though studying its size might not be one of our objectives.

The model above is the same if there are more than two locations, and would still apply if the data were not 'balanced', i.e. if locations were not all sampled on all occasions. With standard statistics packages the inferences can still be made. We return to this in Chapter 19.

Within this general context, significance tests are often used to provide guidance on how complex a model needs to be. Using the chosen model, we then estimate the properties that correspond to our objectives, and give a measure of precision to indicate our level of confidence in reporting the results.

Earlier, one limitation was that the data had to come from a distribution that was approximately normal, but this is no longer the case. Parametric methods are now very flexible in dealing with well-behaved data, even when not normally distributed and this often provides a more attractive framework for data analysis than the simple tests that are sometimes all that are currently used. For example, when looking at inferences about proportions, instead of using a simple chi-square test to examine relationships in a two-way contingency table, the use of log-linear models can be used with both two-way tables (like a chi-square test) and with more complicated tables of counts. We discuss this more general framework in Section 20.5.

16.9 Consequences for Analysis

The ideas of statistical inference introduced in the previous sections have important consequences for the analysis. Of course they underlie specific analysis methods, but they also have more generic consequences that we highlight here.

16.9.1 Choosing the right unit of analysis

The precision of a comparison depends on the sample size and the variability. Though the variability of what? Imagine a study designed to look at above-ground biomass accumulation in secondary forest. Ten 1 ha plots of each of two ages (10 years since clearance, and 25 years since clearance) are selected, and the biomass measured in each. The biomass is measured by randomly placing five quadrats of 5×5 m in each plot. The biomass of each tree in each quadrat is estimated by allometric methods. Now we have observations at three levels in a hierarchy: 20 plots, 100 quadrats and hundreds of trees. The sample size depends on whether we think of plots, quadrats or trees; the variability between them will also differ.

So which unit should we use if we want to use the ideas of the previous sections? In this case it is variation between the primary sampling units, the plots, which is relevant to comparing the two forest types. Why? Because we wish to see if the difference between the 10-year and 25-year plots is larger than random plot-to-plot variation. Variation between quadrats is part of the within-plot variation, and this is usually less than the between-plot variation. There will also be other factors that lead to these larger plots differing from each other. Hence, if we used the 100 quadrat observations as our sample and the quadrat-to-quadrat variation, we would overestimate the precision of the comparison.

There are methods for looking at the multiple levels of variation present in many studies, and using them correctly to make inferences (Section 20.6). However there is a simple way to get valid estimates of precision. Take the hierarchical data and summarize it to the right level before trying to calculate standard errors or to perform *t*-tests. In the example here, this means first getting an estimate for biomass in each of the 20 plots (probably by summing the biomass of all trees in each quadrat, then averaging the biomass of the 5 quadrats per plot to get an estimated biomass for each plot). The formal statistical procedures, of calculating the variance, the standard error of the difference in means and so on, are then based on the 20 plot values.

16.9.2 Unreplicated studies

Once the importance of analysis at the right level of the hierarchy is understood, it becomes clear that many apparently large studies are actually unreplicated. In each of two villages, 200 households were interviewed to elicit information on levels of innovation in resource management. The overall objective was to look at how the presence of an extension officer in the village influences innovation. One village had an extension officer the other did not. The presence/absence of the extension officer is thus a feature of the village, and the study has effectively only two observations, one for each village. Seen in this way it is clear that we cannot estimate the standard error of the 'extension officer effect'. Despite the 200 interviews per village, the study is unreplicated and, however interesting, should be seen as a case study that will generate insights and hypotheses but will not produce any results that can be relied on to have some applicability in other villages. The good news is that the statistical analysis problem disappears!

Ecologists have long recognized this problem and refer to it as 'pseudo replication'.

16.9.3 Linking the case study to the replicated parts

In Sections 6.4 and 7.9, we considered case studies that were embedded in a larger baseline survey. For example, we might have information on 30 villages in a particular district, as well as detailed information on two of these villages from participatory and other studies. The analysis can proceed in three stages.

1. The analysis of the case studies.
2. The analysis of the baseline information. This is a general analysis, but we have the additional objective of seeing how the units (villages) that were studied in more detail compare with the general set of 30 villages.
3. The two analyses are put together to assess whether there are fruitful ways in which the baseline information can assist us in drawing more general conclusions from the case studies.

The first two stages are standard, so we look briefly at some ways that we might put the two analyses together. For example, in Section 20.3.1 we will look at a method of analysis called cluster analysis. Say the 30 villages divided neatly into four clusters. The idea is that each cluster contains villages that are similar. We could now see in which clusters our two case studies are located. It would be more plausible to generalize the results from the case study villages, to others in the same cluster, compared to those in different clusters.

Cluster analysis is intended for unstructured data, whereas we usually have well-defined structure to our data. In such cases the analysis of the baseline information should be in relation to the known structure. For example, suppose two key factors are the presence (or absence) of an extension agent in the village, and the distance of the village from a bank, or other avenue for credit. Suppose also that we find that propensity for innovation, depends crucially on the presence of an extension agent, and linearly on the distance from a bank. This is found from the baseline study and is one of the subjects covered in more detail in the case studies.

Firstly, consider that we have just two case study villages. Both have extension agents and one is near, and the other is a middling distance from a bank. We suggest that any generalization of the case study information to other villages would mainly be informally stated. We would have more confidence in statements made for other villages with extension agents, feeling that those without are likely to have a lower propensity for innovation. Similarly for generalization to other villages that are further from a bank than either case study village.

On the other hand, suppose that there were six case study villages and they were chosen in a factorial way, as described in Section 6.5.2, so there were two villages close to a bank, two at middle-distance, and two far from a bank. At each distance, one of these villages had an extension agent, and the other did not. A sample of six villages is still small in its own right, but now the third stage of the analysis can be undertaken with much more confidence. It might now be possible to make formal adjustments to the results, so they do apply to the larger set, and possibly to give some rough measures of precision, that have been derived from the second stage of the analysis.

16.9.4 Using the structure

In this chapter we have used three examples to illustrate the importance of recognizing the structure of the data when doing the analysis. In Section 16.4, the fourth point was headed *missing variables*, and illustrated how the conclusions could be changed if a key variable was included in a table of results. In Section 16.6.4 we showed that failure to recognize the *paired* nature of our data could affect our ability to detect important treatment effects. And in Section 16.9.1, we discussed the importance of recognizing the correct units for the analysis.

One failing in many analyses is that researchers are sometimes inconsistent in allowing for structure. Sometimes this is partly due to the statistical software that may make it easy to include all of the structure with one method of analysis, but not with another. This may indicate a weakness in the software, or that the suggested method of analysis is not appropriate.

For example, in Section 20.3.1 we look briefly at cluster analysis, a multivariate method of analysis. This can be a powerful tool to discover whether there are groups in your data. Perhaps these may be sets of villages that are similar, and can therefore be treated the same way when giving extension advice. While this is useful in some circumstances, the method is meant for data where there is no inherent structure. Sometimes the method is merely a complicated way to rediscover the structure in your data that you already knew, perhaps that neighbouring villages are more similar than those further apart!

16.9.5 Calculating sample sizes

Finally, we return to an aspect of design: the choice of the sample size. This subject is mentioned here, because we need the results from previous sections of this chapter to outline the ideas.

The formula that is key to understanding the concepts associated with determining the sample size was given in Section 16.5.2. It is that the standard error (s.e.) of the sample mean ($\bar{x}$) in estimating the population mean (μ), is given by s.e. = $s/\sqrt{n}$. Then we saw that the 95% confidence interval for μ is given roughly by

$$\bar{x} \pm 2 \times s/\sqrt{n}$$

From this formula, we see that information must be given on two key aspects to be able to choose the required sample size (n). First, we need an idea of how variable the observations will be, and this is reflected in the standard deviation (s). Second, we need to state how small a difference it is important to detect, namely how wide do we require the resulting confidence interval to be.

The simplicity of hypothesis tests allows them to provide a basis for the evaluation of many sample size calculations. This involves the power of the test, i.e. the probability of correctly rejecting the null hypothesis when it is false. If your sample size is sufficient, then you will have a high power to detect a difference that you care about, and low power for a meaningless difference.

Modern statistical packages incorporate extensive facilities for sample size or power calculation. They also provide help on their use and interpretation. By using a software package to try different calculations, it is possible to acquire a 'feel' for power or sample size calculations.

As an example, we take the paired *t*-test considered earlier, in Section 16.6.4. We make the assumption that our aim is to choose the sample size, i.e. how many observations we need for a similar study. We suppose that the standard deviation will be about the same as before, that is about $s = 1.1$ and that we would like to detect a mean difference of more than 1 unit in sediment load with probability 0.95, i.e. we look for a power of 0.95. Putting these conditions into a statistics package gives a required sample size of 18 observations.

If this is too many, and we find that we can only have 10 observations, we can then keep our difference of 1 unit and would find that the power is 0.72. Or we can ask for what difference the power will be 0.95, giving a value of a mean difference of 1.4. These results can then provide a basis for a discussion on the study to be conducted.

A study with low power may have little ability to discern meaningful results. It should be reconsidered, so that it is large enough to establish important effects, or abandoned if it cannot be expected to do so. When a study becomes too large, it wastes resources; while one that is too small is also wasteful if it does not generate clear findings. Study size calculations are closely related to decisions on expending resources, so it is important to get them right.

In this example we have used a paired *t*-test to emphasize that the variability is usually the residual standard deviation of the data, because it is the *unexplained* variation of the data that relates to the precision of the study. Hence if the study needs to be large, because there is a lot of unexplained variation, then consider whether the study can be designed in such a way that more variation can be explained. This can be by sensible stratification (blocking) or by measuring the variables that explain as much variation in your response as possible. (These ideas are covered in Section 5.7.) In our example, if the samples had not been paired, we would need over 2000 measurements instead of 18, to give the same 95% power to detect a mean difference of one unit!

When using results such as the example above it is important to remember that the calculations of sample size or power relate to a single hypothesis. Most studies have a number of objectives and significance testing is usually only a small part of the analysis. In general, the same type of calculation should therefore be done for a number of the key analyses to ensure that the sample size is sufficient for all the objectives of the study. Thus the proper planning of a data collection study requires that the main analyses are foreseen and planned for, before the data collection is allowed to start.

Chapter 17

Analysis of Survey Data

17.1 Preparing for the Analysis

In this chapter we discuss themes concerned with the analysis of survey data. Some of the issues are also relevant for other types of study. In particular, they are important when analysing survey data collected using questionnaires.

In Sections 17.1 and 17.2, we consider preparatory ideas including weighting, coding and data in the form of ranks and scores. In Section 17.3, we look at issues concerned with tabulation, multiple responses, profiles and indicators.

17.1.1 Data types

The scale on which data are recorded influences the analyses that should be used. On a nominal scale, the categories recorded are described in words, rather than numerically. If a single one-way table resulting from a simple summary of nominal (also called categorical) data contains frequencies as in Table 17.1a, it is difficult to present this same data in any other form. We could report highest frequency first as opposed to placing them in alphabetic order, or reduce the information, for example if one distinction is of key importance compared to the others (Table 17.1b).

Table 17.1a. A simple frequency table.

Christian	Hindu	Muslim	Sikh	Other
29	243	117	86	25

Table 17.1b.

Hindu	Non-Hindu
243	257

On the other hand, where there are ordered categories, the sequence will only make sense in one way, or in exactly the opposite order (Table 17.2a).

Table 17.2a. A table with ordered categories.

Excellent	Good	Moderate	Poor	Very bad
29	243	117	86	25

Table 17.2b. Another table with ordered categories.

Excellent	Good	Moderate	Poor	Very bad
79	193	117	36	75

We could reduce the information in Table 17.2a by combining categories, but we can also summarize the information numerically. For example, accepting a degree of arbitrariness, we might give scores to the categories from 1 for very bad to 5 for excellent, and then produce an 'average score': a numerical indicator, for the sample. This is an analogue of the arithmetic calculation that could be done, if the categories were numbers (e.g. family sizes). This gives an average score of $(1\times25+2\times86+3\times117+4\times243+5\times290)/500 = 3.33$.

The same average score of 3.33 could arise from differently patterned data, e.g. from rather more extreme results as in Table 17.2b. As with any other indicator, this 'average' only represents one feature of the data and several summaries will sometimes be needed.

17.2 Data Structure

17.2.1 Simple data structure

The data from a single-round survey, analysed with limited reference to other information can often be thought of as a 'flat' rectangular file of numbers. These numbers could be made up of counts/measurements, codes, or a combination of these. In a structured survey with numbered questions, the flat file has a column for each question, and a row for each respondent, a convention common to almost all standard statistical packages. If the data form a perfect rectangular grid with a number in every cell, analysis is made relatively easy. However, there are many reasons why this will not always be the case and flat file data will be incomplete or irregular.

Surveys often involve 'skip' questions where sections are left blank if irrelevant, for example, details of spouse's employment do not exist for the unmarried. These arise legitimately, but imply that different subsets of people respond to different questions. 'Contingent questions', where not everyone 'qualifies' to answer, often lead to results that seem inconsistent. If the overall sample size is just adequate, the subset who 'qualify' for a particular set of contingent questions may be too small to analyse in the detail required.

If any respondents fail to respond to particular questions (item non-response), there will be holes in the rectangle. Non-informative non-response occurs if the data is missing for a reason unrelated to the true answers, such as the interviewer turned over two pages instead of one. Informative non-response means that the absence of an answer itself tells you something. For example, you are almost sure that the missing income value was withheld, as it may be the highest in the community, and the individual did not wish to disclose this information. A small amount of potentially informative non-response may be possible to ignore if there is plenty of data. However, if data are sparse or if informative non-response is frequent, the analysis should take account of what can be concluded from knowing that there are informative missing values.

17.2.2 Hierarchical data structure

Another complexity of survey data structure arises if the data are hierarchical. A common type of hierarchy is often seen in questionnaires where a series of questions is repeated. For example questions may initially be asked concerning each child in a household, and this is then combined with a questionnaire for the entire household, and also with data collected at the community level.

For analysis, we can create a rectangular flat file, at the 'child level', by repeating relevant household information in separate rows for each child. Similarly, we can summarize relevant information about the children in a household, to create a 'household level' analysis file. The number of children in the household is usually a desirable part of the summary. This 'post-stratification' variable can be used to produce sub-group analyses at household level, separating out households with different numbers of child members. The way the sampling was done can have an effect on interpretation or analysis of a hierarchical study. For example if children were chosen at random, households with more children would have a greater chance of inclusion, and a simple average of the household sizes would be biased upwards. It should be corrected for selection probabilities.

Hierarchical structure becomes important, and harder to handle, if the data are collected at more than one level. For example, a study on targeting safety net interventions for vulnerable households, used data at five different levels. The top level was government guidance on the allocation of resources. Then there were District Development Committee interpretations of the guidance, Village Task Force selections of safety-net beneficiaries and the households. Finally there were individuals whose vulnerabilities and opportunities were affected by targeting decisions taken at higher levels in the hierarchy. To reflect such a hierarchical structure, a relational database is more suitable than a spreadsheet. It defines and retains the inter-relationships between levels, and creates many analysis files at different levels. These issues are described in Chapter 12. Any of the analysis files may be used as discussed below; but we will be looking only at one facet of the structure, and several analyses will have to be brought together for an overall interpretation. A more sophisticated approach, using multilevel modelling, is described in Section 20.6 and provides a way to look at several levels together.

17.2.3 Stages of analysis

The steps of analysis were described in Chapters 15 and 16, and apply equally to the analysis of surveys. Typically, the exploratory analysis overlaps with data cleaning. It is the stage where anomalies become evident. For example, individually plausible values may lead to a point that is way-out when combined with other variables on a scatterplot. In an ideal situation, exploratory analysis would end with the confidence that one has a clean dataset, so that a single version of the main data files can be finalized and 'locked' and all published analyses derived from a single consistent form of 'the data'. In practice, later stages of analysis often produce additional queries about data values.

Such exploratory analysis will also show up limitations in contingent questions, for example we might find we do not have enough married women to separately analyse their income sources by district. The exploratory analysis should include the final reconciliation of analysis ambitions with data limitations. This phase can allow the form of analysis to be tried and agreed, developing an analysis plan and program code in parallel with the final data collection, data entry and checking. Purposeful exploratory analysis allows the subsequent stage of deriving the main findings to be relatively quick, uncontroversial, and well organized.

The next step is to generate the main findings. This may or may not include 'formal' modelling and inference. The second stage will ideally begin with a clear-cut clean version of the data, so that analysis files are consistent with one another, and any inconsistencies can be clearly explained. This stage is described further in Section 17.3. It should generate the summary findings, relationships, models, interpretations and narratives. It should also include the first recommendations that research users will need to begin utilizing the results. In addition, time must be allowed for 'extra' but usually inevitable tasks such as follow-up work, to produce additional more detailed findings.

A change will be made to the data each time a previously unsuspected recording or data entry error emerges. It is then important to correct the database and all analysis files that have already been created, that involve the value to be corrected. This means repeating analyses that have already been done, using the erroneous value. If the analysis was done by 'mouse clicking' and with no record of the steps, this can be very tedious. This stage of work is best undertaken using software that can keep a log: it records the analyses in the form of program instructions that can readily and accurately be repeated.

17.2.4 Population description as the major objective

In the next section we look at the objective of comparing results from subgroups, but a more basic aim is to estimate a characteristic like the absolute number in a category of proposed beneficiaries, or a relative number such as the prevalence of HIV seropositives. The estimate may be needed to describe a whole population or sections of it. In the basic analyses discussed below, we need to bear in mind both the planned and the achieved sampling structure.

Suppose 'before' and 'after' surveys were each planned to have a 50:50 split of urban and rural respondents. Even if we achieved a 50:50 split, these would need some manipulation if we wanted to generalize the results to represent an actual population split of 70:30 of urban:rural. Say we wanted to assess the change from 'before' to 'after' and the achieved samples were in fact split 55:45 and 45:55. We would have to correct the results carefully to get a meaningful estimate of change.

Samples are often stratified or structured to capture and represent particular segments of the target population. This may be much more sophisticated than the urban/rural split in the previous paragraph. Within-stratum summaries serve to describe and characterize each of these parts individually. Overall summaries, which put the strata together, are needed to describe and characterize the whole population, when required by the objectives.

It may be sufficient to treat the sample as a whole and produce simple unweighted summaries. This is usually if we have set out and managed to sample the strata proportionately, and there are no problems due to hierarchical structure. Non-proportionality arises from various distinct sources, in particular:

- Sampling is often disproportionate across strata by design. For example, the urban situation is more novel, complex, interesting, or accessible, and gets greater coverage than the fraction of the population classed as rural.
- Particular strata sometimes cause continual trouble with high levels of non-response, so that the data are not proportionate to stratum sizes. This can happen even when the original plan has taken this into consideration.

If we ignore non-proportionality, a simple summary over all cases is not a proper representation of the population. The 'mechanistic' response to 'correct' both the above cases is: (i) to produce within-stratum results (tables,etc.); (ii) to scale the numbers in them to represent the true population fraction that each stratum comprises; and (iii) to combine the results.

There is often a problem with doing this where non-response is an important part of the disproportionality. The reasons why data are missing from particular strata often correspond to real differences in the behaviour of respondents, especially those omitted or under-sampled. We see reports such as 'We had very good response rates everywhere except in the north. In the north there is a high proportion of the population who are nomadic, and we failed to find many of them'. Scaling up the data from settled northerners would not take into account the different lifestyles and livelihoods of the missing nomads. If you have missed almost a complete category, it is honest to report partial results, making it clear which categories are not covered and why.

One common 'sampling' problem arises when a substantial part of the target population is unwilling or unable to cooperate, so that the results only represent a limited subset: those who volunteer or agree to take part. Of course, the results are biased towards those subsets who command sufficient resources to afford the time, or those who habitually take it upon themselves to represent others. We would be suspicious of any study which appeared to have relied on volunteers, but did not look carefully at the limits this imposed on the generalizability of the conclusions.

There may be instances where there is a low response rate from one stratum, but you are still prepared to argue that the data are representative. However, where you have disproportionately few responses for one stratum, the multipliers used in scaling up to 'represent' the stratum will be very high, so your limited data will be heavily weighted in the final overall summary. If there is any possible argument that these results are untypical, it is worthwhile to think carefully before giving them extra prominence in this way.

17.2.5 Comparison as the major objective

One sound reason for disproportionate sampling is that the main objective is a comparison of subgroups in the population. Even if one of two groups to be compared is very small, say 10% of the total number in the population, we now usually want an approximately equal number of observations from each subgroup, to describe both groups with roughly equal precision. There is no point in comparing a very accurate set of results from one group with a very vague, ill-defined description of the other. The same broad principle applies whether the comparison is a wholly quantitative one, looking at the difference in means of a numerical measure between groups, or a much looser verbal comparison, for example an assessment of differences in pattern across a range of cross-tabulations.

If for a subsidiary objective we produce an overall summary giving 'the general picture' of which both groups are part, the 50:50 sampling may need to be re-weighted to 90:10 to produce a quantitative overall picture of the sampled population.

The main difference between true experimental approaches and surveys is that experiments usually involve a relatively specific comparison as the major objective, while surveys often do not. Many surveys have multiple objectives though they may not be clearly set out. Along with the likelihood of some non-response, there may not be a sampling scheme that is optimal for all parts of the analysis. Therefore, various different weighting schemes may be needed in the analysis of a single survey.

17.2.6 When weighting matters

Several times we have discussed how survey results may need to be scaled or weighted to allow for or 'correct for' inequalities in the sample representation of the population. Sometimes this is of great importance, at other times it is not. A fair evaluation of survey work ought to consider whether an appropriate trade-off has been achieved between the need for accuracy and the benefits of simplicity.

If the objective is formal estimation, such as finding the total population size from a census or a sample of communities, we are concerned to produce a strictly numerical answer that should be as accurate as circumstances allow. We should then correct as best we can for a distorted representation of the population in the sample. If groups being formally compared run across several population strata, we should try to ensure the comparison is fair by similar corrections, so that the groups are compared on the basis of consistent samples. In these cases we have to face up to problems such as unusually large weights attached to poorly-responding strata, and we may need to investigate the extent to which the final answer is dubious because of sensitivity to results from such sub-samples.

Survey findings are often used in 'less numerical' ways, where it may not be so important to achieve accurate weighting. For example, 'whatever varieties they grow for sale, a large majority of farm households in Sri Lanka prefer traditional red rice varieties for home consumption because they prefer their flavour'. It is important to keep the analysis simple and thus more cost effective, whether the information will be used as stand-alone information, or used as a basis for further investigation. This also makes the process of interpreting the findings more accessible to those not directly involved in the study. Weighting schemes also depend on good information to create the weighting factors though this may be hard to pin down.

Where we have worryingly large weights, attached to small amounts of doubtful information, it is natural to put limits on, or 'cap', the high weights, even at the expense of introducing some bias, i.e. to prevent any part of the data having too much impact on the result.

The ultimate form of capping is to express doubts about all the data, and to give equal weight to every observation. The rationale, though often not clearly stated, is to minimize the maximum weight given to any data item. This lends some support to the common practice of analysing survey data as if they were a simple random sample from an unstructured population. For 'less numerical' usages, this may not be particularly problematic as far as simple description is concerned. Of course it is wrong, and it may be very misleading to follow this up by calculating standard deviations and making claims of accuracy about the results, which their derivation will not sustain!

17.2.7 Coding

We recognize that purely qualitative researchers may prefer to use qualitative analysis methods and software, but where open-form and other verbal responses occur alongside numerical data it is often sensible to use a quantitative tool. From the statistical viewpoint, basic coding implies that we have material, which can be put into nominal-level categories. Usually this is recorded in verbal or pictorial form, on audio- or videotape, written down by interviewers or self-reported. We advocate computerizing the raw data, so it is archived. The following refers to extracting codes, usually describing the routine comments, rather than those that are unique which can be used for subsequent qualitative analysis.

By scanning the set of responses, themes are developed which reflect the items noted in the material. These should reflect the objectives of the activity. It is not necessary to code rare, irrelevant, or uninteresting material.

In the code development phase, a sufficiently large range of the responses is scanned, to be reasonably sure that commonly occurring themes have been noted. If previous literature or theory suggests other themes, these are noted too. Ideally, each theme is broken down into unambiguous, mutually exclusive and exhaustive categories so that any response segment can be assigned to just one category, and assigned the corresponding code value. A 'codebook' is then prepared where the categories are listed and codes assigned to them. Codes do not have to be consecutive numbers. It is common to think of codes as presence/absence markers, but there is no intrinsic reason why they should not be graded as ordered categorical variables if appropriate, such as on a scale of: fervent, positive, uninterested/no opinion, negative.

The entire body of material is then reviewed and codes are recorded. Additional items may arise during coding and need to be added to the codebook. This may mean another pass through material already reviewed, to add these additional codes.

From the point of view of analysis, no particular significance is attached to particular numbers that are used as codes, but it is worth bearing in mind that statistical packages are usually excellent at sorting, selecting or flagging, for example, 'numbers between 10 and 19' and other arithmetically defined sets. If these all referred to a theme such as 'forest exploitation activities of male farmers', they could easily be bundled together. It is of course impossible to separate items given the same code; so deciding the right level of coding detail is essential at an early stage in the process.

When codes are analysed, they can be treated like other nominal or ordered categorical data. The frequencies of different types of response can be counted or cross-tabulated. Since they often derive from text passages, they are particularly well-adapted for use in sorting listings of verbal comments into relevant bundles for detailed non-quantitative analysis.

17.2.8 Ranking and scoring

A common means of eliciting data is to ask individuals or groups to rank a set of options. The researchers' decision to use ranks in the first place means that results are less informative than scoring, especially if respondents are forced to choose between some nearly-equal alternatives and some very different ones. A British 8-year-old offered fish and chips, or baked beans on toast, or chicken burger, or sushi with hot chillies, might rank these 1, 2, 3, 4 but score them 9, 8.5, 8, and 0.5 on a zero to ten scale!

Ranking is an easy task where the set of ranks has no more than four or five choices. It is common to ask respondents to rank, say, their best four from a list of ten, with 1 = best, etc. Accepting a degree of arbitrariness, we would usually replace ranks 1, 2, 3, 4, and a string of blanks by pseudo-scores 4, 3, 2, 1, and a string of zeros. This gives a complete array of numbers that we can summarize, rather than a sparse array where we don't know how to handle the blanks.

Where the instructions are to rank as many as you wish from a fixed long list, it is advisable to replace the variable-length sets of ranks with scores. These scores might be developed as if respondents each had a fixed amount, e.g. 100 beans, to allocate as they saw fit. If four items were chosen these might be scored 40, 30, 20, 10, or with five chosen 30, 25, 20, 15, 10, with zeros again for unranked items. These scores are arbitrary as 40, 30, 20, 10 could instead be any number of choices such as 34, 28, 22, 16 or 40, 25, 20, 15. This reflects the rather uninformative nature of rankings, and the difficulty of *post hoc* construction of information that was not elicited effectively in the first place.

Having reflected and having replaced ranks by scores we would usually treat these like any other numerical data, with one change of emphasis. Where results might be sensitive to the actual values attributed to ranks, we would stress sensitivity analysis more than with other types of numerical data; for example repeating analyses with (4, 3, 2, 1, 0, 0, ...) pseudo-scores replaced by (6, 4, 2, 1, 0, 0, ...). If the interpretations of results are insensitive to such changes, the choice of scores is not critical.

Formal methods of analysing ranks and scores, which do not depend on treating the allocated numbers as measured numerical quantities, are mentioned in Section 20.4.

17.3 Doing the Analysis

17.3.1 Approaches

Data listings are readily produced by database and statistical packages. They are generally on a case-by-case basis, so they are particularly suitable in exploratory analysis as a means of tracking down odd values or patterns to be explored. For example, if material is in verbal form, listings can give exactly what every respondent was recorded as saying. Sorting these records, according to who collected them, may show differences in aptitude, awareness or the approach to data collection by each field worker. Data listings can be an adjunct to tabulation: in EXCEL, for example, the 'drill down' feature allows one to look at the data from individuals who appear together in a single cell.

Graphical methods of data presentation can be valuable, where messages can be expressed in a simple and attention-grabbing form. Packages offer many ways of making results bright and colourful, without necessarily conveying more information. A few basic points are covered in Chapter 21.

Where the data are at all voluminous, it is advisable to tabulate the 'qualitative' but numerically-coded data, i.e. the binary, nominal or ordered categorical types mentioned above. Tables can be very effective in presentations if stripped down to focus on key findings, and are crisply presented. In longer reports, a carefully crafted and well-documented set of cross-tabulations is usually an essential component of summary and comparative analysis, because of the limitations of approaches that avoid tabulation.

Large numbers of charts and pictures can become expensive, but also repetitive, confusing and difficult to use as a source of detailed information.

With substantial data, a purely narrative full description will be so long-winded and repetitive that readers will have great difficulty getting a clear picture of what the results have to say. With a briefer verbal description, it is difficult not to be overly selective. The reader then has to question why such an effort went into collecting data that merits little description, and should also question the impartiality of the reporting.

At the other extreme, some analysts will skip or skimp the tabulation stage and move rapidly to complex statistical modelling. Their findings are also to be distrusted. The models may be based on preconceptions rather than evidence, or they may fit badly and conceal important variations in the underlying patterns.

In producing final outputs, data listings seldom get more than a place in an appendix. They are usually too extensive to be assimilated by the busy reader, and are unsuitable for presentation purposes.

17.3.2 One-way tables

The most straightforward form of analysis, and one that often supplies much of the basic information needed, is to tabulate results question by question as one-way tables. Sometimes this can be done by recording the frequency or number of people who 'ticked each box' on an original questionnaire. Of course, this does not identify which respondents produced particular combinations of responses, but this is often a first step where a quick and/or simple summary is required.

17.3.3 Cross-tabulation: Two-way and higher-way tables

At the most basic level, cross-tabulations break down the sample into two-way tables showing the response categories of one question as row headings, and those of another question as column headings. Some examples were given in Section 15.6.4.

For example, if each question has five possible answers, a two-way table breaks the total sample down into 25 subgroups. If the answers are further sub-divided by sex of respondent, there will be one three-way table, 5×5×2, probably shown on the page as separate two-way tables for males and for females. The total sample size is now split between 50 categories, and the degree to which the data can sensibly be disaggregated will be constrained by the total number of respondents represented.

There are usually many possible two-way tables, and even more three-way tables. The main analysis needs to involve careful thought as to which ones are necessary, and how much detail is needed.

Even after deciding that we want some cross-tabulation with categories of 'question J' as rows and 'question K' as columns, there are several other decisions to be made. The number in the cells of the table may be the frequency or number of respondents who gave that combination of answers. This may be represented as a proportion or a percentage of the total. Alternatively, percentages can be scaled so they total 100% across each row or down each column, to make particular comparisons clearer. The contents of a cell can also be a statistic derived from one or more other questions (e.g. the proportion of the respondents falling in that cell who were economically active women). Often such a table has an associated frequency table to show how many responses went into each cell. If the cell frequencies represent small sub-samples, the results can vary wildly, just by chance, and should not be over-interpreted.

Where interest focuses mainly on one 'area' of a two-way table it may be possible to combine the rows and columns that do not need to be separated out (e.g. ruling party supporters vs. supporters of all other parties). This simplifies interpretation and presentation, as well as reducing the impact of chance variations where there are small cell counts.

We do not usually want the cross-tabulation for 'all respondents'. We may want to have separate tables with the same information for each region of the country, (known as segmentation), or for a particular group on whom we wish to focus such as 'AIDS orphans', which is known as selection.

Because of varying levels of success in covering a population, the response set may end up being very uneven in its coverage of the target population. Then simply combining the respondents can misrepresent the intended population. It may be necessary to show the patterns in tables, subgroup by subgroup to convey the whole picture. An alternative, discussed in Section 17.2, is to weight the results from the subgroups to give a fair representation of the whole.

17.3.4 Tabulation and the assessment of accuracy

Tabulation is purely descriptive, with limited effort made to assess the 'accuracy' of the numbers tabulated. We caution that confidence intervals are sometimes very wide when survey samples have been disaggregated into various subgroups. If crucial decisions rely on a few numbers it may well be worth putting extra effort into assessing and discussing their reliability. In contrast, if the intended uses for various tables are not numerical, nor very crucial, it is likely to cause unjustifiable delay and frustration to attempt to put formal measures of precision on the results.

The most important considerations in assessing the 'quality', 'value' or 'accuracy' of results are usually not those relating to 'statistical sampling variation', but those concerned with the following aspects:

- Evenness of coverage of the target (intended) population.
- Suitability of the sampling scheme reviewed in the light of field experience and findings.
- Sophistication and uniformity of response elicitation and accuracy of field recording.
- Efficacy of measures to prevent, compensate for, and understand non-response.
- Quality of data entry, cleaning, and metadata recording.
- Selection of appropriate subgroups in analysis.

If any of the above aspects raises important concerns, it is necessary to contemplate the interpretation of 'statistical' measures of precision such as standard errors. Uneven effects introduce biases, whose size and ability to be detected should be appraised objectively, and reported with the conclusions.

Inferential statistical procedures can then be used to guide generalizations from the sample to the population when the survey is not badly affected by any of the above. Inference may be particularly valuable in determining the appropriate form of presentation of survey results. Consider an adoption study, which examined socio-economic factors affecting adoption of a new technology. Households are classified as male or female headed, and the level of education and access to credit of the head is recorded. At its most complicated, the total number of households in the sample would be classified by adoption, gender of the head of the household, level of education and access to credit, resulting in a four-way table.

Now suppose that from chi-square tests we find no evidence of any relationship between adoption, and education or access to credit. In this case, the results of the simple two-way table of adoption by gender of the head of the household would probably be appropriate. If on the other hand, access to credit were the main criterion affecting the chance of adoption and if this association varied according to the gender of the head of the household, the simple two-way table of adoption by gender would no longer be appropriate and a three-way table would be necessary. Inferential procedures thus help in deciding whether the presentation of results should be in terms of one-way, two-way or higher dimensional tables.

Chi-square tests are limited to examining association in two-way tables, so have to be used in a piecemeal fashion for more complicated situations like that above. A more powerful way to examine tabulated data is to use log-linear models. They are a special case of the more general models that are described in Section 20.5.

17.3.5 Multiple response data

Surveys often contain questions where respondents can choose a number of relevant responses, as shown in the example below:

'If you are not using an improved fallow on any of your land, please tick from the list below, any reasons that apply to you:

1. *Do not have any land of my own.*
2. *Do not have any suitable crop for an improved fallow.*
3. *Cannot afford to buy the seed or plants.*
4. *Do not have the time/labour.'*

There are three ways of computerizing these data.

The simplest is to provide as many columns as there are alternatives. This is called a 'multiple dichotomy', because there is a yes/no (or 1/0) response in each case, indicating that the respondent ticked/did not tick each item in the list.

The second way is to find the maximum number of ticks from anyone and then have this number of columns, entering the codes for ticked responses, one per column. This is known as 'multiple response' data. This is a useful method if the question asks respondents to put the alternatives in order of importance, because the first column can give the most important reason, and so on.

A third method is to have a separate table for the data, with just 2 columns. The first identifies the person and the second gives their responses. There are as many rows of data as there are responses. There is no entry for a person who gives no reasons. Thus, in this third method the length of the columns is equal to the number of responses rather than the number of respondents.

If there are 'follow-up' questions about each reason, the third method above is the obvious way to organize the data, and readers may identify the general concept as being that of data at another 'level', i.e. the reason level. More about organizing and managing this type of data is in Chapter 12.

Essentially such data are analysed by building up counts of the numbers of mentions of each response. Apart from SPSS, few standard statistics packages have any special facilities for processing multiple response, and multiple dichotomy data. Almost any package can be used with a little ingenuity, but working from first principles is a time-consuming business.

17.3.6 Profiles

Questions put to respondents in a survey need to represent 'atomic' facets of an issue, expressed in concrete terms and simplified as much as possible, so that there is no ambiguity and so they will be consistently interpreted by respondents.

Cross-tabulations are based on reporting responses to individual questions and are therefore narrowly issue-specific. A different approach is needed if the researchers' ambitions include taking an overall view of responses from individuals or small groups, for example in relation to their livelihood. Cross-tabulations of individual questions are not a sensible approach to a 'people-centred' or 'holistic' summary of results.

Usually, even when tackling issues less complicated than livelihoods, the important research outputs are like 'complex molecules', which bring together responses from numerous questions to produce higher-level conclusions described in more abstract terms. For example several questions may each enquire whether the respondent follows a particular recommendation, whereas the output may be concerned with overall 'compliance': the abstract concept behind the questioning. A profile is a description, synthesizing responses to a range of questions. It may describe an individual, cluster of respondents or an entire population.

One approach to discussing a larger concept is to produce numerous cross-tabulations reflecting actual questions and to synthesize their information content verbally. However, this tends to lose sight of the 'profiling' element: if particular groups of respondents tend to reply to a range of questions in a similar way, this overall grouping will often be difficult to detect. If you try to follow the group of individuals who appear together in one corner cell of the first cross-tabulation, you cannot easily track whether they stay together in a cross-tabulation of other variables.

Another type of approach may be more constructive: to derive synthetic variables (indicators) which bring together inputs from a range of questions, say into a measure of 'compliance', and to analyse them, by cross-tabulation or other methods (see Section 17.3.8 below). If we have a dataset with a row for each respondent and a column for each question, the derivation of a synthetic variable just corresponds to adding an extra column to the dataset. This is then used in the analysis just like any other column. A profile for an individual will often be the set of values of several indicators.

17.3.7 Looking for respondent groups

Profiling is often concerned with acknowledging that respondents are not just a homogeneous mass, and distinguishing between different groups of respondents.

Cluster analysis is a data-driven statistical technique that can draw out and then characterize groups of respondents whose response profiles are similar to one another. The response profiles may serve to differentiate one group from another if they are somewhat distinct. This might be needed if the aim were, say, to define target groups for distinct safety net interventions. The analysis could help clarify the distinguishing features of the groups, their sizes, and their distinctness or otherwise. Unfortunately, there is no guarantee that groupings derived from data alone will make good sense in terms of profiling respondents. Cluster analysis does not characterize the groupings; you have to study each cluster to see what they have in common. Nor does it prove that they constitute suitable target groups for any particular purpose.

An example of cluster analysis is in Section 20.3.1. It is an exploratory technique, which may help to screen a large volume of data, and prompt more thoughtful analysis, by raising questions such as:

- Are there any signs that the respondents do fall into clear-cut subgroups?
- How many groups do there seem to be, and how important are their separations?
- If there are distinct groups, what sorts of responses do 'typical' group members give?

17.3.8 Indicators

Indicators are summary measures. Magazines provide many examples: a computer magazine, giving an assessment of personal computers, may give a score in numerical form, such as '7 out of 10', or a pictorial quality rating, such as that shown in Fig. 17.1.

Fig. 17.1. A pictorial form of quality rating.

Very good	Good	Moderate	Poor	Very poor
★	☆	○	O	●

This review of computers may give scores (indicators) for each of several characteristics, where the maximum score for each characteristic reflects its importance. For example for one model: build quality (7/10), screen quality (8/20), processor speed (18/30), hard disk capacity (17/20) and software provided (10/20). The maximum score over all characteristics in the summary indicator is in this case (10+20+30+20+20) = 100, so the total score for each computer is a percentage. With the example above, (7+8+18+17+10) = 60%. The popularity of such summaries demonstrates that readers find them accessible, convenient and to a degree useful. This is either because there is little time to absorb detailed information, or because the indicators provide a baseline from which to weigh up the finer points.

Many disciplines are awash with suggested indicators, from simple averages to housing quality measures, social capital assessment tools, or quality-adjusted years of life. New indicators should be developed only if others do not exist or are unsatisfactory. Well-understood, well-validated indicators, relevant to the situation in hand are quicker and more cost-effective to use. Defining an economical set of meaningful indicators before data collection ought to imply that at analysis, their calculation follows a pre-defined path, and the values are readily interpreted and used.

Sometimes it is legitimate to create new indicators after data collection and during analysis. It is expected in genuine 'research' where fieldwork allows new ideas to come forward, for example if new lines of questioning have been used, or if survey findings take the researchers into areas not already covered by existing indicators.

When it is necessary to create new indicators, how does the analyst proceed? It is important not to create unnecessary confusion. An indicator should synthesize information and serve to represent a reasonable measure of some issue or concept. The concept should have an agreed name so that users can discuss it meaningfully, such as 'compliance' or 'vulnerability to flooding'. A specific meaning is attached to the name, so it is important to realize that the jargon thus created needs careful explanation to 'outsiders'. Consultation or brainstorming leading to a consensus is often desirable when new indicators are created. Indicators created 'on-the-fly' by analysts as the work is rushed to a conclusion, are prone to suffer from their hasty introduction. This can lead to misinterpretation, and over-interpretation by prospective users. It is all too easy for a small amount of information concerning a limited part of the issue, to be taken as 'the' answer to 'the problem'!

As far as possible, creating indicators during analysis should follow the same lines as when the process is done *a priori*, namely:

- Decide on the facets to include, to give a good feel for the concept.
- Tie these to the questions or observations needed to measure these facets.
- Ensure balanced coverage, so that the right input comes from each facet.
- Work out how to combine the information gathered into a synthesis that everyone agrees is sensible.

These are all parts of ensuring face (or content) validity as in the next section. This should be done in an uncomplicated way, so that the user community are aware of the definitions of measurement.

There is some advantage in creating indicators when datasets are already available. You can look at how well the indicators serve to describe the relevant issues and groups, and select those that are most effective. Some analysts rely heavily on data reduction techniques such as factor analysis or cluster analysis as a substitute for contemplating the issues. We argue that an intellectual process of indicator development should build on, or dispense with, more data-driven approaches. Principal component analysis is data-driven, but readily provides weighted averages. An example is in Section 20.3.2. These weighted averages should be seen as no more than a foundation for useful forms of indicator.

17.3.9 Indicator validity

The basic question behind the concept of validity is whether an indicator measures what we say or what we believe it does. This may be a rather basic question if the subject matter of the indicator is visible and readily understood, but the practicalities can be more complex in ordinary, but sensitive areas such as measurement of household income. When considering issues such as the value attached to indigenous knowledge, the question can become very complex. Numerous variations on the validity theme are discussed extensively in social science research methodology literature.

Validity takes us into issues of how different people interpret the meaning of words, during the development of the indicator and its use. It is good practice to try a variety of approaches with a wide range of relevant people, and carefully compare the interpretations, behaviours and attitudes revealed, to make sure there are no major discrepancies of understanding. The processes of comparison and reflection, then the redevelopment of definitions, approaches and research instruments, may all be encompassed in what is sometimes called triangulation: using the results of different approaches to synthesize robust, clear, and easily interpreted results. Survey instrument or indicator validity is a discussion topic, not a statistical measure. However, there are two themes with which statistical survey analysts regularly need to engage, those of content and criterion.

Content validity looks at the extent to which the questions in a survey and the weights the results are given in a set of indicators, serve to cover in a balanced way, the important facets of the notion the indicator is supposed to represent.

Criterion validity looks at how the observed values of the indicator tie up with something readily measurable to which they can relate. Its aim is to validate a new indicator with reference to something more established, or to validate a prediction retrospectively, against the actual outcome. If we measure an indicator of 'intention to participate' or 'likelihood of participating' beforehand, then for the same individuals ascertain whether they did participate, we can check the accuracy of the stated intentions, and the degree of reliance that can be placed on the indicator.

As a statistical exercise, criterion validation has to be done through sensible analyses of good-quality data. If the reason for developing the indicator is that there is no satisfactory way of establishing a criterion measure, criterion validity is not a sensible approach.

17.3.10 Putting it into practice

Readers may be surprised that in a chapter on analysis of surveys we have not explained how to do the chi-square tests, *t*-tests and regressions beloved by survey analysts. Our intention is not to downplay the role of more sophisticated methods in survey analysis. The model-based analyses described later are certainly useful, particularly in trying to understand the contribution of numerous quantities that may all be associated with a response, and in understanding multiple layers of variation in a hierarchical design.

Our reasons for emphasizing the points made in the chapter are as follows:

1. We discuss in this chapter the elements of good practice most commonly ignored by survey analysts; however we do not cover the implementation of standard inferential statistical methods.

2. The theoretical background to the formal methods of analysis is the same for survey analysis as for those of other types of study. This is discussed in Chapters 16, 19 and 20, and is also well described in many other books.

Much of the remaining statistical analysis is mechanical. Once the items in this chapter are understood, including taking sensible decisions of coding and weighting the data, suitable computer software can provide the tools for the rest of the analysis.

Chapter 18

Analysis of Experimental Data

18.1 Strategy for Data Analysis

The analysis of experimental data may seem simple, at least in contrast to the analysis of survey data, considered in Chapter 17. With experiments there is usually a main response variable, and initial tables will be created with this variable for each level of the treatment factor. In traditional experiments, such as crop or animal experiments, this response is often yield. In the broader, natural resource setting it could be an index of sustainability, calculated from many of the measurements. Indicators were described in Section 17.3.

In the first part of this chapter we outline the steps to follow on processing the data for a simple experiment. As with the example in Section 15.6, one aim is to show how easily experimental data can be processed. In practice, almost all experiments involve some complications; we look at common complications in the second part of the chapter.

As a general rule an experiment can be described in terms of its treatments, its layout and its measurements. It is often helpful to view the analysis in terms of these three components, and we try to highlight this within the chapter.

18.2 The Essentials of Data Analysis

In the same way that it takes time to design and to carry out a good experiment, it also takes time to conduct an effective data analysis. The first issue is data entry into the computer and ensuring it is in a suitable format for analysis. The data for analysis may also have to be derived from the raw data. For instance, they may have to be summarized to the 'right' level, e.g. individual plant height to mean height per plot, or transformed, for example from kg/plot to t/ha. This can be performed in the statistics package or in the database environment (e.g. ACCESS or EXCEL) that was used for the data entry. The choice is up to the user. The topic was described earlier in Chapters 11 and 12.

The analysis should then unfold as an iterative sequence, rather than a straight jump to the analysis of variance. There are three main parts:

- Relating the analysis to the stated objectives of the study, which is crucial to addressing the researchers questions.
- Exploratory and descriptive analysis where preliminary answers to the objectives can be realized via simple tables or graphs of the treatment means or other summaries.
- ANOVA and formal inference methods to check on the adequacy of the previous steps and to add precision to the findings of the exploratory analysis.

Sometimes some of the steps above need revisiting. For example, the first use of ANOVA is usually largely exploratory and often prompts a fresh look at the objectives of the study.

Sections 18.2.2 to 18.2.4 are concerned with the first two parts of this process. These are straightforward and many trials can be reported, at least initially, following these parts of the process.

The last step allows the researcher to attach *p*-values to the comparisons of interest, and to quantify the uncertainty in the assertions about treatment effects. Since formal analysis is necessary for published reports, we will consider how to tackle any analysis, no matter how messy, by following the principles of statistical modelling. This is essential for getting the most out of data, particularly where the data structures are complex, as in large on-farm trials. Modern software is now sufficiently user-friendly for researchers to complete some of these analyses, once they understand the general principles of modelling. In this chapter we only set the scene for a modelling approach to data analysis. Chapter 19 is devoted entirely to the topic.

18.2.1 Objectives of the analysis

The first stage is to clarify the objectives of the analysis. These are based on the objectives of the study and therefore require the experimental protocol. The design and analysis should be linked at the planning stage of a study, with the objectives of an experiment being given in the protocol in such a way that they define the treatments to be used and the measurements that are needed.

Some experiments have a single treatment factor, such as ten varieties of maize, plus one or two control treatments representing the varieties in current use. An objective might then be to decide whether to recommend some of the new varieties, compared to the ones currently in use. This recommendation could be based on one or more criteria such as high yield, or acceptable taste, and these correspond to the measurements taken.

Many trials have a factorial treatment structure, with two or more factors, such as variety, level of fertilizer and frequency of weeding. In such trials there are usually more objectives: some that relate separately to the individual factors and some to the more complicated recommendations that are needed if the factors do not act independently of one another. Where treatment factors are quantitative, such as amount of fertilizer, the objectives are often to recommend a level that produces an economic increase in the yield, compared to the cost.

The above examples emphasize the role played by the objectives of the trial in determining the analysis. In principle the objectives in the protocol and the objectives of the analysis will be the same. However the objectives of the analysis may be different for a number of reasons.

- The objectives of the study may have been stated in a vague way.
- There may be other unstated objectives that can be added, once the data are available.
- It may not be possible to meet some of the original objectives because the required data were not measured or because of unexpected complications during the trial.
- The objectives of the analysis may evolve as the analysis progresses.

In realizing the objectives, it is often necessary to clarify the response variables of interest. For instance, if the objective of the trial is to look at the disease resistance of a crop, and individual plants in a plot are assessed on a scale from 0 (no disease) to 5 (very badly diseased), how is disease resistance to be assessed? Is it, for instance, by the proportion of plants in a plot that are disease free; or by the mean or maximum disease score of the plot? This should normally be thought out beforehand and specified in the experimental protocol.

We strongly recommend that this preparatory stage include the preparation of the initial dummy tables and graphs that the scientist feels will meet the objectives of the analysis. They will often use the treatment means, because the treatments were chosen to satisfy the objectives. The specification of these tables/graphs should define the particular measurements that will be used. Thus, in the first example above, where new varieties of maize were being compared with ones in current use, our dummy table may look like Table 18.1.

Table 18.1. Dummy table.

Variety	Mean yield (kg/ha)	Ranking of taste
Control variety 1	...	...
Control variety 2	...	...
	...	...
Variety A	...	...
Variety B	...	...
Variety ... etc., in decreasing order of mean yield	...	...

In the example where one treatment factor was the level of fertilizer, a graph of mean 'profit' (i.e. income from yield minus cost of fertilizer) on the y-axis, versus amount of fertilizer on the x-axis, might be suggested as one presentation.

18.2.2 Understanding variability

Having a clear understanding of the variability in a set of data will help with its analysis and interpretation. Here we discuss variability in a traditional setting; but the ideas extend to any natural resources experiment.

The variability observed in data from plots or animals in an experiment can be due to the treatments which were applied, or to the layout (e.g. in which block the plot is sited: a shady area as opposed to a sunny area), or because any two plots experiencing the same experimental conditions give different yields (i.e. the plot-to-plot variability). Statistical analysis is concerned with explaining the variability in the data, and determining whether, for instance, treatment effects are larger than would be expected of random variability.

It is useful to use the following general form, introduced in Section 16.8, to describe the data for any particular measurement in an experiment

data = pattern + residual

Pattern is the result of factors such as the experimental treatments and other characteristics often determined by the layout. *Residual* is the remaining unexplained variation, which we will also seek to explain further (Section 18.2.6).

Identifying how much of the pattern is due to the treatments, is an important part of the analysis, because it relates directly to the objectives of the study. The residual is also worth a close look as it gives us an understanding of the variability that is not due to the pattern.

In the preliminary analyses we want to 'get a feel' for which features (treatments, blocks, covariates) are responsible for creating these patterns, and what patterns they cause. Looking for strange values, checking if variability is different in different places or on different treatments are our explorations of the residual.

At later, more formal stages of analysis, our investigations of pattern are concerned with quantifying and testing hypotheses concerning the treatments, and other features. Investigations of the residual then focus on: (i) checking assumptions underlying the analysis; and (ii) checking whether we have missed any part of the pattern.

All data analyses depend on the above framework, and we return to it later when discussing more formal methods. The type of measurement also determines the method of analysis. Data such as yields can be analysed by methods in the class of linear models, of which ANOVA is one. Other types of data such as insect counts, number of diseased leaves per plant or preference scores, use methods classed as generalized linear models, but the ideas of expressing variability in terms of pattern and residual still apply. We return to this later in Chapter 20.

18.2.3 Preliminary analysis

The simple methods described here apply similarly to all types of data. Assuming that some tables and graphs have already been identified from the objectives, it is easy to 'fill in the numbers' and observe the patterns of response.

Boxplots, scatterplots and trellis plots are all useful tools of exploratory methods of data analysis. They give insights into the variability of the data and the treatment pattern, and highlight any outliers that need further clarification. Suspected patterns, such as block effects or a fertility gradient can also be explored, and sometimes unexpected patterns emerge that the researcher may want to investigate further.

Figure 18.1 shows a simple example of an exploratory analysis, where the main objective of the trial was to suggest the high yielding varieties. It shows that the yields in all three replicates are consistent except for Variety 7, which has yields of 2.4, 2.0 and 0.6 t/ha in the three replicates. Clearly something is odd. Examination of other variables confirmed that the problem was not a typing error, but it was too late to return to the field to confirm the data. One possibility is to omit this variety from the formal analysis and report its results separately.

Fig. 18.1. Exploratory plot of yield, against variety for three replicates.

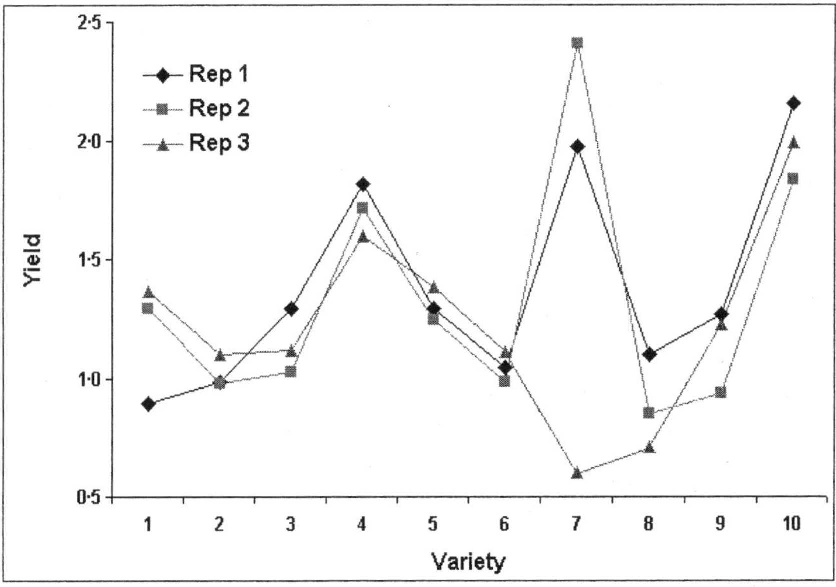

In simple experiments, these preliminary phases may only take a short time. In contrast, in a large on-farm study, most of the analysis consists of: (a) summarizing the interview data as simple tables; and (b) looking at treatment comparisons in subgroups of farms. The researcher also needs to identify all the elements of pattern, or possible elements, that will be incorporated into the statistical modelling that forms the next stage of analysis. These can only be identified by looking at many two- or three-way tables or graphs. The preliminary work therefore constitutes the bulk of the analysis, though much of it may go unreported at the end of the day.

Once the appropriate descriptive statistics have been tabulated and/or graphed, the researcher should be able to draw some, albeit subjective, conclusions about the study objectives. It is therefore possible at this stage to write a draft report of the study.

We are advocating here that a researcher can learn a lot about the data by paying close attention to the objectives and producing some summary tables and graphs. So why not stop here? What is the use of ANOVA and any more formal methods of analysis?

There are two main limitations to the exploratory analysis. The first is that the data are usually only examined in 'slices'. For instance, we can look at the pattern due to treatment (and the residual variability) ignoring blocks, and then look at the pattern due to blocks; however it is difficult to look at several components of the pattern at the same time. This is particularly important when there are many factors or when data structures are complicated, as in large on-farm trials where the farms may be in different agro-ecological zones and farmers have different management practices. We need a method of apportioning variability across different sources all at once; this is the role of ANOVA or, more generally, linear models.

Second, the conclusions drawn from the summary tables are still subjective. Some measure of precision needs to be attached to the observed effects. For this we need the formal methods of analysis.

18.2.4 The role of analysis of variance (ANOVA)

Most scientists know that ANOVA is a key tool in the analysis of experimental data, yet many do not fully understand how to interpret it. Here we illustrate what it provides, in addition to the tables of means that have been used above.

As a descriptive tool

In the previous section, we introduced the idea that for any measurement we can consider

data = *pattern* + *residual*

as a summary of the variation in our data. The analysis of variance is a technique with which we can look at the components of the pattern all at once, (e.g. blocks and all treatment factors together) and identify those that are important. It also provides an opportunity to look at the residual, the part of the data that cannot be explained by the pattern in the experiment.

The ANOVA table contains elements called sums of squares (SS), degrees of freedom (DF), and mean squares (MS), for both the components of the pattern and the residual. See Table 18.2 for an example.

The sums of squares allow us to see what proportion of the variation in the data can be explained by the different parts of the pattern. The residual sum of squares shows us what remains unexplained, whilst the residual mean square, estimates the variance of the units. This should be as small as possible so that differences between the treatments can easily be detected.

Table 18.2. ANOVA table for a randomized block experiment with 9 treatments laid out in 4 blocks.

Source	DF	SS	MS
• Blocks	3	1.078	0.359
• Treatments	8	37.781	4.723
Residual	24	5.646	0.235
Total	35	44.505	

Each observation has a contribution to make to the residual sum of squares, and one large aberrant observation can have a large impact on this. These individual contributions are called *residuals*, and they should be inspected for their influence on the estimate of random variability. Most statistical packages will automatically indicate which observations have large residuals, and these should warrant further investigation. The reader should be aware though, that if the study is of a reasonable size, there would usually be one or two reasonably large residuals just by chance. Thus, the first use of ANOVA is as an exploratory tool.

Exploring the treatment structure further

In trials where there is more than one treatment factor, each line in the ANOVA table corresponds to a table of treatment means. Therefore, the ANOVA table can be used as a sort of 'passport' to the respective tables of means. This can help us decide which tables to use to present the results. For example, in a two-factor experiment, with variety and fertilizer, we may have proposed to present the means for each factor separately. The ANOVA table may indicate that a large part of the pattern in the data is due to the interaction between variety and fertilizer, thus suggesting that we should instead interpret and present our findings using the two-way table of means.

Another step in the analysis of variance is to identify which parts of the treatment pattern relate to the different objectives, and then to examine the corresponding sum of squares and effects. For qualitative treatment factors such as variety or intervention, this is often through the use of treatment contrasts. For quantitative factors, like spacing or amount of phosphorus, it is often through consideration of a line or curve that models the effect of the changing level of the factor. These graphs would have normally been generated earlier, but we now know that they arise from a treatment effect that has explained a reasonable proportion of the variability in the trial.

Statistical inference

So far we have discussed the ANOVA table without the distraction of F-probabilities and t-values; we now add the ideas of statistical inference, which were reviewed in Chapter 16. These include the ideas that the standard error of an estimate is a measure of its precision, and that a confidence interval for a treatment mean is an interval likely to contain the mean, not an interval that contains most of the measurements for that treatment.

The first step in formal statistical inference in the analysis of experimental data is usually based on the F-probabilities that are given in the final column of the ANOVA table above. They enable us to test certain hypotheses about the components of the pattern part of the data. The p-values (significance levels), and the magnitude of the relevant mean squares in the table, help us to decide which components of the pattern to look at in more detail.

The p-values must not be used too strictly. The F-test is usually used to investigate if there are treatment differences between any of the levels. Usually there are, so the null hypothesis is untrue. What we want to find out instead is whether any of the differences that relate to the objectives can be estimated with reliability, given the 'noise', or residual variability that is in the data.

The second step, which is the main component of statistical inference, is based on the standard errors that are associated with the important differences in treatment effects. We recommend that tables of treatment means be accompanied by the standard error of a difference, rather than the standard error of treatment means, since it is usually the difference between treatments, which is of interest. Values for any important treatment contrasts should be accompanied by their standard errors, and normally reported in the text. Chapter 21 gives some guidance on the presentation of graphs.

For simple experiments, we now have all the tools that are required to conduct a full analysis of the data and write a report.

18.2.5 Statistical models

The essentials of the statistical model have already been given as

$$data = pattern + residual$$

This description is now formalized through specific examples. The first example is of simple regression, where yields may be related to the quantity of nitrogen in the soil by

$$yield = a + b.nitrogen + residual$$

if there is a linear relationship between yield and nitrogen.

This is an example where the pattern in the yield is a result of its relationship with a variate, but the idea is the same if we include factors in the model. Therefore, in a simple randomized complete block trial of varieties, we might consider the model as *yield = constant + block effect + variety effect + residual*

A third possibility is that we have a mixture of both variates and factors in the model, for example

yield = a + b.nitrogen + block + treatment + residual

To add another piece of terminology, we can now estimate the unknown *parameters* in the model; for example, the slope of the regression line would be a parameter in a simple regression model. In the models with factors, the estimates of the treatment effects are particularly important, because they relate to the objectives of the trial.

Once users are familiar with the idea of simple linear regression, these models with factors are all simple to fit with current statistics software. Most packages have an ANOVA facility for the analysis of data from simple experiments, and a regression facility for this more general approach. Care must be taken not to confuse the term ANOVA, used to denote the analysis of a simple experiment and the ANOVA table, which is given by both approaches as an initial summary of the data.

What does the user gain and lose by this modelling approach, compared to thinking in terms of the treatment means as a summary of the effect of each treatment? If an experiment is simple, such as a simple randomized block design, then the modelling approach is identical to using the simple approach. Thus the estimates of the treatment effects are the treatment means. In such cases, the ANOVA facility in statistical software is usually used as it provides a clearer display of the results than the regression approach.

The modelling approach comes into its own, when the experiment is 'unbalanced' or there are several missing values. For example, in a simple on-farm trial, all farmers may have the same three treatments and so each farmer constitutes a 'replicate'. In a more general case, perhaps one or two treatments are in common and farmers then choose other varieties they would like to try, on an individual basis. There may be a total of eight varieties overall, and an initial examination of the data indicates that the model for the yields of

yield = constant + farmer effect + variety effect + residual

is sensible to try. This now needs the more general approach and the estimates of the variety effects would not necessarily be the simple treatment means, but would automatically be 'adjusted means', where the adjustments are for the farmer differences. It should be clear that an adjustment may be needed to compensate for the possibility that a particular variety might have only been used by the farmers who tended to get better yields. As some treatments were used by all farmers, those yields provide information to make the adjustment. We consider the modelling approach in more detail in Chapter 19.

18.2.6 Have I done my analysis correctly?

Thinking in terms of the modelling approach, it is also useful to examine both the assumptions of the proposed model and to see if it can be improved. In examining a model, we examine its two parts; the 'pattern' component of the model, and the 'residuals'?

We often look at the residuals (i.e. the individual residuals introduced in Section 18.2.2) to check if they still contain some extra part that we could move into the pattern. In the example above, the model assumes that a 'good' variety produces a high yield for all farmers. However one of the 'good' varieties may not be as successful on farms in which striga is present. If the presence or absence of striga is measured for each farm, then the striga-pattern could be added to the model; if there were only a few farmers with striga, we could split the data and look separately at the non-striga farms.

It is also important to check the residuals in a model to see if they satisfy the assumptions of the analysis. One assumption is that the residuals are from a normal distribution. This is not usually critical, but it often helps to check if there are odd observations that need special examination. Residuals can also be examined to check whether the data are equally variable, because the ANOVA and regression approaches both assume there is only one measure of spread for the whole experiment. Unequal variance can be handled in a number of ways, such as splitting the data into homogeneous subsets, or transforming or finding a more realistic model that does not assume it is constant.

18.3 Complications in Experiments

Researchers sometimes state that their pre-computer training in statistics is sufficient, because they only do simple trials. However, in our experience it is rare to encounter a trial that does not have some unforeseen complication. Indeed, perhaps a trial with absolutely no complications is too good to be true!

The three main components of a trial are the treatments that are applied, the layout of the experiment and the measurements that are taken. Describing which of these components has become more complicated is useful in categorizing the types of solution.

One of the most common complications in experimental data arises because aspects of the trial are at multiple levels. For example in an on-farm trial, there are normally some measurements at the farm level from interviews and others at the plot level from yields.

A related problem that is also common in many trials is that of repeated measurements, in other words measurements made repeatedly over time on the same unit. Examples might be milk yield or weight gain measured on the same animals on several occasions over a period of time. Depth and distance are two other ways in which measurements can be repeated on the same unit.

Both of these complications are to some extent planned, as they are part of the design of the experiment, but there are also unexpected complications that can arise. Planned complications are discussed here, before giving guidance on some of the common 'surprise' complications.

18.4 Multiple Levels

18.4.1 Introduction

In a simple randomized complete block experiment with treatments applied to the plots and a single measurement (e.g. maize yield) made on each plot, the analysis is simple. Treatments and measurements are at the plot level and so the treatment effects are assessed relative to the variation between plots.

In many experiments, however, the experimental material has a multilevel structure. To explain this, consider the situation where the experimental unit, to which treatments are applied, is a plot containing a number of trees. If measurements are taken for individual trees within each plot, then the data have been collected at a lower level to that of the treatments.

Such experiments are sometimes wrongly analysed, because the different levels of the data are ignored, giving the impression that there is much more information than there really is. Failure to recognize the multilevel structure results in the between-plot and within-plot variations being mixed into a single 'variance' that has no clear meaning. This can result in wrong conclusions being made about the treatments, since the between-plot variation is often much larger than the within-plot variation.

A split-plot experiment is one in which the treatments are applied at two levels, (i.e. some to the main plots and others to the subplots). Data are usually collected at the lowest (subplot) level. Although this is a common experimental design, the complication is that the analysis still requires the two different levels of variation to be correctly recognized. Many common statistics packages produce, or can be made to produce, the correct analysis of variance table, but do a poor job in presenting the full results for this design. Very few supply all the standard errors needed to compare the different treatment means.

For those familiar with large variety trials, another example of multiple levels is a lattice design, where there are more treatments than can be accommodated in a complete block design. Here the multiple levels are in the layout component. The treatments and measurements are all at the plot level, but the layout consists of replicates, blocks within replicates and then plots within blocks. Again, these multiple levels must be recognized in the analysis.

Many on-farm trials have a similar hierarchy to the examples above. The layout structure might have many levels: perhaps village, then farming group within village, then household, then field, and finally plot within field.

In on-farm trials, each farmer is often the equivalent of a 'block' or 'replicate'. The treatments are applied to plots within the farm and some measurements are taken at the plot level. We also interview the farmer, or observe characteristics of the whole field. These measurements at the farm level, are higher than that at which the treatments were applied. Some measurements, as in a participatory study, could be from a higher level still, perhaps from discussions at the farming-group level.

Other studies may have structures that are more complex, with 'treatments' applied at more than one level, and data collected at a number of levels. An experiment on community forest management may have community level treatments (with and without training) and plot level treatments (with and without grazing control), with measurements taken on communities, plots, households within communities and quadrats within plots. Such studies are often not thought of as experiments, yet the same analysis concepts and methods apply.

18.4.2 Dealing with multiple levels

Once the correct multiple-level structure is recognized, some of the analysis complications can be overcome. There are two obvious strategies, which we consider in turn:

- Eliminate the different levels and conduct each analysis at a single level.
- Accept the different levels and analyse the data as they stand, as in a split plot analysis.

Trials where soil measurements are made within each plot are an example where the observations are made at a lower level to that at which the treatments were applied. For example, five cores may be made within each plot, but then be reduced to the plot level by mixing the soil before analysis. This is the same idea as scoring ten plants for disease within each plot, and then calculating a summary statistic, such as the mean disease score, before the main analysis.

This is the appropriate solution for many studies where measurements have been taken within the plot (or within the unit to which treatments were applied). The summary is now at the plot level, where the treatments were applied and the analysis proceeds. When the data are individual measurements of height, weight or yield, then usually a mean or total value for the plot is all that is required.

Sometimes, depending on the objectives, more than one summary value will be calculated. For instance, disease severity might be assessed in terms of the mean disease score, the number of plants with a disease score more than three, and the score of the most diseased plant.

In general we recommend that the raw data at the within plot level are the data that are recorded and computerized, rather than only the summaries at the plot level. This is partly for checking purposes, and also for ease of calculation. The computer is a tool to assist the researcher in the analysis, and should be used for as much data management and calculation as possible.

On-farm trials often have measurements at the plot level, where the treatments were applied, and at the farmer level, with data from interviews. Plot yields are used for the objectives related to the observed performance of the treatments, and the interview measurements relate to objectives concerned with the farmers' priorities and views. In addition, there may be farm-level data collected by remote sensing (e.g. average ground cover, distance from road) and village-level data such as effectiveness of the extension agent. Here there is no problem: we analyse each set of measurements at its own level.

The difficulty in analysing on-farm trials arises because we often need to combine information across the two levels. For example, there may be large variability in the effect of the treatments (plot level), and we believe some of this variability could be explained by the different planting dates, recorded at farm level. The solution is easy where there are just two treatments. Then we can just calculate the treatment difference for each farmer, which is a single value, and relate it to the other farm-level data.

This solution can be extended to situations where there are more than two treatments per farm, by applying the same approach to the important treatment contrasts in turn and relating them to the other farm level measurements. However, this is a 'piecemeal' solution and not particularly suitable if there are a reasonable number of treatments, or the treatments have a factorial structure. The alternative is to use a multilevel, or mixed model analysis which can handle the data over the different levels. This approach is similar to a split-plot analysis, where main plots are the farms, except that it can deal with data that are unbalanced. It is discussed in Section 20.5.

18.5 Repeated Measures Made Easy

Many experiments include repeated measurements, either in time or in space. In an animal experiment, we may record the weight of the animal each month. We may harvest the fruit from trees each season, or record the disease score each week and so on.

These are situations where the treatments are applied at the plot or animal level and measurements are then made at a lower level, at time points within each plot or animal. It is tempting to think of this as a multiple-levels problem, and indeed the same 'solution' is often possible as was described in the previous section for measurements taken on plants within plots.

The important difference between the ten plants per plot and the repeated measurements is that the ten plants are normally selected at random. If instead they are deliberately taken along a transect, and their position noted, then they become repeated measurements in space.

Thus, the new feature is that these measurements are 'in order' and it is usually important to take account of the ordering in the analysis. In the preliminary analysis, the first step is usually to graph the raw data, (i.e. to produce a time-series graph showing the development of the measurement for each experimental plot or animal). For simple summaries of the experimental treatments, the graphs are often conveniently grouped so that all the lines corresponding to the same treatment are on the same graph. An example is shown in Fig. 18.2.

The figure depicts milk consumption over an eight-week period for individual calves all being fed the same food supplement. It shows that milk consumption increases in a linear fashion over time until about weeks six or seven, after which time it starts to level off or decrease. We can also see that consumption becomes slightly more variable at the later times.

Fig. 18.2. Repeated measurements on animals, plotted as time series.

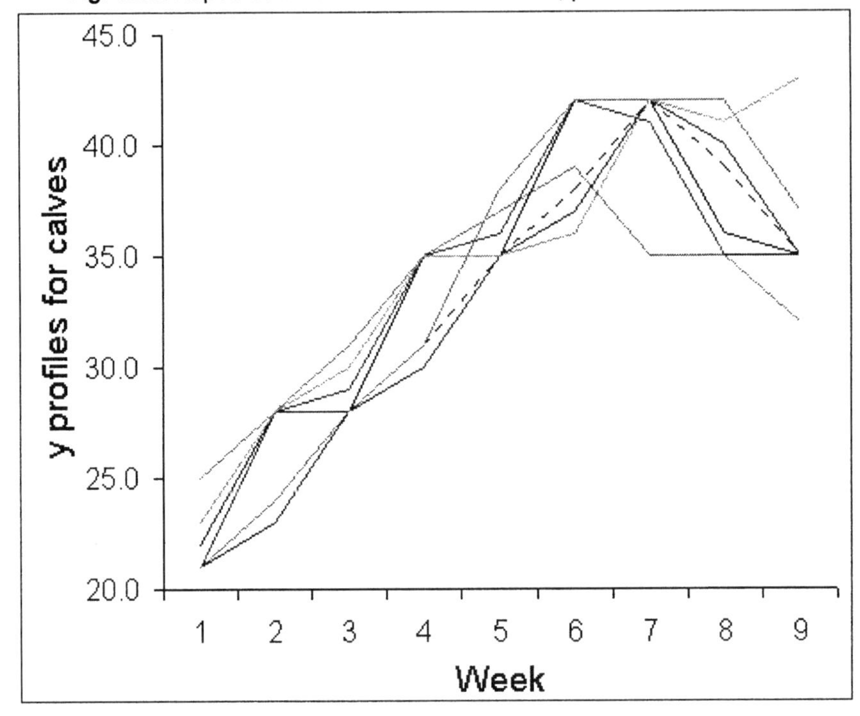

Such graphs, together with the objectives of the trial, will normally indicate the summaries that should be taken of the repeated measures, at the unit (animal or plot) level. Thus, as in the previous section, we eliminate the problem by moving the repeated measures data up to the level of the experimental unit. The reason this is usually sufficient is that the objectives are usually directly related to the different treatments and they are applied at this higher level.

As with the multiple levels situation, there is no one 'standard' summary with repeated measurements. It may be the value measured at a particular time point, or the difference between the value at two points, or something more complicated. For instance, had our example above related to live-weight gain in animals, then a reasonable summary might have been the slope of the regression line, over the first six weeks of the study.

More advanced procedures are available for the analysis of repeated measurements, which extract more information from the data. They model both the variations at, and correlations amongst, the different times. They are useful if there is no clear summary feature that can be analysed and particularly if there are missing values in the dataset. They are not discussed here, because the simple approaches are sufficient for most situations.

Repeated measures in space can be handled in a similar way. They arise when multiple measurements are taken in each plot but these are not all equivalent. For example, a plot may consist of a fruit tree with surrounding crops, with the treatments being different management regimes applied to the tree. Crops and soil are measured at different distances from the tree in each plot. This data structure is not the same as taking a number of randomly placed measurements in each plot, as we are specifically interested in each distance. For example, the objective may be to compare the distances over which tree management has some impact on soil water. However, these repeated measures cannot be treated as if they came from different plots. The simplest solution is to produce a summary for each plot, such as the distance to which the tree reduces soil water, or the difference in yield between measurements close to and far from the tree. These plot levels summaries can then be analysed to compare treatments.

18.6 Surprise Complications

Table 18.3 lists some common complications that can arise in experimental research. They are categorized as complications relating to the treatment structure, the layout or the measurements. For example, farmers may not apply the treatments exactly as specified, there may be missing plots, or there may be some zero or otherwise odd values.

Table 18.3. Complications in experiments.

Type	Complication	Example
Treatment	Levels modified. Applied to different units.	Farmers applied 'about' 100 kg of mulch. Seed shortage of one variety, so control applied to multiple plots in reps 3 and 4.
Layout	Post-hoc blocking useful. Missing values.	Observations indicated that blocking should have been in other direction. Some farmers left trial. Stover yields missing on 3 out of 36 plots in mulching trial. Some plants missing within the row.
Measurement	Zero values. Strange values. Censored values. Different variability. Affect treatments.	Some trees did not survive. Yields for one farmer had different pattern to the others. All trees counted, circumference only measured on large trees. Soil nitrogen values showed more plant-to-plant variation in some treatments. a) Some plants heavily affected with striga. b) Some farms have streak virus.

We describe six alternative strategies for coping with such complications.

Ignore the problem

This is sometimes the correct strategy. For example in a study on leucaena where an objective was to estimate the volume of wood, all trees were counted, but the circumference was not recorded on the small trees. This could perhaps be ignored, because the small trees contribute little to the volume. Alternatively, our second strategy is to modify the objectives, or the definition of a treatment or measurement. With the same example, we could modify our objective to one of estimating the volume of wood from trees with a diameter of more than 10 mm.

Abandon an objective

This does not imply that the whole analysis need be abandoned, because other objectives may still be obtainable. In some cases, the problem can be turned to your advantage in that new objectives can be studied. For example, in an on-farm study, two of the treatments involved requesting the farmers to apply 100 kg mulch. In the actual trial, the quantity applied varied between 11 kg and 211 kg. This might permit a study to be made of the effects of applying different quantities of mulch. However, we find that such complications are usually more useful in reassessing the design for a subsequent trial. Note that in the new trial the appropriate question might not be 'How do we force the farmers to apply 100 kg?' but 'Why are the quantities applied SO variable?'

Use a 'quick fix'

With missing values, a common quick fix is to use the built-in facilities in many statistics packages, namely to estimate them from the remaining data and then proceed roughly as if the experiment did not have the problem. A quick fix in the problem of the small stems in the leucaena trial mentioned earlier, might be to assume they are all 5 mm in diameter.

Quick fixes are not risk-free even if they are built into a statistics package and we caution against using the automatic facilities for estimating missing values if more than a tiny fraction are missing. It is quite easy to use our next strategy, at least to check that the approximate analysis is reasonable.

Use a more flexible approach to the analysis

That often means we move from the ANOVA to the more general linear, or regression modelling approach, because the latter does not require the data to be balanced. Hence, it can adjust for missing values properly. Similarly this approach can be used in a trial where *post-hoc* blocking would be useful, because the initial blocking was in the wrong direction.

Sometimes the more flexible approach implies that a transformation of the data would give a more reasonable model. Alternatively, where some treatments are more variable than others, the solution is to analyse the data separately for the two groups of treatments.

Do a sensitivity analysis

For example, with transformations or odd observations, for which there is no obvious explanation, it is usually very quick to do two or three alternative analyses. These correspond to different models. If there is no difference in the conclusions then the results are not sensitive to the particular model. It is then often useful to report this investigation as justification for using the model that is simplest to interpret in relation to the objectives. Where the conclusions do change, you know that more work is needed to identify the appropriate model.

Build a model to try to solve the problem

For example, with the leucaena stems it might be possible to build a 'one off' model for stem diameters of leucaena trees. Often the construction of this model is not part of the original objectives and may not be needed; the quick fix is sufficient. Yet, it may be interesting and of general use later. Thus, it adds a methodological objective that is of general use and may be incorporated in the design of a subsequent study.

18.7 What Next?

In this chapter we have emphasized that the key to a successful analysis is that it should relate directly to the objectives of the experiment, so that as far as possible these are satisfied, given the data. The analysis should be driven by a sense of curiosity about the data, and the approach should be flexible.

In Sections 18.3 to 18.6 we have given guidance on how to deal with some complications that can arise in data analysis. These could be dealt with using sophisticated methods, but often a simple solution is all that is necessary. Though we have not discussed complications where more advanced approaches are necessary, there are two instances where such methods will often be required. The first concerns the type of data, and the second the issue of multiple levels.

Trials typically include measurements of different types. Often there are measurements of yield. There may also be counts of weeds per plot (a disease score may be on a scale of 1 to 9 per plot), and seed germination may be recorded as a binary variable (yes or no).

Any of these types of measurements may be summarized with the simple methods described in the preliminary analysis, but we now use tables and graphs of counts or percentages (e.g. of farmers who valued a particular variety) instead of summary tables of mean yields.

The standard ANOVA and the regression modelling, assume the data are from a normal distribution, and are no longer directly applicable for these types of data. Hence, researchers often transform the data to make the assumption of normality more reasonable. However, there is a more modern alternative approach. The simple modelling ideas described in Section 18.2.5 are often called 'general linear modelling' where the 'general' implies that it applies to balanced and unbalanced data. The analyses can be extended to 'generalized linear modelling' where the 'generalized' is because it applies to other distributions besides the normal. We look at general linear models in more detail in Chapter 19 and generalized linear models in Section 20.4.

The second main area, which requires methods that are more complex, is where the data are at multiple levels and the simple approaches are not sufficient. Multiple levels cannot be handled by the modelling approach described earlier, because this is limited to information at a single level.

The standard ANOVA will handle multiple levels only for balanced data, as in the case of the split-plot experiment. Therefore, just as the regression modelling approach generalizes the single-level ANOVA, we need an equivalent to the split-plot analysis to generalize to situations where multiple-level data are unbalanced.

An example is the on-farm trial described earlier where we now want to relate some of the farm level measurements (such as striga infestation), to the plot level yields. However, there are unequal numbers of farms with and without striga, and not all farmers grew the same three varieties. We want to study the interaction table in the same way as we would in a standard split plot analysis, with one factor (striga) on the main plots and the other (variety) on the subplots, but the data structure is unbalanced.

Multiple levels cannot be handled by the modelling approach described in Section 18.2, but need their own type of modelling. This is called mixed-models or multilevel modelling, and is sometimes known by the names used by statistical software, such as REML in GENSTAT or PROC MIXED in SAS. We return to these ideas in Section 20.5.

Chapter 19

General Linear Models

19.1 Introduction

In this chapter we complete our DIY Toolbox with a discussion of general linear models. They are important because they provide a common framework for many statistical methods, and are the basis for others that are introduced in Chapter 20. The value of those further methods can be appreciated more easily once these general linear models are understood.

19.2 A Simple Model

We start by reviewing the 'simple linear regression model'. The data used throughout this chapter are from an experiment comparing nine soil fertility regimes in a randomized block design with four replicates. Crop yield was measured as a response, together with various indicators of soil quality.

Exploring the data is straightforward: we are interested in the way yield is affected by changes in inorganic nitrogen so it is natural to graph the data. The scatter diagram in Fig. 19.1 shows results much as expected: a clear relationship between yield and inorganic nitrogen, but with a fair amount of scatter. Yield increases with increasing nitrogen (N), but for any given level of N, there is a wide range of yields, of about ±1 t/ha.

The description of *data = pattern + residual* is now formalized by defining a specific model. This means describing precisely both the pattern and the residual.

The pattern could be described by many different curves, which are consistent with the data. Some considerations when choosing the shape of the curve are described in Chapter 20. For these data, a straight line shows the results clearly. This is chosen, as it is biologically sound, the simplest mathematically, and as there is nothing in the data to suggest we need anything else. Thus, we can write

yield = *a* + *b*.(*inorganic nitrogen*) + *residual*

The value of the parameters *a* and *b*, are at this stage unknown. The residual is the difference between the pattern and the data, and can be found once *a* and *b* are known.

Fig. 19.1. Relationship between yield and inorganic nitrogen.

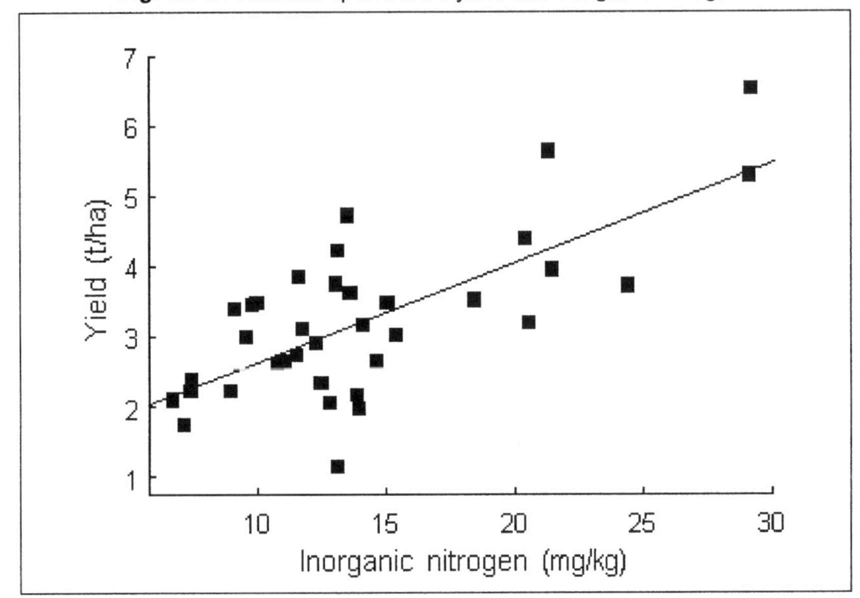

The next step is to find suitable values of *a* and *b*. There are many ways of choosing these, but the most common is to choose them as the values which give the 'best fit', defined as minimizing the sum of squared residuals or, the variance of the residuals. The fitting will be done with suitable statistical software. One reason this modelling approach is important, is the ease with which it is implemented in good statistical software. To show this and the similarities in fitting a range of models, we show the dialogue of 'general linear regression' in the GENSTAT software system. In Fig. 19.2, GENSTAT needs to know the *response variate* of the data (i.e. the left-hand side of the model), namely the yield and the pattern, in this case just the inorganic nitrogen. Notice that the intercept is included automatically.

Fig. 19.2. GENSTAT dialogue for general linear regression.

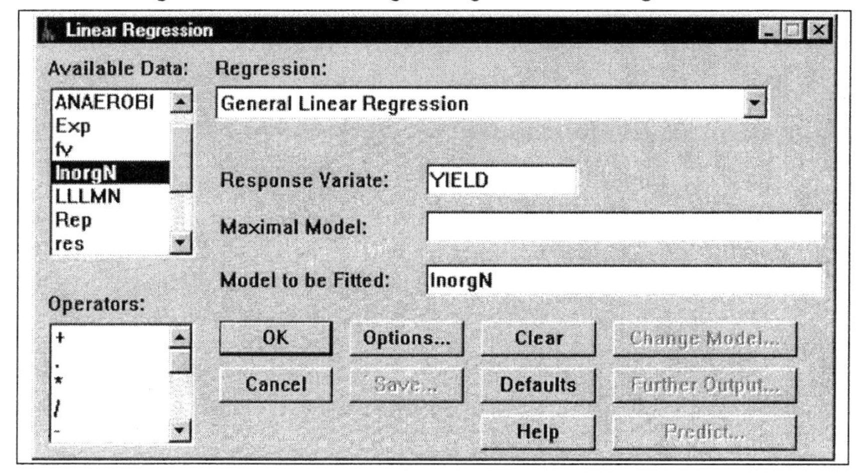

The estimates that are produced are shown in Fig. 19.3.

Fig. 19.3. Part of the output showing the parameters of the equation.

```
          estimate  s.e.
Constant  1.2280    0.3660
InorgN    0.1419    0.0241
```

The results include the parameters of the equation of the line, as shown in Fig. 19.3. The value of a is estimated as 1.23 t/ha and that of b as 0.14 t/ha (per mg/kg of inorganic nitrogen). Thus, the pattern part of the data is estimated to be

yield = 1.23 + 0.14 × *(inorganic nitrogen)*

The residuals can now be calculated, and we need to check that there is no more pattern to the residuals. They are plotted in various different ways, but this does not reveal any further pattern. The residuals can be summarized, by giving their variance or standard deviation (0.81 t/ha) and perhaps their distribution, revealed in a histogram (see Fig. 19.4).

Our look at the residuals suggests the description of the pattern is acceptable, so we continue by interpreting the model. It suggests that a technology that can increase the inorganic nitrogen content of the soil by 10 mg/kg should give an increase in yield of about 10×0.14 = 1.4 t/ha.

The a parameter mathematically represents the yield at zero inorganic nitrogen. However, we should be cautious when interpreting this value, since it is an extrapolation beyond the range of the data. The minimum value observed was 6.6 mg/kg and we cannot be sure that the straight line is a good representation for the relationship from here back to zero; indeed biological considerations would suggest this would be most unlikely. For similar reasons it rarely makes sense to use a simple linear model without a constant a, even if the biology of the situation is such that the line has to go through the origin. This is one reason why the constant is automatically included in the model definition. The other is so that the baseline or null model, with no explanatory variables against which others are compared, is

response = *a* + *residual*, and not
response = 0 + *residual*.

Fig. 19.4. Histogram of residuals.

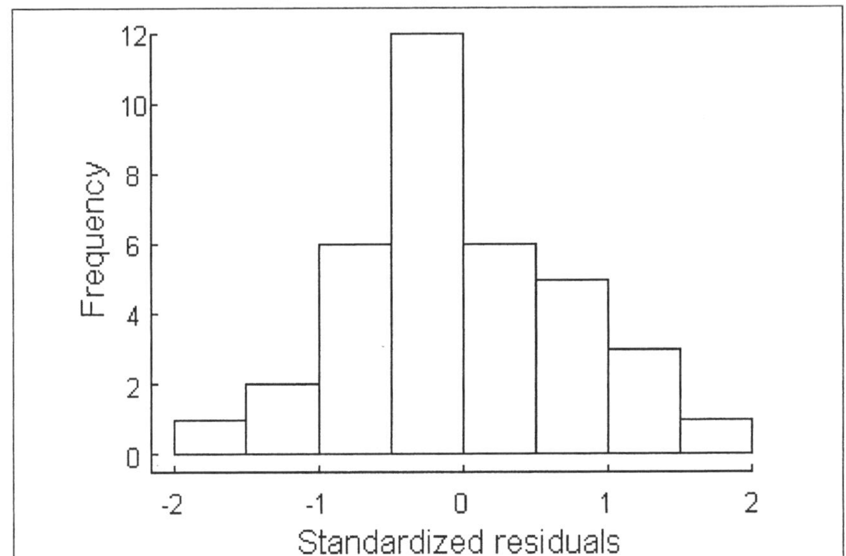

19.3 Further Information about the Model

The full output produced by the software on fitting this simple linear regression model is shown in Fig. 19.5. The analysis of variance table can be interpreted as described in Sections 16.7 and 18.2.4. The total variation in yield (s.s. = 44.51) is broken down into two components that are explained by the regression line (s.s. = 22.46) and the variation in the residuals (s.s. = 22.05). Thus the pattern accounts for about half (49%) the overall variation in the data, which is sometimes referred to as r^2 and this is stated in the analysis of variance table in Fig. 19.5.

The mean square (m.s. = 0.6485) for the residuals is the variance of the residuals, so $\sqrt{0.6485} = 0.805$ is their standard deviation. The residuals appear to have roughly a normal distribution, so we expect most of them to lie in the range ±2 standard deviations or ±1.6. This is confirmed by looking at the list of residuals or at the graph in Fig. 19.6. In Fig. 19.5, one residual has been picked out as being larger than expected. The graphs (e.g. in Fig. 19.6), suggest that this just happens to be the largest negative residual, but it is not so much larger than the rest to cause suspicion.

Fig. 19.5. Output from the dialogue shown in Fig. 19.2.

```
***** Regression Analysis *****

Response variate: YIELD
 Fitted terms: Constant, InorgN

*** Summary of analysis ***

            d.f.    s.s.    m.s.     v.r.   F pr.
Regression    1    22.46  22.4579   34.63  <.001
Residual     34    22.05   0.6485
Total        35    44.51   1.2716

Percentage variance accounted for 49.0
Standard error of observations is estimated to be 0.805
* MESSAGE: The following units have large standardized residuals:
 Unit  Response  Residual
 8      1.138     -2.45
* MESSAGE: The following units have high leverage:
 Unit  Response  Leverage
 15     3.720     0.123
 19     5.290     0.228
 28     6.542     0.230

*** Estimates of parameters ***

          estimate  s.e.    t(34)  t pr.
Constant  1.2280    0.3660  3.36   0.002
InorgN    0.1419    0.0241  5.88   <.001
```

The results in Fig. 19.5 also include three points with 'high leverage'. These are points that could have a large influence on the estimates of the pattern. They are also the three points with the highest values of inorganic nitrogen. The reasons they are influential can be seen in the scatterplot (Fig. 19.6); they are on the right of the scatterplot where there are relatively few data points. They will therefore tend to 'pull' the fitted line towards themselves more than other points will. This is not necessarily a problem, but is something to be aware of. It may be uncomfortable to have your conclusions possibly dependent on just one or at best, a few observations. In the case of a single explanatory variable, influential points can be seen on a scatter diagram. In the more general case they can be difficult to spot, which is why the numerical diagnostics produced by the software have been developed.

The output in Fig. 19.5 also includes information to help in formal inference. The most useful are standard errors of parameter estimates. The estimate of the slope parameter has a standard error (s.e.) of 0.024. Thus the estimated increase in yield with an increase of 10 mg/kg inorganic nitrogen has a standard error of 0.24. An approximate 95% confidence interval for this increase is therefore 1.4 $\pm$ 2×0.24 or 0.9 to 1.9 t/ha. In the ANOVA table, results of an F-test are presented and in the list of parameter estimates, the results of t-tests are shown. The F-test and t-test for inorganic nitrogen both test the null hypothesis of no linear relationship between yield and inorganic nitrogen (i.e. b = 0). It is clear from the graph that this null hypothesis is not of interest, so the test results do not add anything useful. The *t*-test for the constant, tests the hypothesis that $a = 0$. Unless there is a specific scientific reason for looking at this value, it is also not of interest.

Fig. 19.6. Graph of residuals against fitted values.

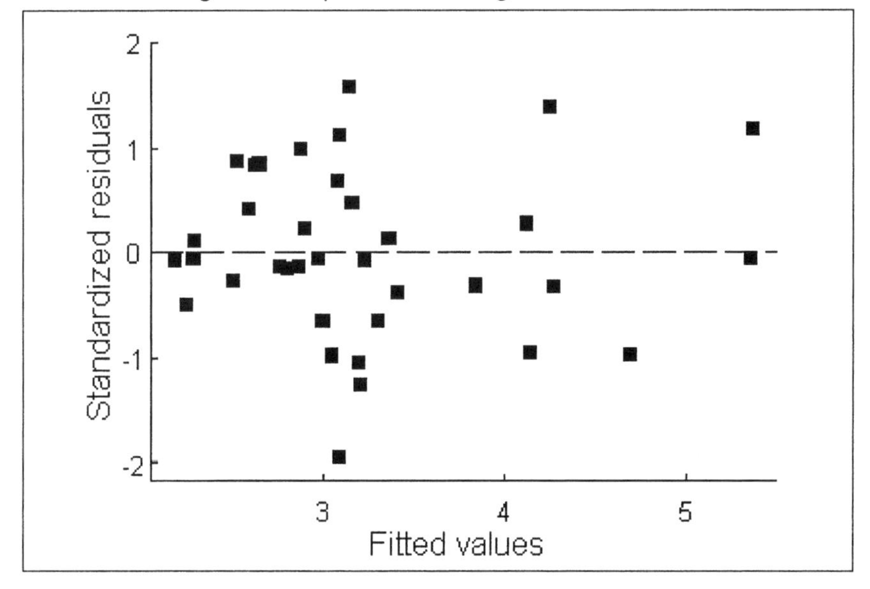

These inference results can be mathematically justified if certain conditions are true:

1. The model structure (a straight line plus independent residuals) is appropriate for the data.
2. The variance of the residuals is constant.
3. The residuals are more or less normally distributed.

Assumptions 1 and 3 have already been checked and seen to be reasonable. Assumption 2 is checked by looking at the variation of the residuals, relative to anything that might change them. A common problem occurs when the variation of residuals increases with the mean, and this can be checked by plotting residuals versus fitted values (i.e. the values on the fitted line for any value of inorganic nitrogen). Figure 19.6 does not reveal any strong pattern in variance across the range of means.

19.4 Steps in Statistical Modelling

The steps that have been followed in the previous section apply for any statistical model and are therefore worth emphasizing. They are:

1. Explore the data to detect patterns and relationships.
2. Choose a possible model, based on the patterns seen, knowledge of the design that generated the data, and any information about the underlying scientific processes involved.
3. Fit the model (i.e. estimate the unknown parameters).
4. Check that the model is fitting well. If it is not, go back to step 2.
5. Interpret the fitted model to satisfy the analysis objectives.

19.5 A Model with Factors

Yield data from the experiment was analysed in the previous section by looking at its relationship with soil inorganic nitrogen. However there is another way of looking at the data, corresponding to another of the objectives, and that is to determine which of the fallow treatments give good yields. The design of this trial is regular, as each block has each treatment once. Hence, simple analysis of variance, with calculation of treatment means, will give an analysis that meets the objectives. The corresponding analysis of variance table was given in Section 18.2.4. Here we do an identical analysis using the general linear modelling framework.

First, what is the form of the model to be fitted? Exploring the data, for example by boxplots, suggests that certainly there are treatment differences. The design of the trial allows for block (replicate) differences so these should be included. We will assume additive block and treatment effects, which have not been explicitly justified but is the same assumption as is used in the ANOVA approach. The model can be written as

yield = constant + replicate effect + treatment effect + residual
(i.e. *data = pattern + residual*)

The formulation of this model is different from the previous regression model. The term 'treatment effect' means that the yield is increased or decreased by a fixed amount, depending on which treatment was applied. The term thus represents a series of quantities describing the effect of each different treatment. It does not represent a slope of a regression line. The same is true of the 'replicate effect'. This can be represented as

$$yield_{ij} = constant + rep_i + treat_j + residual$$

Where $yield_{ij}$ is the crop yield for the i'th replicate of treatment j, rep_i are the four replicate effects, and $treat_j$ are the nine treatment effects. The GENSTAT dialogue for fitting this model is shown in Fig. 19.7; the resulting output is shown in Fig. 19.8.

Fig. 19.7. Dialogue for ANOVA using the general linear modelling approach.

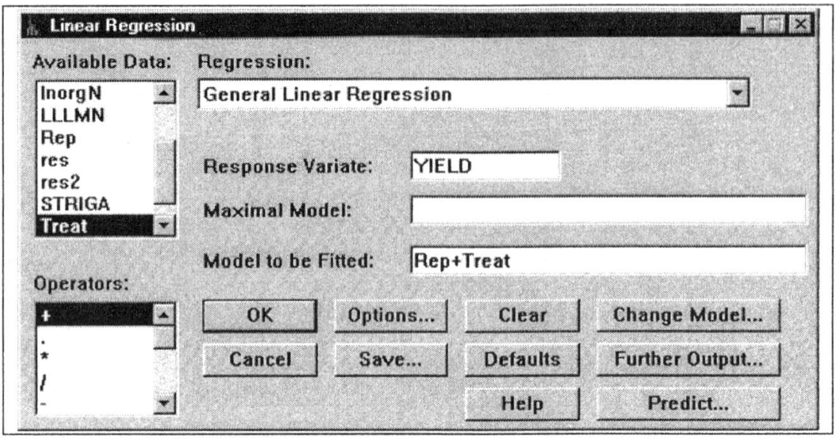

Linear Regression

Available Data: InorgN, LLLMN, Rep, res, res2, STRIGA, Treat

Regression: General Linear Regression

Response Variate: YIELD

Maximal Model:

Model to be Fitted: Rep+Treat

Operators: + . * / -

OK | Options... | Clear | Change Model... | Cancel | Save... | Defaults | Further Output... | Help | Predict...

Fig. 19.8. Output from the dialogue in Fig. 19.7.

```
***** Regression Analysis *****

Response variate: YIELD
 Fitted terms: Constant + Rep + Treat

*** Summary of analysis ***

             d.f.   s.s.     m.s.    v.r.   F pr.
Regression   11     38.859   3.5327  15.02  <.001
Residual     24      5.646   0.2353
Total        35     44.506   1.2716

Percentage variance accounted for 81.5
Standard error of observations is estimated to be 0.485
* MESSAGE: The following units have large standardized residuals:
 Unit  Response  Residual
  6    2.057     -2.14
 10    4.716     -2.36

*** Accumulated analysis of variance ***

Change     d.f.  s.s.     m.s.     v.r.
+ Rep       3    1.0783   0.3594   1.53
+ Treat     8   37.7811   4.7226  20.07
Residual   24    5.6463   0.2353

Total 35 44.5058 1.2716
```

In constructing the accumulated ANOVA table in Fig. 19.8 we have taken the total variation in the yield, and looked to see how much of that variation was due to the fact that plots were in different replicates. We see that 1.08 of the total sum of squares (44.51) was due to the differences between replicates. This left a remaining unexplained variation of 43.43. We looked to see how much could be explained by the different effects of treatment. It turned out that 37.78 could be attributed to the treatments, leaving 5.65 unexplained.

Would it have made a difference if we had fitted treatment effects first, and then looked to see how much of the remaining variation could be explained by replicates? The answer to this question is no, not in this case. When data structures are orthogonal, the order of fitting is unimportant. The sums of squares are the same no matter what order the terms are fitted. This is not so when data structures are non-orthogonal, then the order of fitting does matter and consideration needs to be given as to how that is done.

The results from the model fitting now give parameter estimates, rather than treatment means (see Fig. 19.9). These correspond to the terms fitted in the model.

Fig. 19.9. Further output for the general linear modelling approach.

```
*** Estimates of parameters ***

                         estimate   s.e.     t(24)    t pr.
Constant                  5.370     0.280    19.18    <.001
Rep 2                     0.280     0.229     1.23    0.232
Rep 3                     0.018     0.229     0.08    0.937
Rep 4                     0.406     0.229     1.78    0.088
Treat 2 C.calo           -2.961     0.343    -8.63    <.001
Treat 3 G.sepium         -1.702     0.343    -4.96    <.001
Treat 4 L.leuco          -1.868     0.343    -5.45    <.001
Treat 5 F.congesta       -2.087     0.343    -6.08    <.001
Treat 6 C.siamea         -3.440     0.343   -10.03    <.001
Treat 7 groundnut        -2.466     0.343    -7.19    <.001
Treat 8 nat.fallow       -2.734     0.343    -7.97    <.001
Treat 9 M.only           -3.571     0.343   -10.41    <.001
```

Here we see a *Constant* parameter, three *Rep* parameters, and eight *Treat* parameters. There is no estimate for *Rep 1* or *Treat 1*: those values are not printed as they are fixed to be zero.

Table 19.1. Using the parameter estimates to predict the yields (fitted values).

Treatment	Replicate				Estimated yield
1	1	5.370			= 5.370
1	2	5.370	+ 0.280		= 5.650
1	3	5.370	+ 0.018		= 5.450
1	4	5.370	+ 0.406		= 5.776
1	*Mean*				= *5.546*
2	1	5.370		- 2.961	= 2.409
2	2	5.370	+ 0.280	- 2.961	= 2.689
...	...	...	...	...	...
...	...	...	...	...	...

The parameter estimates can be used to estimate, or predict, the yield to expect for each treatment in each replicate, i.e. using the model pattern

yield = constant + replicate + treatment

The estimated yields for the treatments in each replicate are shown in Table 19.1.

But what has happened to our mean values? When presenting results we normally use treatment means. With a general linear model, it is possible to construct mean values using the parameter estimates, and most software packages will calculate them automatically. For the nine treatments in our example, the estimates are shown in Fig. 19.10.

Fig. 19.10. Estimated adjusted means for each treatment.

```
Response variate: YIELD

                Prediction  s.e.
 Treat
 1 S.sesban     5.546       0.243
 2 C.calo       2.585       0.243
 3 G.sepium     3.844       0.243
 4 L.leuco      3.678       0.243
 5 F.congesta   3.459       0.243
 6 C.siamea     2.106       0.243
 7 groundnut    3.080       0.243
 8 nat.fallow   2.812       0.243
 9 M.only       1.975       0.243
```

These estimates are sometimes called 'adjusted mean values', because they are averaged over the levels of the other factor or factors in the model; in this example this is only the replicate. Referring back to the parameter estimates and the estimated treatment yields for each replicate, we can see how these mean values are calculated (the value for Treat 1 is included in Table 19.1). In this example, the adjusted mean yields and their standard errors (s.e.) are the same as the ordinary treatment means and standard errors. This only happens with orthogonal data. If the data were non-orthogonal, the adjusted means would be different from the ordinary means.

Consider now a comparison of two treatments. Suppose one objective was to estimate the mean difference in yield between Treat 2 and Treat 1. How do we estimate this difference? One way is to look at the difference between the adjusted means i.e. 2.585-5.546 (see Fig. 19.10) which gives an estimate of -2.961 t/ha.

Alternatively, we can look at the way the adjusted means are constructed from the parameter estimates. The adjusted means for Treat 1 and Treat 2 are:

constant + (Rep1 + Rep 2 + Rep 3 + Rep 4)/4 + Treat 1
constant + (Rep1 + Rep 2 + Rep 3 + Rep 4)/4 + Treat 2

Hence, the difference in adjusted means is simply (Treat 2 - Treat 1). The parameter estimate for Treat 1 is zero, so the difference is just the parameter labelled Treat 2, which is -2.961 (from Fig. 19.9, and reproduced in part as Fig. 19.11). Hence, the estimate in the line starting Treat 2 is the estimate of the difference between Treat 1 and Treat 2. Similarly, Treat 3 estimates the difference between treatments 1 and 3, and so on. The difference between Treat 2 and Treat 3 would be estimated by (Treat 2 - Treat 3).

Fig. 19.11. Estimates of treatment effects, extracted from Fig. 19.9.

```
*** Estimates of parameters ***

                    estimate  s.e.   t(24)   t pr.
Treat 2 C.calo      -2.961    0.343  -8.63   <.001
Treat 3 G.sepium    -1.702    0.343  -4.96   <.001
```

The parameter estimate information for Treat 2, shown in Fig. 19.11, says a lot about the comparison between Treat 1 and Treat 2. Not only does it give the estimate of the difference, it also gives us the standard error of this difference (s.e. = 0.343), and a *t*-test for testing the hypothesis that there is no difference between these two treatments. A *t*-value of -8.63, compared against the *t*-distribution with 24 degrees of freedom, is highly significant ($p<0.001$).

The equivalent estimate for (Treat 2 - Treat 3) is also found by subtraction, i.e. -2.961 - (-1.702) = -1.259 (from Fig. 19.11). The standard error and *t*-value are a little harder to obtain, but can be given by the software.

19.6 Analysing Other Designs

The previous section introduced a more complex way of doing something that we could do simply before! What is the point? A little experience shows us that the ANOVA facilities in statistical software often fail. Whenever the design is 'incomplete', or 'unbalanced' in some way: for example, we do not have every treatment occurring equally often in each block, the simple analysis of variance will give a message such as that shown in Fig. 19.12.

Fig. 19.12. Simple ANOVA cannot be used for unbalanced data.

```
Design unbalanced - cannot be analysed by ANOVA
Model term Treat (non-orthogonal to term Rep) is unbalanced, in the
Rep.*Units* stratum.
```

However, the general linear modelling approach works for these examples in exactly the same way as before. In the following example, the data from the first year of the trial is analysed again but this time the second row of data (Treat 3 from Rep 1) has been removed. The output from fitting the model has of course changed (Fig. 19.13) but can be interpreted in exactly the same way as before.

Notice that the s.e. of the Treat 3 parameter is larger than the others. This should be expected as we have less information now about Treat 3 than about the other treatments.

19.7 One Further Model

In Section 19.2, we fitted a simple linear regression model with a variate on the right-hand side of the equation, to explain the relationship between a response and a continually varying 'explanatory' variable. In a second example (Sections 19.5 and 19.6), we fitted a model with factors on the right-hand side, to estimate the difference between factor levels; in this case treatments. The modelling framework is so general that we can combine these ideas. Here we show an example that involves both an explanatory variable and a factor.

Fig. 19.13. General linear model output for unbalanced data.

```
Response variate: YIELD

 Fitted terms: Constant + Rep + Treat

*** Summary of analysis ***

             d.f.   s.s.    m.s.    v.r.   F pr.
Regression   11    38.263  3.4784  16.41  <.001
Residual     23     4.875  0.2119
Total        34    43.137  1.2687

Percentage variance accounted for 83.3
Standard error of observations is estimated to be 0.460
* MESSAGE: The following units have large standardized residuals:
 Unit  Response  Residual
 5     2.057     -2.03
 9     4.716     -2.57

*** Accumulated analysis of variance ***

Change     d.f.  s.s.     m.s.    v.r.
+ Rep       3    1.6990  0.5663   2.67
+ Treat     8   36.5637  4.5705  21.56
Residual   23    4.8746  0.2119
Total      34   43.1374  1.2687

*** Estimates of parameters ***

                         estimate  s.e.    t(23)   t pr.
Constant                  5.280    0.270   19.56  <.001
Rep 2                     0.400    0.226    1.77  0.090
Rep 3                     0.138    0.226    0.61  0.548
Rep 4                     0.526    0.226    2.33  0.029
Treat 2 C.calo           -2.961    0.326   -9.10  <.001
Treat 3 G.sepium         -1.971    0.355   -5.56  <.001
Treat 4 L.leuco          -1.868    0.326   -5.74  <.001
Treat 5 F.congesta       -2.087    0.326   -6.41  <.001
Treat 6 C.siamea         -3.440    0.326  -10.57  <.001
Treat 7 groundnut        -2.466    0.326   -7.57  <.001
Treat 8 nat.fallow       -2.734    0.326   -8.40  <.001
Treat 9 M.only           -3.571    0.326  -10.97  <.001
```

Still using the same trial, consider the following question. The yield is strongly determined by the inorganic nitrogen in the soil. The yield also varies between different treatments, which can change the level of inorganic nitrogen. But, does the change in inorganic nitrogen account for all the effect of the treatments on the yield, or are the treatments having some effect in addition to the increase in inorganic nitrogen?

To investigate this question we start with the simple regression analysis:

yield = a + b.(inorganic nitrogen)

The residuals from this model are variations in yield which are not explained by a simple straight-line relationship with inorganic nitrogen. A boxplot of these residuals by treatment (Fig. 19.14) suggests that there are large differences. The sesbania plots (Treat 1), for example, have positive residuals (higher yield than expected from the model) and the maize-only plots (Treat 9) have negative residuals.

Fig. 19.14. Boxplots of standardized residuals for each treatment.

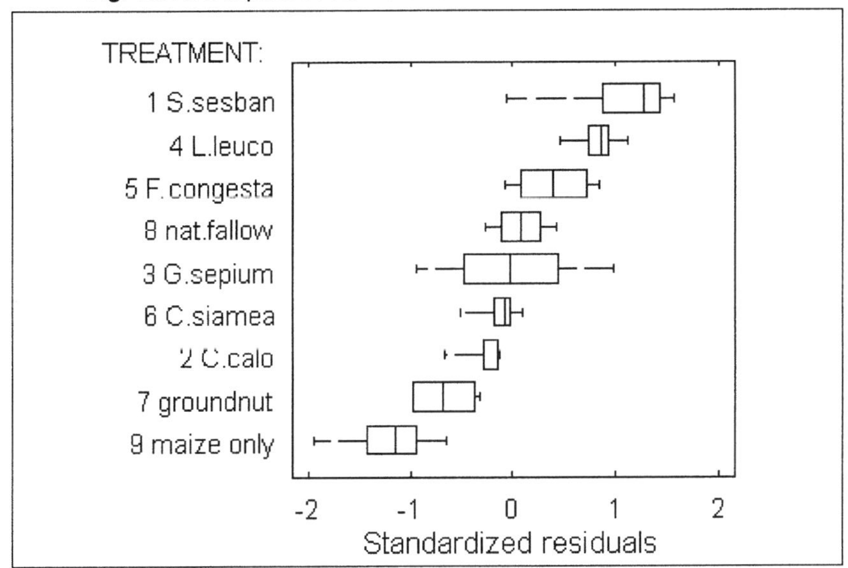

Fig. 19.15. GENSTAT dialogue for a regression model with both variates and factors.

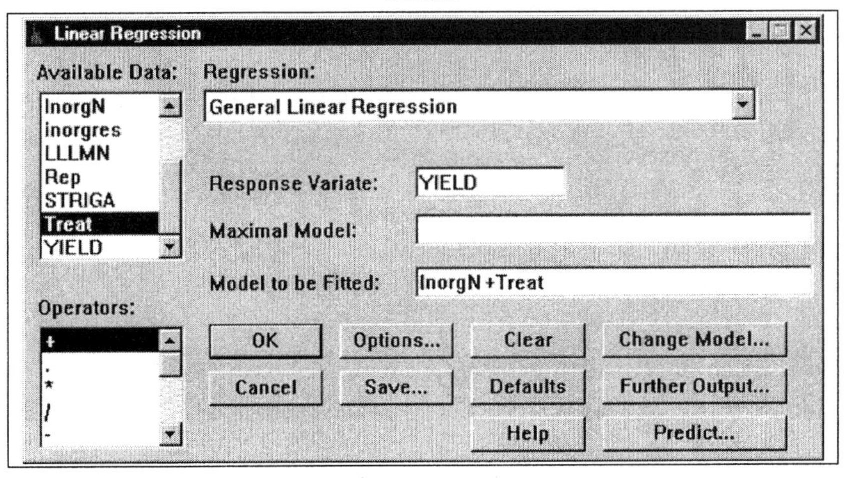

Fig. 19.16. ANOVA table for the model specified in Fig. 19.15.

```
*** Accumulated analysis of variance ***

Change     d.f.   s.s.      m.s.       v.r.
+ InorgN    1     22.4579   22.4579    121.20
+ Treat     8     17.2300    2.1537     11.62
Residual   26      4.8179    0.1853

Total      35     44.5058    1.2716
```

Hence, we need a model that accounts for this additional pattern that was previously not noticed. The dialogue above (Fig. 19.15) shows how a model with a linear effect of inorganic nitrogen and additive treatment effects can be fitted.

Fig. 19.17. Parameter estimates for the model specified in Fig. 19.15.

```
*** Estimates of parameters ***

                        estimate    s.e.
Constant                  3.826     0.578
InorgN                    0.0739    0.0231
Treat 2 C.calo           -2.089     0.408
Treat 3 G.sepium         -1.349     0.324
Treat 4 L.leuco          -0.995     0.408
Treat 5 F.congesta       -1.327     0.386
Treat 6 C.siamea         -2.249     0.480
Treat 7 groundnut        -2.057     0.330
Treat 8 nat.fallow       -1.801     0.421
Treat 9 maize only       -2.878     0.373
```

The analysis of variance table in Fig. 19.16 shows that a substantial part of the residual variation from the simple regression is actually explained as treatment variation (sum of squares of 17.23 with 8 d.f.). There is very strong evidence that the treatments have some effect on yield, which is not accounted for by change in the inorganic nitrogen level.

The parameter estimates in Fig. 19.17 show what these effects are. For instance, for a fixed level of inorganic nitrogen, the yield following a sesbania fallow exceeds that of a natural fallow by (Treat 1 - Treat 8) = (0 - (-1.801)) = 1.801 t/ha with a standard error 0.421.

We could alternatively have fitted the model with the terms in the other order, simply by specifying the model as 'InorgN + Treat' in Fig. 19.15. In other words, these two models are identical, in that they give identical parameter estimates:

yield = constant + b.(inorganic nitrogen) + Treat
yield = constant + Treat + b.(inorganic nitrogen)

However, the accumulated analysis of variance produced when fitting the two models is different. Compare Fig. 19.18 with Fig. 19.16.

Remember how the earlier ANOVA was found: identify how much variation is explained by the first term (this time Treat) and then how much additional variation is explained by the second term (now InorgN). Thus the line in the ANOVA table for inorganic nitrogen in Fig. 19.18 shows how much variation can be attributed to changes in inorganic nitrogen once we have accounted for differences between the treatments. It is not surprising this is small since there is not much variation between levels of inorganic nitrogen within treatments. The previous ANOVA showed how much variation is due to inorganic nitrogen when treatments are ignored. These two quantities are clearly different. The two ANOVA tables are both correct but address different questions and are suitable for different analysis objectives.

Fig. 19.18. ANOVA table when fitting the model 'Treat + InorgN'.

```
*** Accumulated analysis of variance ***

Change     d.f.   s.s.      m.s.     v.r.
+ Treat     8    37.7811   4.7226   25.49
+ InorgN    1     1.9067   1.9067   10.29
Residual   26     4.8179   0.1853

Total      35    44.5058   1.2716
```

There are further steps we could take with this modelling. For example, we have ignored blocks. Should they also be included in the model? Can we explain the treatment effects using, in addition to the inorganic nitrogen, any of the other variables that were measured? No new ideas are needed to pursue these objectives.

19.8 Assumptions Underlying the Model

As with earlier analyses, there are some assumptions underlying this modelling approach. There are two parts to the model, the pattern and the residual components. The assumptions made about both parts can be checked.

The pattern part of the model contains effects of continuous variates and/or factors specifying a series of levels. Graphs can be used to check that a simple relationship is a straight line. Non-linear forms will require a different model. If there are several variates in the pattern, other methods still allow us to check whether the relationship is a straight line. Likewise there are some graphical methods for checking if the block + treatment form of the model is appropriate. Much of the important information about pattern will have been detected during the exploratory analysis of the data.

The assumptions made about the residual part of the model should be checked. They play an important role in the statistical inference. The key ones are:

- Residuals are independent (knowing one residual will not tell you anything about the next). Check by looking at the residuals in field (or perhaps time) order.

- Residuals have a constant variance. Check by looking at the variance of residuals for different groupings of the data (e.g. treatment groups). Try plotting the residuals against the predicted value, since one of the most frequent ways this assumption is violated is for the variance to increase with the mean value. Some software produce these residual plots automatically.
- Residuals have a more or less normal distribution. Look at a histogram (if the dataset is large enough), or better still a probability plot.
- There should be no pattern left in the residuals. Any systematic pattern is a sign that something is missing from the first part of the model.

19.9 The Linear Model Language

So far, the models fitted have been of the form:

response = constant + effects + residual

The effects are specified using a 'language', though the details vary slightly for different statistical software. However, all are similar to the following:

- The effects are specified as a list of terms, usually separated by '+' symbols.
- If the term is a variate it is treated as a regression, i.e. included in the model as a *b.variate* effect where *b* is a parameter to estimate.
- If the term is a factor, a series of constants are included with a separate parameter estimated for each level of the factor.
- Two or more factors can be combined, to give interaction terms that fit a separate constant to each combination of levels of the two factors.
- A factor and variate can be combined, to give a term that fits a separate regression slope for each level of the factor.

This same language is used in specifying the various generalizations to the models that are introduced in Chapter 20.

Chapter 20

The Craftsman's Toolbox

20.1 Introduction

In this chapter, we show researchers some of the more specialist tools that may be of use in analysing data from natural resource management research. We indicate which tools and methods are available and why they are particularly useful in NRM research.

Our aim in this chapter is to raise awareness of these tools. We start with a review of some advanced approaches to exploratory statistics, that can be described as 'modern' in the sense that they are now more readily available because of recent advances in statistical software. We outline two popular multivariate methods, namely cluster analysis and principal components, and then give some general points on statistical modelling. This is followed by a discussion of two important classes of advanced models (generalized linear models and mixed models) that build on the ideas of earlier chapters.

The final section is a reality check. An effective analysis is one that achieves the objectives as simply as possible. Being aware of the craftsman's tools is sometimes useful to confirm that an analysis needs only the DIY toolbox.

20.2 Visualization

Graphical methods are extremely valuable in data exploration because the human brain is excellent at detecting pattern. In a well-drawn graph, we can see the nature of trends and relationships, we can spot similarities and differences between groups, and we notice oddities and surprising aspects of the data. However, simple graphs are not of much value for large and complex datasets. In addition, graphs can lose their appeal if they take too long to construct. In this section we give pointers for the production of useful exploratory graphs for more complex datasets. As these are for exploration rather than final presentation, it is important that they can be produced quickly. Rapid production requires suitable statistical graphics software. Not all statistical analysis software is equally useful when it comes to producing graphs, and spreadsheets are only useful for the very simplest graphical problems. Comments on preparing graphs for final presentation and publication are given in Chapter 21.

It is worth noting here that most software uses colour in graphical displays, which helps to show up possible patterns. Our examples are all in black-and-white and have therefore lost a little of their visual impact. We ask the reader to imagine how much better they would look in colour!

Fig. 20.1. Proportion of crop area devoted to cash crops (vertical axis) plotted against travel time to market for land-use survey data in Kenya.

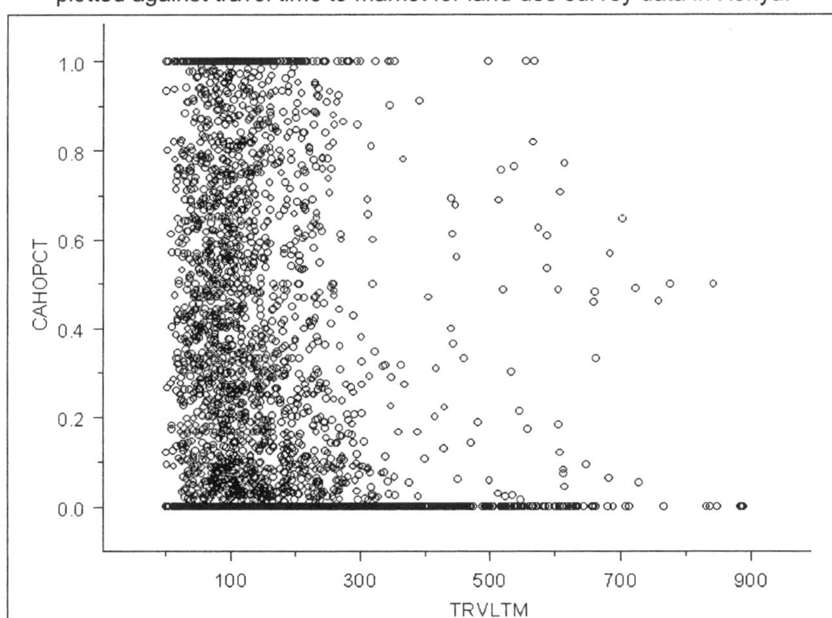

As an example, we start with a simple scatter diagram to look for a relationship between response and explanatory variables. The example in Fig. 20.1 uses data from a survey in Kenya designed to look at factors related to poverty and economic development. The response (vertical) is the proportion of crop area devoted to cash crops. The explanatory (horizontal) variable is an estimate of travel time to the nearest large market. The diagram shows very little, apart from indicating that more could be seen more effectively with a univariate display (such as the large number of observations with zero or all cash crops).

Adding a trend line (Fig. 20.2) shows a rather different picture: there is a clear decrease in the importance of cash crops with increasing distance. This simple act of adding a trend line has revealed an important pattern in an otherwise uninteresting graph.

Fig. 20.2. As Fig. 20.1, with a smooth trend line added.

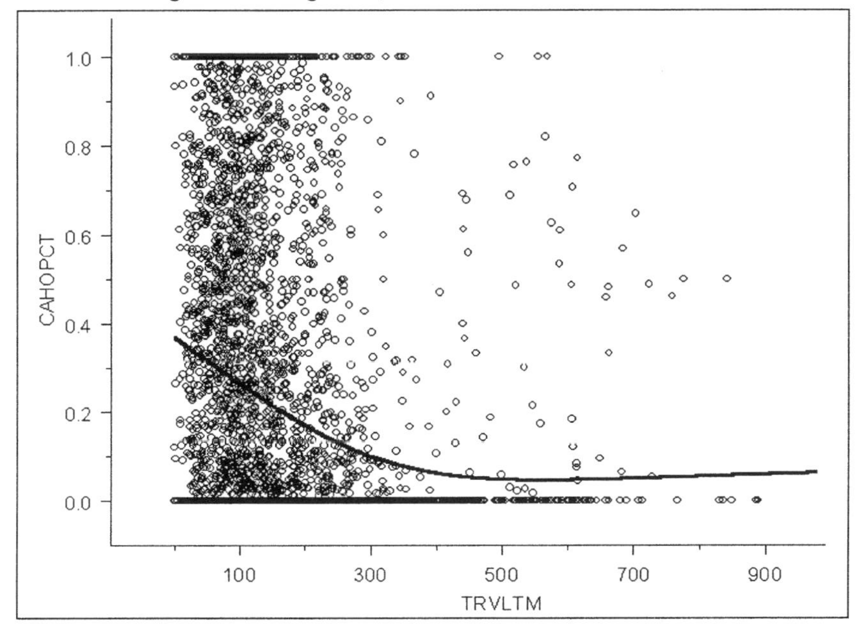

Once we need to look at more than two variables at the same time, there are several options. Simply varying the style or colour of the plotting symbol by the value of a third variable (as shown in Fig. 20.3) may reveal important patterns. However in this case, this leads to a graph that is too cluttered. A better option is to break down the graph into several small repeats of the same pattern, each for different values of the third variable (Fig. 20.4).

This type of 'trellis plot' is flexible and powerful. The conditioning quantity used to define the subplots may be a factor with a different subplot for each level as in this example, or may be a variate that is grouped into classes for the purpose.

Trellis plots can be constructed for more than one conditioning variable. A typical example of a trellis plot is shown in Fig. 20.5, which indicates the variation in soil measurements over several years, at three sites for each of three separate treatments. The graph contains a lot of information and makes many comparisons obvious. For example:

- Patterns for each treatment are similar.
- Patterns for each site are similar.
- Year 92 has the highest level, 93-95 are similar, and 96 shows a slight increase followed by a decrease in 97.
- The year-to-year variation is similar for each site and treatment.

- A straight-line response is a poor description of trends over time.
- Replicate to replicate variation is similar in all sites and treatments, and highest in 92.

This kind of information will make it relatively straightforward to choose a strategy for further analysis and model fitting.

Fig. 20.3. As shown in Fig. 20.2, but with different symbols for observations from each of 12 ethnic groups.

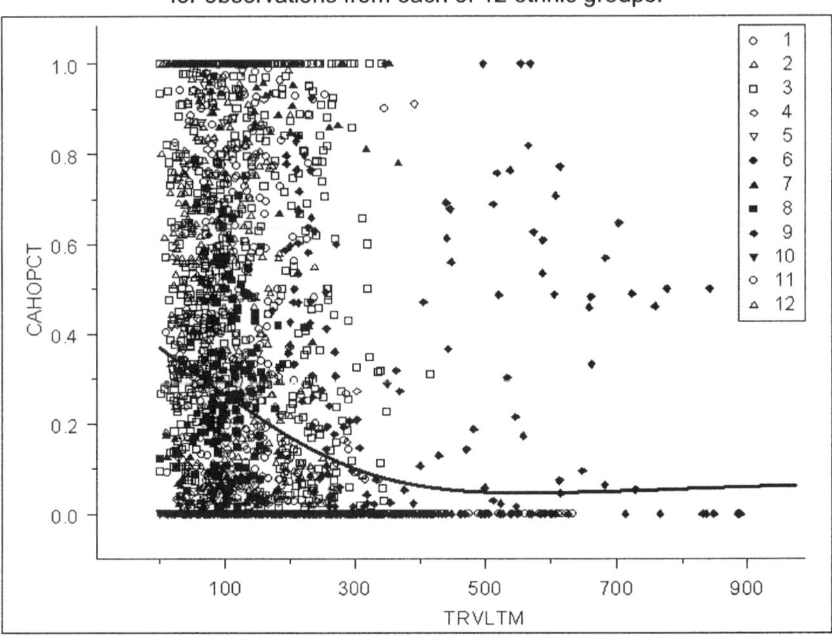

Variation around trends and relationships can be as important as the trends themselves, particularly when looking for outliers and for choosing the specification for the 'residual' part of a model. Scatter diagrams, such as those above, show variation. However, they can be misleading as the eye usually sees the envelope of points (the area delineated by the minima and maxima) rather than the variance defined by the bulk of the points. Figure 20.6 illustrates this point for a large dataset, relating kernel weight of a fruit (*Ricinodendron heudelotii*) to its length. Figure 20.6a shows a scatterplot, with a relationship between kernel weight and fruit length, that could be expected. There is a lot of scatter around the trend line, and it appears greater for lengths around the centre of the range. However, the boxplot in Fig. 20.6b shows this is an illusion. In a boxplot the length of the box depends on the spread, which in this case is constant for much of the range in lengths, but larger towards the extremes.

Fig. 20.4. A trellis plot as shown in Fig. 20.2, but with a separate scatterplot for data from each ethnic group.

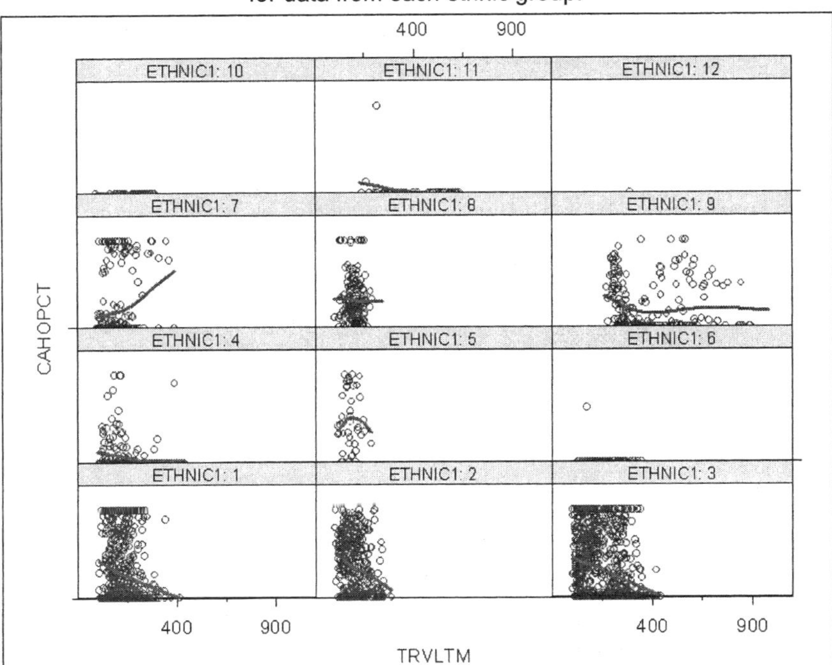

Fig. 20.5. Trellis plot of data from an experiment with 3 treatments at 3 sites with Na (sodium) concentration (vertical axis) plotted against time.

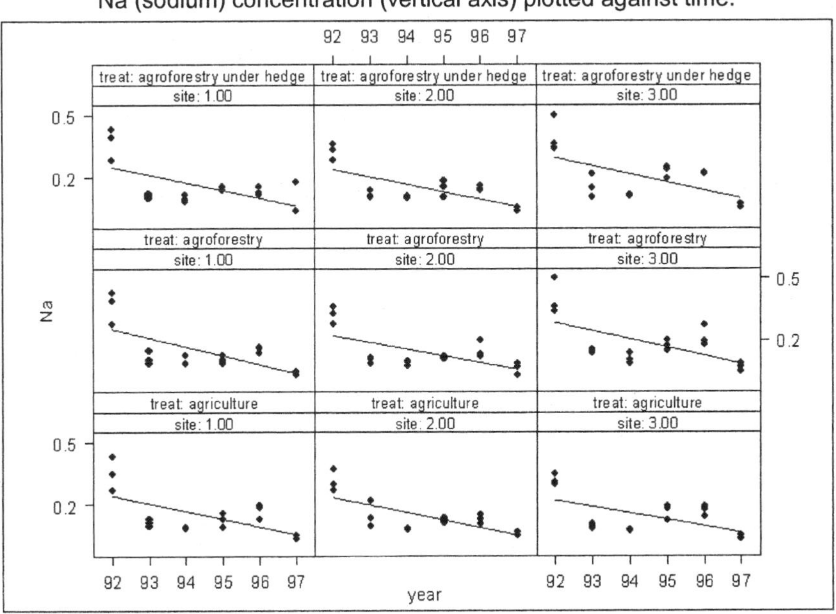

Boxplots can be combined with the idea of trellis plots to look at the way means and variances change with several factors. Figure 20.7 shows multiple boxplots to demonstrate the effect of increasing levels of nitrogen (N) on barley yields at several sites (Ibitin, Mariamine, Hobar, Breda, Jifr Marsow) in Syria. In addition to demonstrating how yields vary across N levels and how the yield-N relationship varies across sites, individual boxplots show the variation in yields for a constant N application at a particular site. This also highlights the occurrence of outliers (open circle) and extremes (*). This graph is rather complex and confusing for publication purposes but is a useful tool in exploratory data analysis.

Fig. 20.6a. Scatterplot of kernel weight (vertical axis) against fruit length for a sample of *Ricinodendron* fruits.

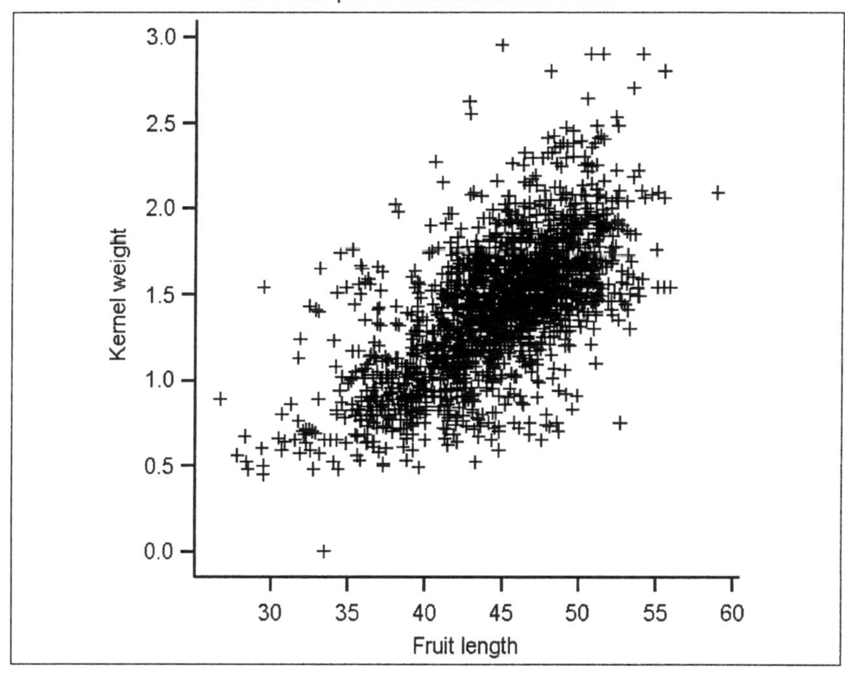

Fig. 20.6b. As in Fig. 20.6a, but with a boxplot drawn for each 2 mm class of fruit lengths.

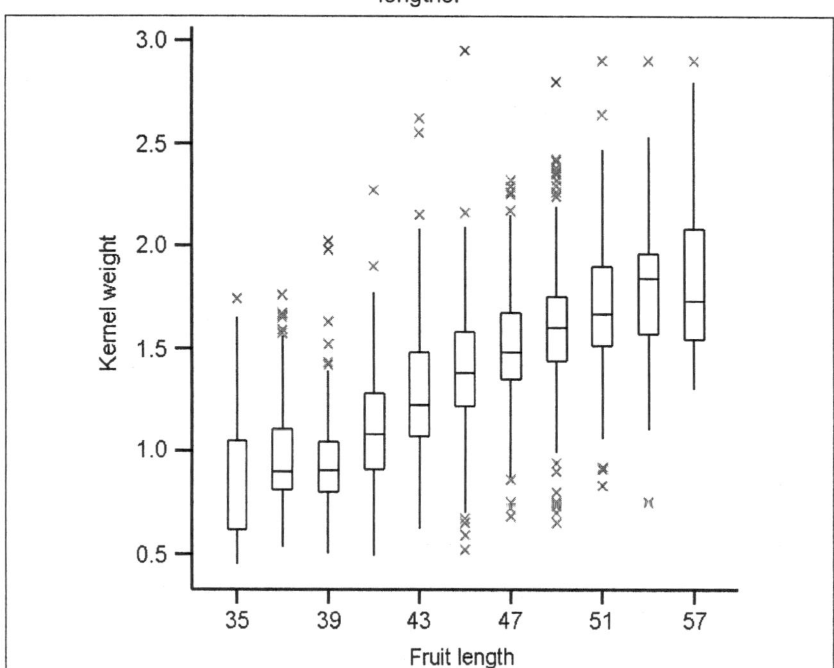

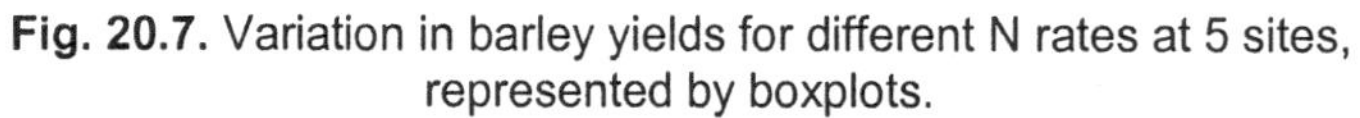

Fig. 20.7. Variation in barley yields for different N rates at 5 sites, represented by boxplots.

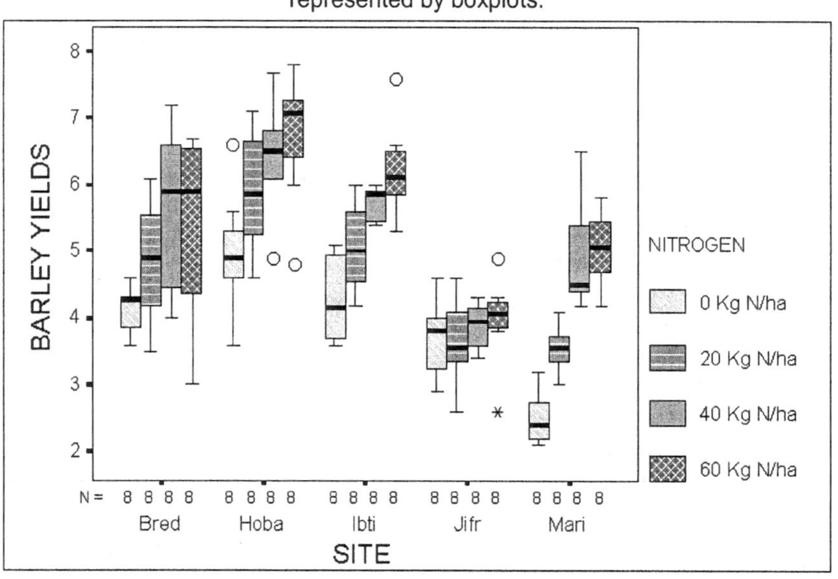

In many resource management studies, the spatial organization of the data is important in interpretation. Locations (e.g. east and north coordinates) and the information derived from them are important structural variables. These can help in graphical analysis, and can be portrayed through mapping the data. The examples below are from a land-use study based on an aerial survey in Kenya. Figure 20.8a shows the extent to which farms grow cash crops (the darker areas indicate where smaller proportions of cash crops are grown, and the lighter areas where higher proportions are grown). This shows that the pattern of cash crop production is clustered. A hypothesis as to the reason for this, is differing travel times to markets, with the expectation that farmers closer to markets commit more resources to cash crop production. Figure 20.8b shows that travel time to market varies, and is spatially clustered, but the clusters do not always coincide with the cash crop patterns. Other factors are also at work.

20.3 Multivariate Methods

Researchers are sometimes concerned that they cannot be exploiting their data fully because they have not used any 'multivariate methods'. This is understandable: we know that resource management problems are complex and require investigation from many angles. Every useful study is 'multivariate' in the sense that more than one variable is measured on each unit. This is true whether the study is a survey or an experiment. However, it does not mean that multivariate statistical methods are necessarily appropriate. In this section, we show some examples of these methods and explain their uses and limitations.

Many of the methods so far described use more than one variable in the analysis. However, they have all assumed for both exploratory and formal analysis, that there is a single response variable. In terms of the generic model

data = *pattern* + *residual*

the data on the left-hand side has consisted of a single variate, even if the pattern has been made up of many, both designed and measured. Multiple response variables are common, for example measuring many components of soil quality, many dimensions of 'wealth', different products from an agroforestry system or the levels of 20 different pests and diseases in a variety trial. Multiple response variables on each study unit also arise when repeated measurements are taken in time or space: measuring each tree in a plot from a field experiment, or measuring water quality every week in a watershed study. However, the existence of this multivariate data does not automatically mean that the statistical methods commonly described as 'multivariate' have to be used.

First, it is perfectly valid to look at a whole series of important response variables in turn: this constitutes a series of univariate analyses. This will tell you of the patterns in these separate responses. If that meets the objectives of the study, then it is sufficient.

Fig. 20.8a. Mapped proportion of crop land under cash crops, from a Kenyan land-use survey (axes in km, measured from a known, arbitrary datum).

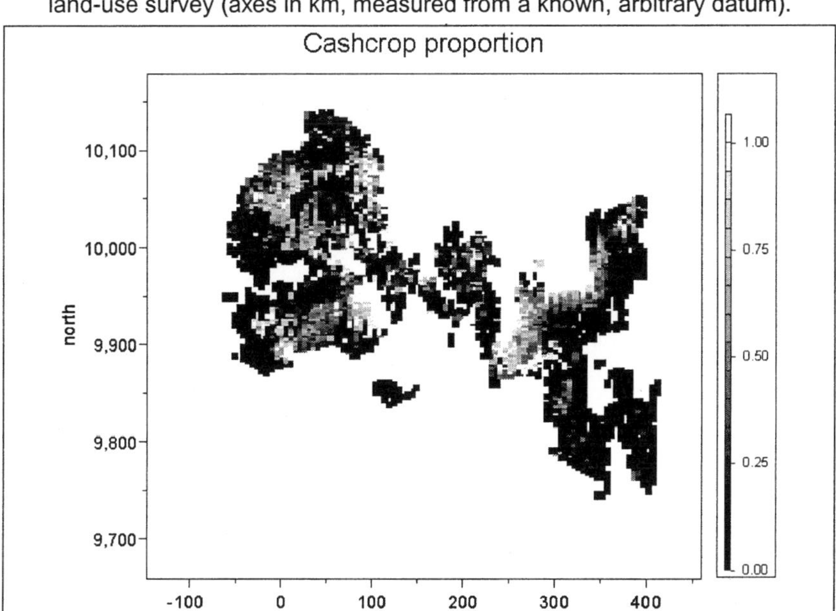

Fig. 20.8b. Mapped travel time to market (hours) for the same scene as in Fig. 20.8a.

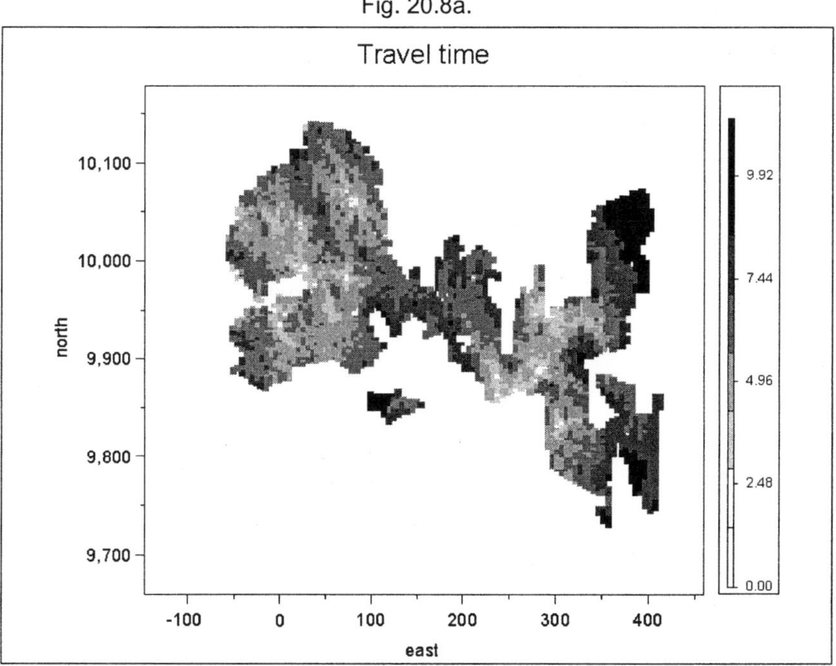

Secondly, it may be sensible to put together your own index or combined response, from the multiple measurements. In some cases this is natural: if each tree in a plot is measured, there is little to be gained from analysing each as a separate variable. If trees have been measured once a month for 5 years, then the average growth rate or final volume, is a sensible constructed response to analyse. The same is true in other situations where we can construct variables such as a total economic output or a 'wealth index'. The method of calculating the former may be more soundly based in theory than the latter, but in both cases the justification for the index is that it reflects a response which meets the study objectives. Indices were discussed in Sections 17.3.8 and 17.3.9.

The analysis sometimes begins with univariate approaches and then concludes with a multivariate method. This is rarely appropriate. Instead, we suggest that the most useful roles of multivariate methods are as exploratory tools to help you to find hidden structure in your data. Two common methods are cluster analysis and principal component analysis. The first aims to find structure among the units, the second among the variables.

20.3.1 Cluster analysis

Consider a baseline survey that has recorded a large number of variables, perhaps 50, on each of 200 farmers in a particular region. Assume most of the variables are categorical and a few are numerical measurements.

Suppose an on-farm trial was used with these farmers to compare a number of weed management strategies (or treatments), giving results that demonstrated a farm by treatment interaction (i.e. not all farmers benefit by the same management strategy). We now want to explore reasons for this interaction. One method may be to investigate whether particular strategies can be recommended for groups of farmers who are similar in terms of their socio-economic characteristics. A first step then is to group the farmers according to their known background variables (this is the part that is multivariate).

Cluster analysis tries to find a natural grouping of the units under consideration, the 200 farmers, into a number of clusters, on the basis of the information available on each of the farmers. The idea is that most people within a group are similar (i.e. they had similar measurements). It would simplify the interpretation and reporting of your data, if you found that the 200 farmers divided neatly into four groups and that farmers within each group had similar responses with respect to the effectiveness of the weed management strategies. The four groups then become the recommendation domains for the weed management strategies.

Cluster analysis is a two-step process. The first step is to find a measure of similarity (or dissimilarity) between the respondents. If the 50 variables are Yes/No answers, then an obvious measure of similarity between 2 respondents is the number of times they gave the same answer. If the variables are of different types, or on different themes, then the construction of a suitable measure needs more care. In such cases, it may be better to do a number of different cluster analyses, each time considering variables that are of the same type, and then seeing whether the different sets of clusters are similar.

Once a measure of similarity has been produced, the second step is to decide how the clusters are to be formed. A simple procedure is to use a hierarchic technique where we start with the individual farms (i.e. clusters of size 1). The closest clusters are then gradually merged until finally there is only a single group. Pictorially this can be represented as a dendrogram, as shown in Fig. 20.9, which relates to the example following. If three clusters are formed, they would be the sets (1), (7) and (2, 3, 4, 5, 6, 8); or five clusters made up of (1), (7), (4, 5, 8), (2,3) and (6), and so on. These sets are obtained by drawing a line at different heights in the dendrogram, and selecting all units below each intersected line.

If you use this type of method, do not spend too long asking for a 'test' to determine which is the 'right' height at which to make the groups. Remember cluster analysis is just descriptive statistics.

As an example, consider the data in Table 20.1 from eight farms, extracted from a preliminary study involving farmers in an on-farm research programme. The variables were recorded as 'Yes' (+) and 'No' (–) on characteristics determined during a visit to each farm. The objective was to investigate whether, on the basis of these characteristics, the farms form a homogeneous group or whether there is evidence that they can be classified into several groups.

For these data a similarity matrix can be calculated by counting the number of +s in common, arguing that the presence of a particular characteristic in two farms shows greater similarity between those two farms than the absence of that characteristic in both. The matrix of similarities between the eight farms appears in Table 20.2 where it can be seen that farms 4 and 8 had 6 answers in common and therefore have greatest similarity. The dendrogram in Fig. 20.9 was produced based on these similarities.

Table 20.1. Farm data showing the presence or absence of a range of farm characteristics.

	Farm (farmer)							
Characteristics	1	2	3	4	5	6	7	8
Upland (+)/Lowland (–)?	–	+	+	+	+	+	–	+
High rainfall?	–	+	+	+	+	–	–	+
High income?	–	+	+	–	–	+	–	–
Large household (>10 members)?	–	+	+	+	–	+	–	+
Access to firewood within 2 km?	+	–	–	+	+	–	+	+
Health facilities within 10 km?	+	–	–	–	–	–	–	–
Female headed?	+	–	–	–	–	–	–	–
Piped water?	–	–	–	–	–	–	+	–
Latrines present on-farm?	+	–	–	–	–	+	–	–
Grows maize?	+	–	–	+	+	–	+	+
Grows pigeonpea?	–	+	+	+	–	+	–	–
Grows beans?	–	–	–	+	+	–	–	+
Grows groundnut?	–	–	–	–	–	–	–	+
Grows sorghum?	+	–	–	–	–	–	–	–
Has livestock?	+	+	–	–	+	–	+	–

Table 20.2. Matrix of similarities between eight farms.

		Farm							
		1	2	3	4	5	6	7	8
	1	-	1	0	2	3	1	3	2
	2		-	5	4	3	4	1	3
	3			-	4	2	4	0	3
Farm	4				-	5	3	2	6
	5					-	1	3	5
	6						-	0	2
	7							-	2
	8								-

Fig. 20.9. Dendrogram formed by the between-farms similarity matrix.

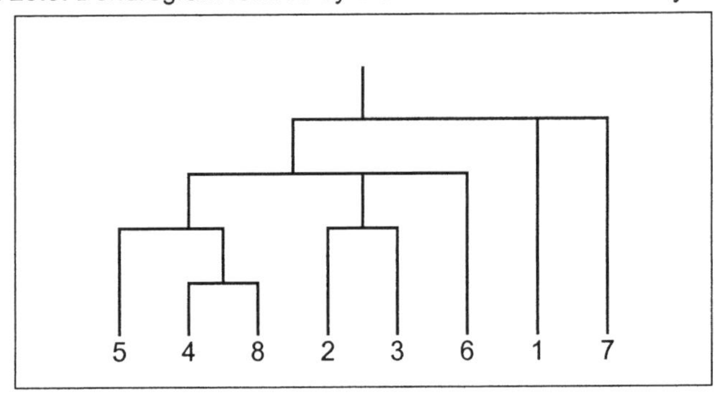

Before leaving this example, we note an alternative strategy for this problem. The cluster analysis approach is 'data driven' in that we look for clusters in the group of 200 farmers and then check whether these 'explain' the farm by treatment interaction. An alternative is a 'theory driven' approach. Based on existing understanding, hypothesize some reasons for the interaction. The data are used to examine these hypotheses and, where necessary, refine the theory. For example, we may hypothesize that reaction to different weeding strategies will be determined by the importance of crop production in the household livelihood. An index of this is constructed from the baseline survey data, using such variables as area cropped, percentage household labour used for cropping and percentage of income from crop production. The relationship between this index and weeding can then be examined. This strategy may well produce results, which are easier to interpret and generalize than the purely data driven approach.

20.3.2 Principal component analysis

If, in the example above, there are inter-correlations between the 50 measurements, then there may be less than 50 pieces of information. The linear combination that explains the maximum amount of variation is called the first 'principal component'. A second component could be found independent of the first, explaining as much as possible of the variability that is left. Suppose that three linear components combined, explain 90% of the variation in your data, then essentially the number of variables that you need to analyse is reduced from 50 down to three!

If you put these ideas concerning cluster analysis and principal component analysis together, then you have reduced your original data of 200 farmers by 50 measurements down to four clusters by three components. A report can now be written, that is much more concise than would have been possible without these methods.

This may seem ideal, but there are catches. The main catch is that few real sets of data would give such a clear set of clusters, or as few as three principal components explaining 90% of the variation in the data. There is also often additional structure in the data that was not used in the analysis.

For instance, the data may have come from a survey in six villages, in which half the respondents in each village were tenant farmers; the remainder owned their land. This information represents potential groups within the data, but which were not included in the cluster analysis. Hence, the cluster analysis will rediscover structure that is known, but was not used in the analysis. It is sometimes constructive to find that the structure of the study is confirmed by the data (e.g. people within the same village are more similar than people in different villages). But that is rarely a key point in the analysis.

This problem affects principal components in a similar way. Much of the variation in the data may be due to the known structure, which has been ignored in the calculation of the correlations that are used in finding the principal components. Another major difficulty with principal component analysis is the need to give a sensible interpretation to each of the components which summarize the data. An illustration is provided below to demonstrate how this may be done, but the answer is not always obvious!

In a fisheries study, 200 respondents were asked to score a number of indicators (on a 1-15 scale), which would show the impact of community-based coastal resource management projects in their area. The indicators were:

1. Overall well-being of *household.*
2. Overall well-being of the fisheries *resources.*
3. Local *income.*
4. *Access* to fisheries resources.
5. *Control* of resources.
6. Ability to *participate* in community affairs.
7. Ability to *influence* community affairs.
8. Community *conflict.*
9. Community *compliance* and resource management.
10. Amount of traditionally *harvested* resource in water.

These ten indicators were subjected to a principal component analysis, to see whether they could be reduced to a smaller number for further analysis. The results are given in Table 20.3 for the first three principal components.

Table 20.3. Results of a principal component analysis.

Variable	Component:	PC1	PC2	PC3
1. Household		0.24	0.11	0.90
2. Resource		0.39	0.63	0.02
3. Income		0.34	0.51	0.55
4. Access		-0.25	0.72	0.17
5. Control		0.57	0.40	0.12
6. Participation		0.77	0.13	0.29
7. Influence		0.75	0.22	0.34
8. Conflict		0.78	0.03	0.18
9. Compliance		0.82	0.12	0.07
10. Harvest		0.38	0.66	0.12
Variance %		33	19	14

Thus, component 1 (PC1) is given by

PC1 = 0.82(Compliance) + 0.78(Conflict) + 0.77(Participation) …
… + 0.24(Household)

By studying the coefficients corresponding to each of the ten indicators, the three components may be interpreted as giving the following new set of composite indicators:

PC1 : an indicator dealing with the community variables;
PC2 : an indicator relating to the fisheries resources;
PC3 : an indicator relating to household well-being.

Fig. 20.10. Biplot of ten land uses, measured on 4759 plots from a land-use survey in Kenya.

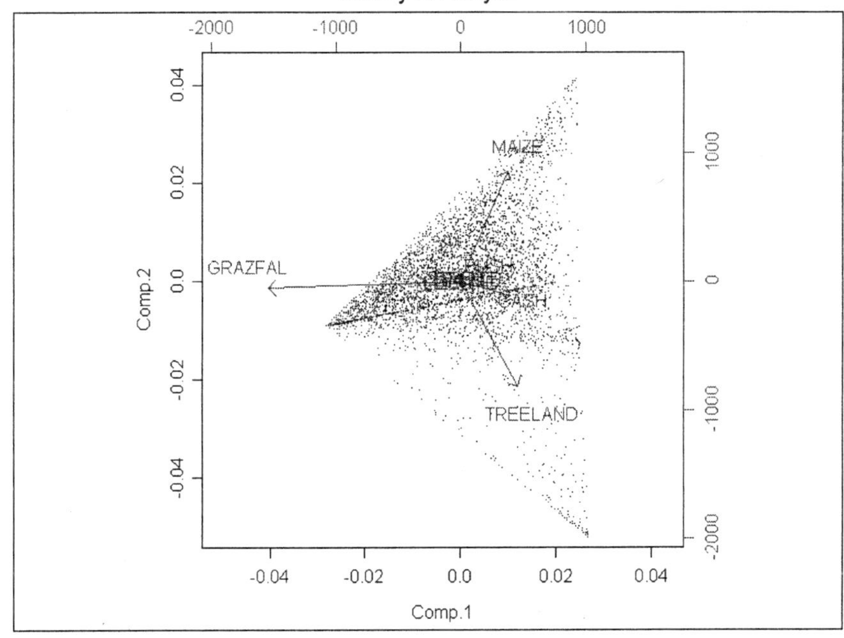

These summary measures may then be further simplified. For example, if we were planning to take the total score in variables 6 to 9 as an indicator of compliance, then principal component analysis might support that this first principal component could be a useful summary for this set of data. Indicators were discussed in Section 17.3.8.

In Section 20.2, we illustrated ways in which data can be displayed graphically for several variables. The methods of multivariate analysis can aid in the construction of useful graphs. For example, for the land-use survey discussed in Section 20.2, the 'biplot' in Fig. 20.10 is a way of representing the results of a principal components analysis, showing both observations (points) and variables (arrows) on the same diagram. The variables used in the analysis are the percentages of land under different uses, and the triangular pattern is an artefact of the variables being percentages. The plot can be used in various ways, revealing locations which are similar, the variables which make them similar, and the variables which are associated. Overall, the graph suggests land use can be considered in the three land use categories of grazing/fallow, trees and maize.

The result is that we can provide three maps of land use (Figs 20.11a, b and c), which capture most of the variation and patterns.

Related methods are often used in community ecology. The multivariate data there consists of a matrix of sites by species, recording the presence or abundance of each species in each site. Often the analysis will have several objectives, concerned with finding relationships between species and between sites. Diagrams such as Fig. 20.10 may help to show these relationships. The analysis can be taken a step further if environmental variables have been recorded at each site. Methods related to regression analysis can be used to explain site patterns, in terms of the environmental variables.

20.3.3 Correlations and structure

The calculation of a correlation matrix is probably the most common multivariate statistical method used in the analysis of both experiments and surveys. The ordinary correlation between two variables measures the extent of linear relationship between them. Identifying relationships between variables is a stated aim of many studies, so it is perhaps natural to look at correlations. When there are a number of response variables, then correlations can be estimated for each pair, resulting in a matrix of correlations. These correlation matrices by themselves, are of limited value for several reasons:

- Correlation coefficients only give a measure of strength of linear relationship. They do not inform about non-linear relationships or other patterns in the data, and can be influenced by outliers. This limitation can be overcome by plotting scatter diagrams to look at the relationship, and only using correlations to summarize patterns that appear linear. Some statistical graphics software will draw a matrix of scatterplots corresponding to the matrix of pairwise correlations.

Fig. 20.11a. Mapped percentage of land under grazing and fallow.

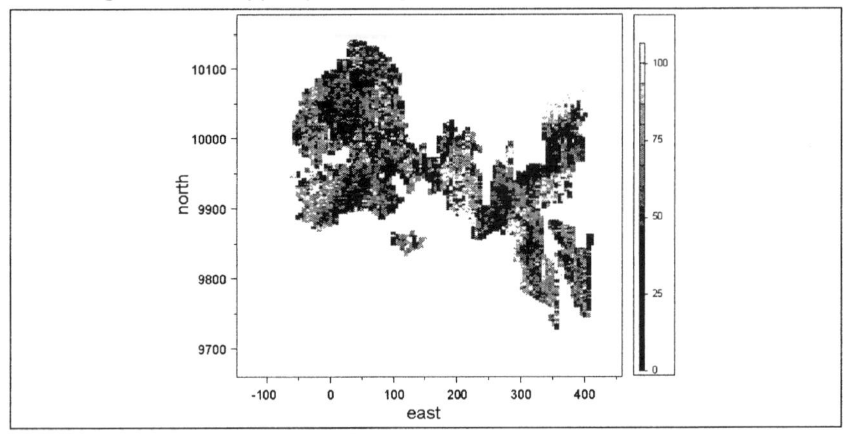

Fig. 20.11b. Mapped percentage of land under trees.

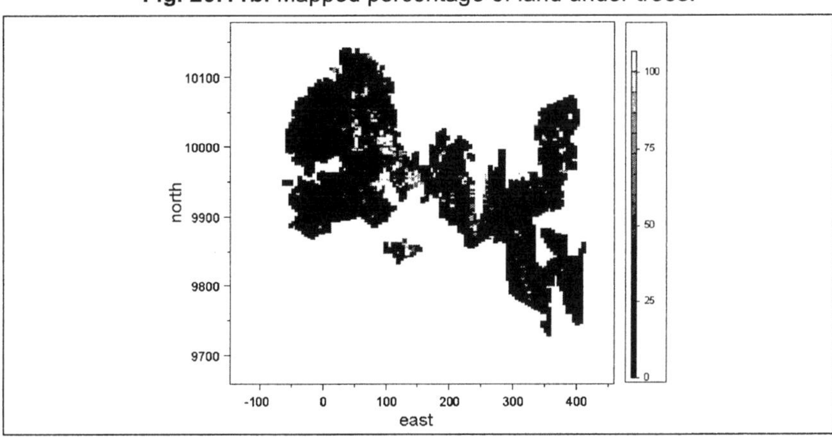

Fig. 20.11c. Mapped percentage of land under maize.

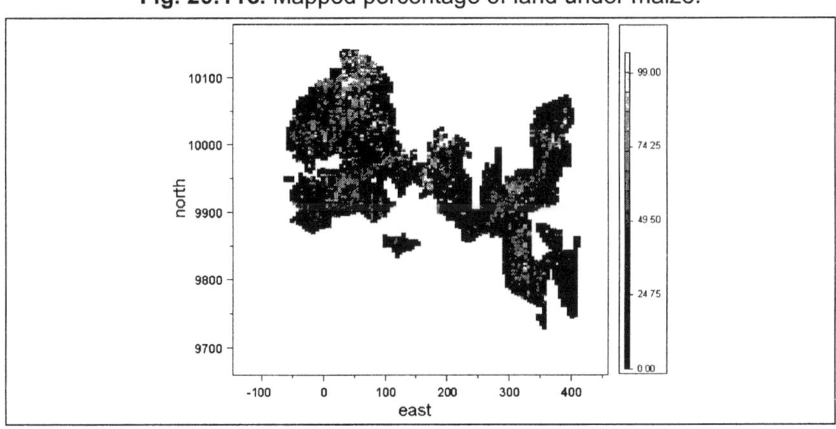

- A correlation matrix of v variables contains $v(v-1)/2$ correlations. This number quickly becomes large as v increases, making it hard to understand the implications of the correlations. The large number of pairwise correlation coefficients may also mislead, as due to sampling variation, one would expect a few large values if there were really no pattern. For example, if $v = 10$, there are 45 correlations and if 'significance' is used to judge which correlations are important, then you are likely to find at least one 'significant' correlation (at $p = 0.05$) among these variables.
- If several important correlations are found, they can be difficult to interpret. If A is correlated with B and B with C then A will be correlated with C as well. The interesting questions are then likely to revolve around the extent to which A and C are correlated, even when B is constant, and so on. These relationships can be disentangled using models of various types, but not through a simple examination of correlation coefficients.
- Correlations are usually calculated ignoring any structure in the dataset. The structure of groupings in the dataset, including that imposed by the study design, may account for the correlations; yet, this will not be apparent from simple correlation coefficients.

Table 20.4. Correlation coefficients from the experiment in Chapter 19.

Source	DF		Correlations			
blocks	7		*y*	*ae*	*an*	*ll*
		ae	0.55			
		an	0.47	0.33		
		ll	0.98	0.57	0.47	
		in	0.60	0.04	0.41	0.53
treat	8					
		ae	0.69			
		an	0.37	0.66		
		ll	0.82	0.85	0.54	
		in	0.88	0.68	0.55	0.87
residual	56					
		ae	0.22			
		an	0.26	0.21		
		ll	0.31	0.16	0.70	
		in	0.33	0.10	0.06	0.10
overall	71					
		ae	0.51			
		an	0.38	0.36		
		ll	0.62	0.47	0.67	
		in	0.66	0.37	0.16	0.35

This last point is illustrated with an example from the experiment analysed in Chapter 19. There were nine treatments in four replicates in eight blocks; four in each of two years. On each plot, a crop yield was measured along with four different measures of available nitrogen. The analyst was interested in the relationships between the four N measures (*ae*, *an*, *ll* and *in*) and yield (y), so the correlation coefficients were calculated. These are shown in Table 20.4

('overall'). The three largest correlation coefficients are between *an* and *ll* (0.67), between *y* and *ll* (0.62) and between *y* and *in* (0.66). Typically, these would now be interpreted.

The data are structured, with eight blocks and nine treatments in each block. Hence, we can identify three sources of variation: blocks, treatments and a residual. It is also possible to find a correlation coefficient for each of these types of variation as shown in Table 20.4. The correlation among the residuals shows that the three larger overall correlations are each quite different. The variables *ll* and *an* have a high correlation among residuals. Hence, plots with high *an* also tend to have a higher *ll,* once all block and treatment effects have been accounted for. This is not true for the correlations of *y* with *ll* and *in*. These large overall correlations of *y* with *ll* and *in* are due to relationships at the block and treatment level of variation, instead. Thus, a treatment with a high *in* mean tends to have a higher mean yield. These different types of correlation could be displayed graphically by drawing scatterplots of block means, treatment means and residuals from an analysis of variance. The central message is to be aware of the differing correlations that can exist due to structure, particularly hierarchical structure. If you are not sure which is the correlation to use, then you should probably not be using any of them!

20.4 General Ideas on Modelling

20.4.1 Statistical model building

There is a common approach to statistical modelling, however complex or novel the nature of the model being used. The analyst should be going through the steps of:

- Choosing a suitable model.
- Fitting the model.
- Assessing the fit.
- Using the model.

The first step is probably the hardest. A suitable model depends on the nature of the observations, the design used to collect them and the objectives: this often being ignored. There is no single suitable model for a dataset, but many; each appropriate for different objectives. Consider a model to help estimate scales of variation in tree growth over a landscape. A suitable model might have variance components describing variation between individuals, niches, and fields, or larger landscape elements. Such a model would show the relative importance of these different levels of variation but would not help understanding of why there is this variation. That is a different objective, for which we might hypothesize that the variation is driven by environment, management, and genetics. A model that incorporates these would help test the hypothesis and estimate the relative importance of each. These are two different models, each valid and useful for analysing the same response variable, tree growth.

A model should also be based on a sound understanding of the science involved. We have described models as:

data = *pattern* + *residual*

It is often in the 'pattern' that we can incorporate some theoretical understanding. For example, in studying decay of organic matter over time, the simplest realistic model is probably not one that says the remaining quantity (y) decreases linearly with time (t):

$$y = a - b.t + residual$$

but one that predicts an exponential rate of decay

$$y = a \exp(-kt) + residual$$

This formula shows what you would expect, if the material decays at a constant rate, with a proportion k disappearing in each time unit. A more elaborate theory postulates that organic matter is composed of fast and slow decaying components. The amount remaining may then be modelled by:

$$y = a \exp(-k_1t) + b.\exp(-k_2t) + residual$$

Investigating these two models would help lend support to that theory. The rate of decay (k), is thought to depend on the lignin (Lig) and nitrogen (N) content of the organic material. Hence, after estimating k for many materials, and measuring their Lig and N content, we can try to look at the dependence. A simplistic approach would be to use linear regression with a model such as:

$$k = c + d.\text{Lig} + e.\text{N} + residual$$

However, there are reasons based on the biology of decomposition for expecting the rate to be better predicted by the ratio of Lig to N. Hence a more useful form of the model may be:

$$k = c + d.\text{Lig/N} + residual$$

This idea of using some theoretical insights to help formulate the model is just as important in social, as in biophysical applications. It is not unusual for the analyst of a social survey to hope to gain insights into some processes by simply piling dozens of 'x-variables' into a multiple regression model, then being disappointed when little emerges that is of interest.

In this example of decay of organic matter, the modelling is envisaged as a two-step process (Fig. 20.12). In the first step (A), the residual material is related to time and the rate constant (k). In the second (B), the rate constant is related to Lig and N content. An alternative approach would be to fit the model in one step. There are some reasons why the single step approach may be useful. For example, uncertainty in the estimates of k from y and t can be correctly carried forward into the study of relationships with Lig and N. However, breaking the modelling down into its component steps has the advantage of simplicity and of allowing investigation of the form of model at each step. It is straightforward to check that the exponential decay curve is suitable for each different material. Once we have the whole set of k parameters estimated they can be suitably plotted with Lig and N, oddities examined, and so on.

Fig. 20.12. Schematic diagram of model relating decomposition of biomass (y) through time (t) to lignin (Lig) and nitrogen (N) contents.

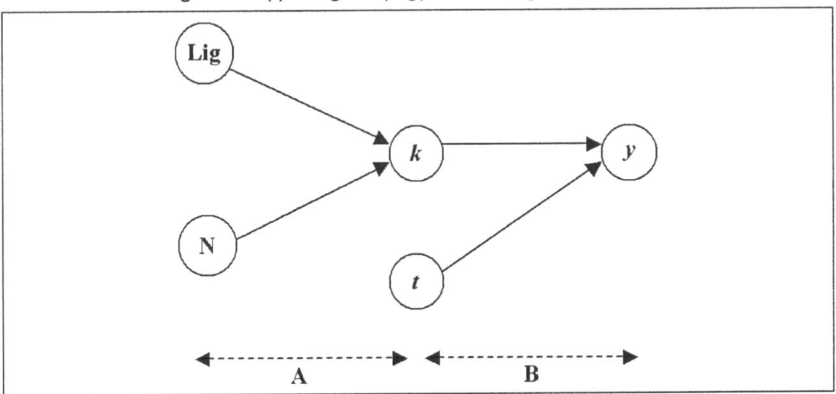

We have referred to the importance of keeping the structure in the analysis. Most data collection strategies, whether from experiments or surveys, result in datasets with a known structure: observations grouped in various ways, and variation both between and within these groups. Ignoring the structure when formulating the model, will give misleading results. An explanation for this, in the case of hierarchical data is given in Section 20.6.

'Fitting the model' means using the available data to estimate the unknown parameters. With the increasing range of models built into good statistical software, fitting models is usually the easy part. Criteria for choosing a 'fit' vary for a number of reasons, as do the actual algorithms used for the calculations. Some are computer intensive, being too complex for hand calculation. However, in most cases the user has little to worry about unless the software produces error or warning messages, in which case specialist help will probably be needed.

The step of 'assessing the fit' has been described for regression models in Chapter 19, and is much the same for models that are more complex. We are looking for signs that 'pattern' has been described in the best way possible, and that the residuals do not contain further important patterns. It is also important to have the properties assumed in the model, and that there are no odd observations unduly affecting the results.

Model assessment may be done using the fitted model, to predict additional observations that were not used in the fitting. If the ultimate purpose of the model is to predict, then this gives a realistic test. Note that predicting the observations used to fit the model will generally overestimate the predictive power of the model. This is to be expected: the fitted model has been determined to be one that describes those observations as well as possible. If the sample size is sufficiently large, some of the sample can be retained to test the model.

However, the structure of the data must be considered, when selecting a sample to retain or 'hold out' from the model. If the data are clustered, whole clusters will usually need to be withheld, otherwise the predictive power of the model will be overestimated. Cross-validation is a form of hold out analysis, in which one observation is withheld; the model is fitted using the remaining data, then the withheld observation is predicted. This process is then repeated for all (or a large sample) of the observations in the dataset. This computer-intensive method again needs to be used with care, especially if the sample is clustered.

The statistical model is to be used and interpreted to meet study objectives. Fitting the model and showing it does a good job is never an end in itself. Often the use will be through incorporating the results into a more comprehensive quantitative description of the problem.

The four steps of: model selection, estimation, assessment, and use, have to be followed in an iterative way. Steps one to three may have to be repeated if the assessment (the third) step, shows the model is not adequate. On occasion, the whole sequence may need repeating. For instance, when a model is used, it may become apparent that it produces 'odd' results for some x-values. This may indicate an inappropriate type of model, or that it is being used outside the range of data that have been collected.

20.4.2 The variety of statistical models

The generic model of *data* = *pattern* + *residual*, has been used in Chapters 16 and 18, and particularly in Chapter 19 for the specific case where:

- '*data*' is a continuous response variate;
- '*pattern*' is a linear combination of explanatory variates and factors;
- '*residual*' comprises independent random components with a constant variance (and with an approximately normal distribution).

Table 20.5. Common modelling options available in good statistical software.

Component	Options	Examples
Data	Continuous	Yields, water available, annual income
	Counts	Insects on a plant, children in a family
	Binary (0/1, yes/no)	Technology adoption, tree survival
	Ordered categories	'excellent' 'ok' 'poor'
Pattern	Linear combinations	Treatment+b.Nitrogen
	Non-linear	a.exp(-kt)
	Non-parametric smooth curves	Splines
Residual	Single component, constant variance	Simple regression, simple ANOVA
	Several components	Mixed models
	Correlated observations	Time and space models

There are other options for each of these components, giving a wide range of model types, yet all can be used in a similar way. Some of the generalizations available in good statistical software are shown in Table 20.5.

20.4.3 Non-parametric methods

Normally distributed measurement is the starting point of most statistical analysis. There are situations however where this seems worryingly inappropriate. Measurements may be from a very skew distribution, where an occasional reading is much larger than the usual range and cannot be explained or discounted. Results may be only quasi-numerical (e.g. an importance score between 1 and 10 allocated to several possible reasons for post-harvest fish losses). Fishermen may assign scores each in their own way, some avoiding extreme scores, while others using them. We may then have reasonable assurance as to the rank order of scores given by each individual, but doubtful about applying processes such as averaging or calculating variances for scores given to each reason.

In such cases, it is sensible to consider using non-parametric methods. A simple example is the paired data shown earlier in Section 16.6.4. The ten differences in sediment load are as follows:

3 2 3 1 3 0 3 1 1 1

Earlier we used the *t*-test, but a simple non-parametric test follows from the fact that nine out of the ten values are positive, with the other being zero. If there were no difference between the locations, we would expect about half to be positive and half negative; so this simple summary of the data provides good evidence ($p = 0.004$, on a formal test) against this hypothesis. Just noting whether the observations are positive, zero or negative is also clearly robust against occasional readings being very large: if the first difference were 30, rather than 3, this would not affect the analysis. Non-parametric methods often provide a simple first step and add easily explained support for the conclusions from a parametric analysis.

We advise caution, however, about the overuse of non-parametric methods. Inadequate understanding of the data generating system by the researcher may be the real reason for messy-looking data. A common reason for apparently extreme values, or 'lumpy' distribution of data, is often that the population sampled has been taken as homogeneous, when it is an aggregate of different strata, within which the observations follow different patterns.

The primary focus of most non-parametric methods is on forms of hypothesis testing, whereas the provision of reasonable estimates usually generates more meaningful and useful results. Non-parametric methods are not 'assumption free'. They may make less stringent assumptions about the data than corresponding parametric methods, but the assumptions made are just as critical and may well not be appropriate in practice. The most important one is independence of the observations. If in the example in Section 16.6.4 there is some serial correlation (successive values tending to be related to each other) the non-parametric test of difference between locations is not valid.

Changes in the scope of readily available statistical models, has also reduced the role of traditional non-parametric methods. The increasing flexibility of methods available (for example, not relying on normal distributions) means that it is easier to find models which are realistic for datasets. In addition, we may be able to model problems for which non-parametric methods do not exist, for example to deal with complex structure or correlations between observations.

The idea of fitting models that do not make strict 'parametric' assumptions about the data, has been behind some potentially useful developments. Regression equations that relate a response to a continuously varying explanatory variable are powerful tools. Yet, it is sometimes hard to find a mathematical function that realistically describes the shape of the response that can be seen in the data. Methods have been developed that allow a smooth curve to be fitted without explicitly giving a formula to the smooth curve. An example is shown in Fig. 20.13.

Fig. 20.13. Grain yield plotted against position along the field, from an experiment with 6 treatments (with trends described by a generalized additive model).

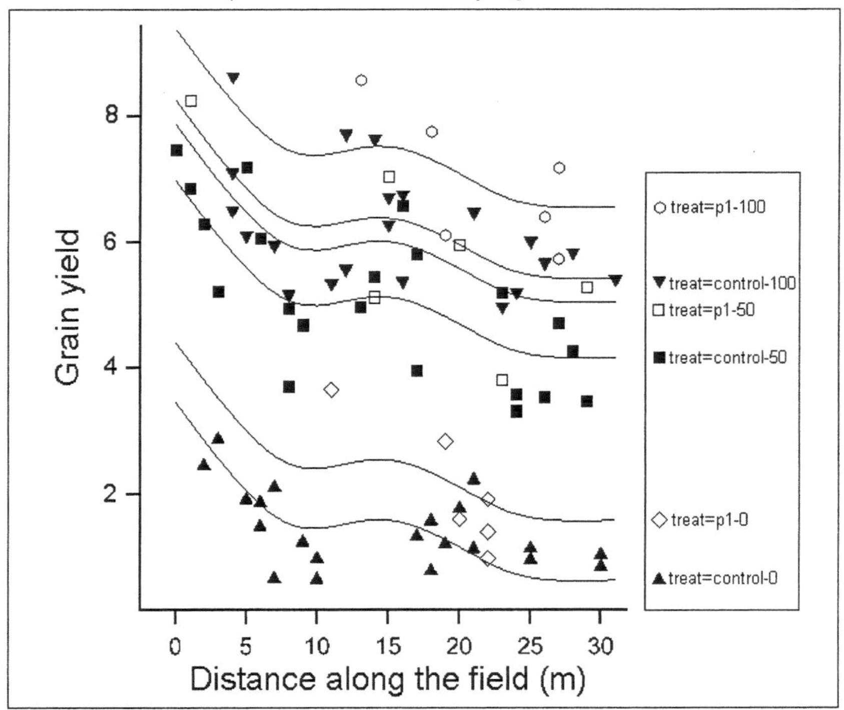

This shows data from an experiment in which there were six treatments and a strong gradient in soil fertility along the field, which had not been accounted for by blocking. Including a term for the trend in the model, gives greatly increased precision of estimates of treatment effects. However, the usual simple models of a straight line (or maybe a quadratic response) do not fit the data. Instead, a smooth 'non-parametric' curve called a 'spline' is fitted.

The 'bootstrap' approach to estimation and inference is another example of modelling that achieves some of the features of non-parametric methods by reducing the stringency of assumptions that need to be made.

20.5 Broadening the Class of Models

The general linear models used in Chapter 19 are powerful and very widely used. Special cases include simple and multiple regression, comparison of regressions, *t*-tests for paired and independent samples, and much of analysis of variance and covariance. However, the methods are still limited in scope. They are only appropriate for some types of response variable (those measured on a continuous scale), and when some strict assumptions about the residual variation (independent, constant variance, approximately normal distribution) are true. In this section, we introduce a more general class of models, called *generalized linear models* (GLM) that are valid for a wider range of response variables, and which relax some of these assumptions. More assumptions will be relaxed in the next section.

The methods are important in data analysis, because they are appropriate for a wide range of problems, and because the basic concepts and skills needed, are much the same as for the linear models, described in Chapter 19. The methods are therefore accessible to non-specialists and have been implemented in many statistical software systems.

The models used so far have been of the form:

data = *pattern* + *residual*

where *data* is a response measured on a continuous scale, *pattern* is a linear combination of explanatory or structure variables, and *residual* is the 'noise' or error, normally distributed with constant mean.

An alternative way to specify the same model is:

1. *data* = a Normal distribution with mean (m) and variance (v);
2. m = *pattern,* a linear combination of explanatory variables;
3. v = a constant.

Each of these can be generalized to give a more widely useful set of models. To illustrate this step, Fig. 20.14 shows a dialogue for one of the analyses described in Section 19.6 (compare Fig. 20.14 with Fig. 19.15). Here the analysis is treated as an example of a generalized linear model.

With these settings (Fig. 20.14), the model and results will be the same as in Chapter 19, because we are using the 'special case' of a normal distribution and an 'identity link'. The extension to generalized linear models allows each of the three components above to be broadened, as follows:

Fig. 20.14. GENSTAT dialogue to specify a generalized linear model.

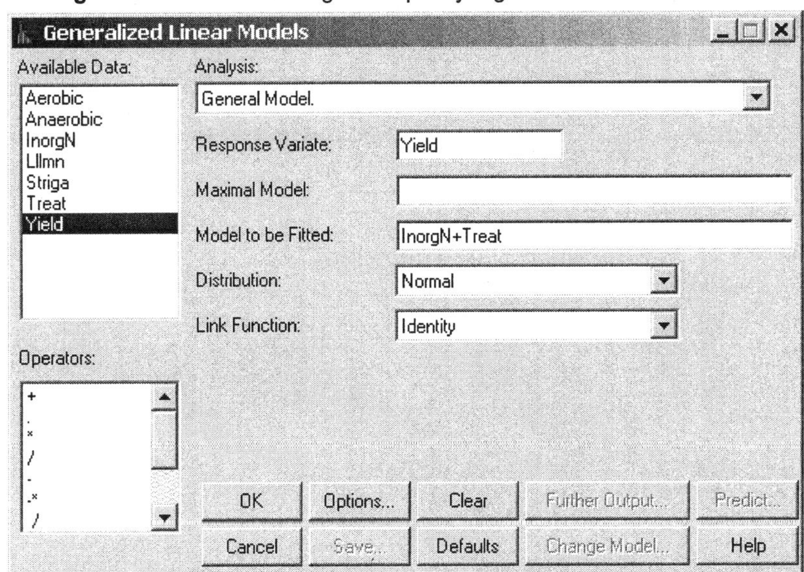

Fig. 20.15. Some of the possible distributions for the data in a GLM.

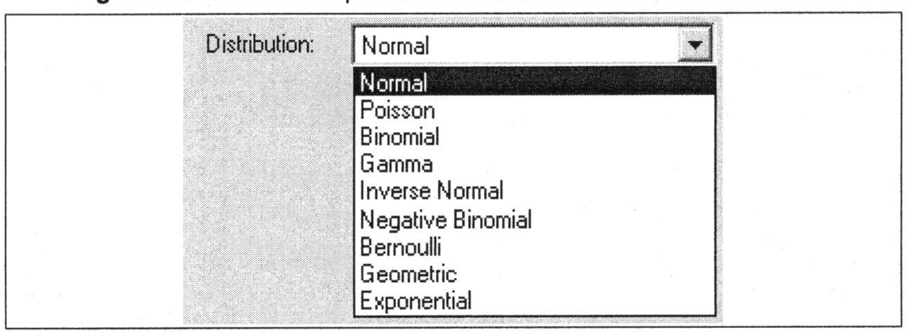

Fig. 20.16. Some of the link functions between the data and the linear model.

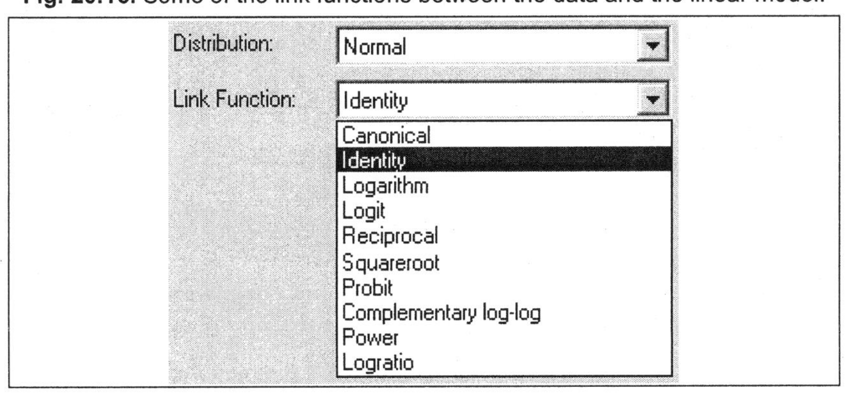

1. The *data* may be from a range of distributions, including the normal, binomial and Poisson (see Fig. 20.15).
2. Some transformation of the mean, rather then the mean itself, is related to the explanatory variables. This is called the link function (see Fig. 20.16).
3. The variance may be constant, or may change systematically with the mean.

We give an example with binary (Yes/No) data. The ideas extend to other examples of non-normal data such as counts, where a Poisson distribution may be assumed.

Example: Logistic model for binary data

Two special cases of generalized linear models are particularly common. The first is to model frequencies in a contingency table. It is called log-linear modelling and generalizes the chi-square test, which is limited to two-way tables.

The second, called logistic modelling, is illustrated here. It is used to model a type of data that is common to many surveys and experiments. On each unit, a simple one/zero or yes/no response is measured. This type of data occurs whenever the result is one of two options, for example adopt/reject, alive/dead, like/do not like, degraded/intact. There is a special name, Bernoulli, for the simple distribution we use here. We assume that observations are independent of each other and each takes the value 1 with probability p, and the value 0 otherwise. So, the Bernoulli is the special case of the binomial distribution, where each trial is considered separately. We need to take each unit separately, because our objective is to model how the probability of success is affected by the pattern in our data. For example, does having a bank nearby make the probability of using a new technology more likely.

The data in our example are from a survey of farmers who have had the opportunity to test a new technology, namely improved fallow, on their land. One of many responses was whether they had planted an improved fallow in the 1997 season. Of the 1480 respondents, 23% tried fallowing that season. It is expected that the use of fallowing may be related to farm size, household wealth (one of three groups based on assets) and to ethnic group (Luo or other). Models that include these variables are defined, fitted and assessed. Looking first at farm size, we would like to assess the extent to which the probability of improved fallow is related to the farm size. The simplest response we can imagine is a straight line. However it does not usually make sense to model the probability directly, and we usually model what is called the 'logit of p'. The model is thus:

$$\text{logit(p)} = a + b.\text{farmsize}$$

where p is the proportion responding with 1 (tried improved fallow). The logit is defined by logit(p)=log(p/1-p), so while p goes from zero to 1, logit(p) goes from minus infinity to plus infinity, just like a straight line.

We use the same dialogue as was shown earlier in Fig. 20.14, but specify that our data have a Bernoulli distribution. The software then suggests the logit link as the default (see Fig. 20.17). Thus fitting the model is no harder than for the general linear models, described in Chapter 19.

Fig. 20.17. GENSTAT dialogue specifying a logistic model.

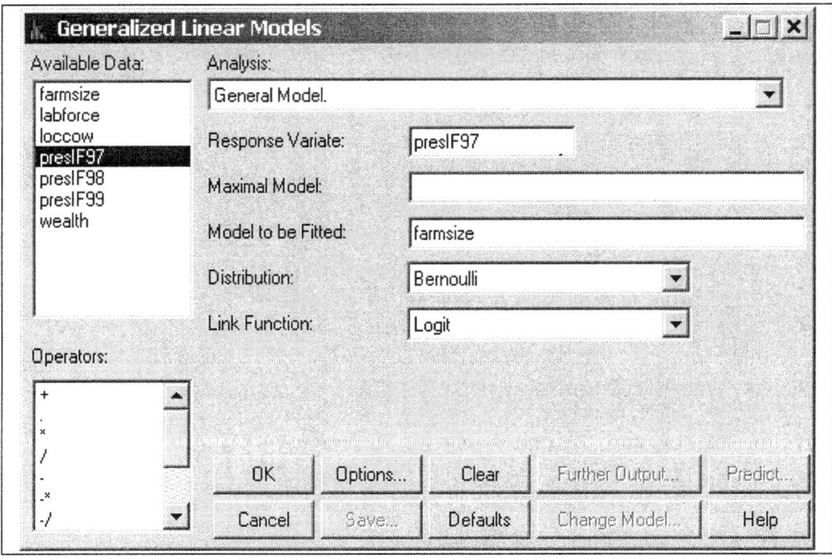

Fig. 20.18. GENSTAT output for the example.

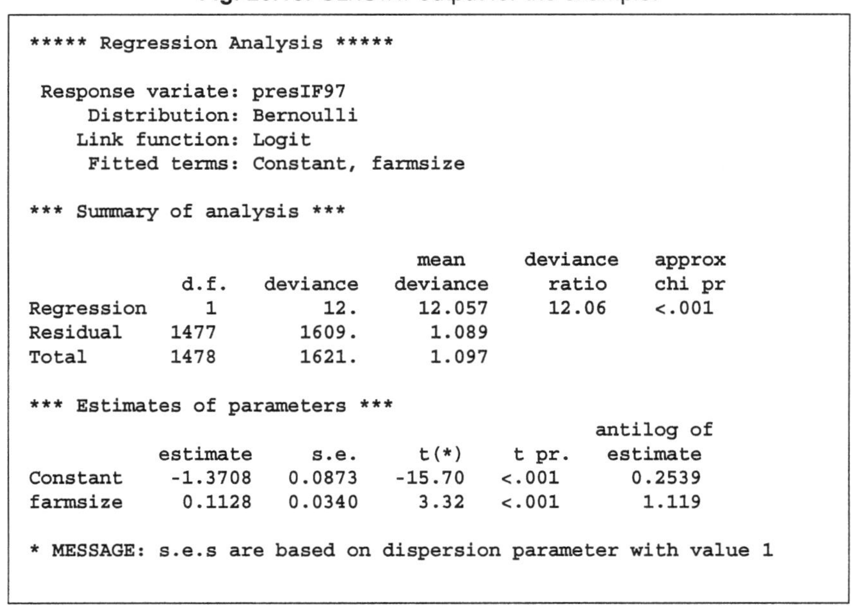

```
***** Regression Analysis *****

 Response variate: presIF97
     Distribution: Bernoulli
    Link function: Logit
     Fitted terms: Constant, farmsize

*** Summary of analysis ***

                                  mean     deviance   approx
             d.f.   deviance   deviance      ratio    chi pr
Regression     1         12.     12.057      12.06    <.001
Residual    1477       1609.      1.089
Total       1478       1621.      1.097

*** Estimates of parameters ***
                                                   antilog of
           estimate      s.e.       t(*)    t pr.   estimate
Constant    -1.3708    0.0873    -15.70    <.001     0.2539
farmsize     0.1128    0.0340      3.32    <.001      1.119

* MESSAGE: s.e.s are based on dispersion parameter with value 1
```

The output associated with the fitted model includes an 'analysis of deviance' table (Fig. 20.18). This corresponds to the analysis of variance

(ANOVA) table for normally distributed data. The residual deviance is analogous to the residual sum of squares in an ANOVA. Chi-square tests replace the usual *F*-tests in the ANOVA, to assess whether the farm size affects the probability of trying the technology. In some specific cases, chi-square tests are not appropriate; we would then use approximate *F*-tests. The analysis of deviance suggests that it does.

The estimates of *a* and *b* are also given in Fig. 20.18. The fitted relationship is more clearly understood by plotting it, as in Fig. 20.19.

Fig. 20.19. Fitted proportion of uptake in 1997 plotted against farm size.

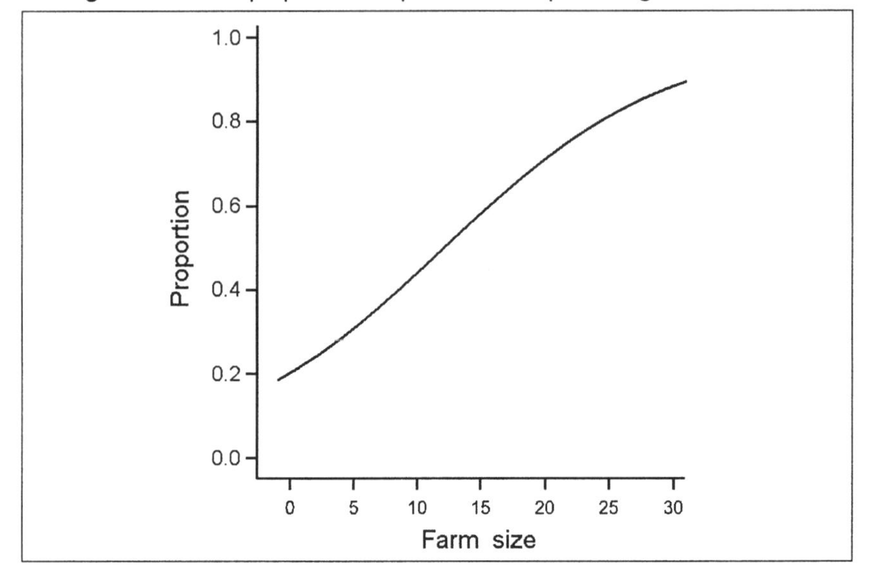

The relationship has a positive estimate of the slope (parameter *b*), indicating that farmers with larger farms are more likely to try the new fallow technology. However the change in proportion over the range 1 to 5 ha, which covers most of the data, is slight, with the proportion only increasing from around 0.2 to 0.3.

The modelling can continue by adding the effects of other variables hypothesized to be important, namely wealth group and ethnicity. The analysis of deviance in Table 20.6 summarizes the results, indicating that there are differences between the wealth groups, but not between ethnic groups.

We can assess the differences between wealth groups by calculating predictions (at an average farm size), as given in Table 20.7. This shows a lower rate of adoption among the poorest group of farmers, but again the difference is small.

The modelling process could be taken still further. We could now omit ethnic group from the model, and add a term to check whether there might be an interaction between the farm size and the wealth group. The resulting analysis of deviance table (not shown) suggests that there is.

Prior to this type of modelling being available, users might have resorted to grouping all the variables and then doing some chi-square tests on the resulting tables of frequencies. The approach above is far better for several reasons, one of which is being able to look at the effect of several explanatory variables simultaneously.

This type of modelling is a powerful way of understanding the multiple effects of several variables that are all related to the 'chance of success'. But don't expect miracles! All the general points about modelling made earlier still apply, so it will not be possible to extract information which is not in the dataset. Note that each observation on a 0/1 or yes/no response contains less information than an observation on a continuous scale. Small sample sizes will therefore not give any clear or useful results.

The logistic regression models illustrated in this example can be extended to other situations. For example, if there are more than two categories of response, such as 'single, married, widowed' or ' bought, inherited, leased', then we can model the proportions responding in each category. It is common for the responses to be orders, such as 'poor, acceptable, very good, excellent'. These 'ordered categorical' responses should be analysed using a method that recognizes the order, and modifications of the logistic regression model that carries this out, are available.

Table 20.6. Analysis of deviance table for the farm study.

Change	DF	Deviance	Mean deviance	Deviance ratio	Approx chi prob.
+ Farm size	1	12.057	12.057	12.06	<0.001
+ Wealth group	2	8.579	4.289	4.29	0.014
+ Luo	1	0.637	0.637	0.64	0.425
Residual	1474	1599.634	1.085		
Total	1478	1620.907	1.097		

Table 20.7. Predicted values for the farm study.

Wealth group	Prediction
Low	0.19
Mid	0.26
Higher	0.25

Standard errors of differences			
Low	*		
Mid	0.027	*	
Higher	0.027	0.028	*
	Low	Mid	Higher

20.6 Mixed Models and More

In Section 20.5, we introduced a set of tools called 'generalized linear models', which although available for a number of years, have only recently become a regular component of researchers' toolboxes, as they are incorporated into standard software. They were characterized as being relevant to many common analysis problems, and provide a common framework that integrates many disparate methods and approaches. In this section, we introduce another tool with similar characteristics. 'Linear mixed models' integrate a number of previously separate models and are increasingly available in statistical software. Most importantly, they are relevant to a number of common problems that arise in analysis of NRM research data. Three topics are described here.

20.6.1 Dealing with hierarchical data

Many studies involve information at multiple levels. A typical example is a survey, or on-farm trial where some information is collected about villages (agro-ecological conditions), some from households in these villages (family size, male or female headed, level of education of household head) and some on farmers' fields (soil measurements, tillage practice, yields). Here there are three levels, namely village, household, and field. The levels are nested or hierarchical, since the fields are within the households and the households are within the villages. We have described earlier (Sections 16.8 and 18.4) why it is important to analyse data at the right level. The main reason is that our inferences depend on the estimated uncertainty, and that depends on the variability at the appropriate level in the hierarchy.

There are two approaches in the 'DIY toolbox' of methods, which, with limitations, can be used in the analysis of hierarchical data.

- Analysis of variance for split plot and similar designs, handles simple multi-level data. However, the formulae and calculation methods only work for 'regular' designs. This approach has often been used for the analysis of on-station trials, but is rarely appropriate for other types of study design.
- The other approach is to 'summarize everything to the right level'. Thus we might do a household level analysis on data defined at the household level, and a separate field level analysis on the variables measured at this level. This approach works well when data can be unambiguously identified and summarized to a single level, but that is not always the case. For example, we might find that most households have a single level of 'quality of weeding' applied to all their fields. If we want to compare different levels of weeding then we are at the household level. However, there may be some households that have varying levels of weeding, on different fields. How do we classify such a household to a weeding level? It may be possible to ignore those few households. Or, we could omit some fields from those households, or do separate analyses for the two groups. However, these strategies may become unfeasible once we start including other variables in the analysis, as we may end up omitting most of the data.

To illustrate mixed modelling, and the way it dovetails with simple split plot analysis, we use data from a study in Kenya that compared three soil fertility treatments on 28 farms. The farms fell into two 'weeding quality' groups. Hence, the treatment factor is at the plot level, while the weeding factor is at the farm level. The standard split-plot analysis of variance is shown in Table 20.8.

Table 20.8. Standard split-plot ANOVA for the soil fertility study.

Source of variation	DF	SS	MS	VR	F prob.
farm stratum					
weeding	1	6.8701	6.8701	0.90	0.350
residual	26	197.5501	7.5981	9.08	
farm.plot stratum					
treat	2	41.9281	20.9641	25.06	<.001
treat.weeding	2	1.9706	0.9853	1.18	0.316
residual	52	43.4946	0.8364		
Total	83	291.8135			

The ANOVA table shows that the simple *data* = *pattern* + *residual* is no longer sufficient. We now have two levels in the study, and therefore we have two residuals. We now need to write something like:

data = *(farm level pattern)* + *(farm level residual)*
+ *(plot level pattern)* + *(plot level residual)*

The key difference is that we now have a 'residual' term at each level. Our analysis is always based on the idea that the residuals are random effects, so we now have more than one random effect. Hence, our model is a mixture of 'patterns' and 'residuals'. This is called a 'mixed model'.

Table 20.9. REML analysis for the soil fertility study.

Estimated variance components				
Random term	Component	s.e.		
farm	2.2539	0.7046		
farm.plot	0.836	0.1640		
Wald tests for fixed effects				
Fixed term	Wald statistic	DF	Wald/DF	Chi-sq prob.
weeding	0.90	1	0.90	0.342
treat	50.13	2	25.06	<0.001
weeding.treat	2.36	2	1.18	0.308

Mixed models can be fitted using a method known as REML, and some of the standard output is shown in Table 20.9. There are some clear similarities with the results from the ANOVA table, as well as new information.

The power of the method becomes apparent if we now include a third factor in the analysis, namely application of manure. This varies both between and within farms, and any attempt to use usual analysis of variance methods will fail. The mixed model approach in Table 20.10 produces correct, and potentially useful results.

Table 20.10. REML analysis with the additional factor of manure.

Estimated variance components				
Random term	Component	s.e.		
farm	1.6999	0.5445		
farm.plot	0.638	0.1334		
Wald tests for fixed effects				
Fixed term	Wald statistic	DF	Wald/DF	Chi-sq prob.
weeding	1.20	1	1.20	0.274
treat	65.74	2	32.87	<0.001
manure	9.61	1	9.61	0.002
weeding.treat	0.24	2	0.12	0.888
weeding.manure	8.71	1	8.71	0.003
treat.manure	10.68	2	5.34	0.005
weeding.treat.manure	4.48	2	2.24	0.106

The approach is very useful for analysing the irregular, multi-level designs common in NRM research. The models are sometimes called Linear Mixed Models, in contrast to the Linear Models described in Chapter 19. Some statistical software packages also have a further extension to Generalized Linear Mixed Models, which extends to hierarchical data the ideas of Generalized Linear Models (described in Section 20.5).

20.6.2 Pattern as random effects

In Section 20.6.1, we showed how the model, *data* = *pattern* + *residual* might have the residual made up of several random components, if it is to realistically represent data from a hierarchical design. There are also situations in which it is also helpful to think of parts of the pattern as random. This has long been done in work on tree and crop varieties. Experiments are carried out with varieties or families as treatments, where one objective is to estimate the variance between varieties, rather than a mean for each variety. This information is needed for example to characterize the variation that exists in landraces, as well as to estimate various genetic parameters such as heritability. An example of the results is shown in Table 20.11.

Table 20.11. Treating the family as a random effect.

Estimated variance components Random term	Component	s.e.
rep	4.54	1.94
rep.block	1.54	1.18
rep.block.plot	0.00	BOUND
rep.block.plot.tree	86.54	3.20
family	9.43	3.43

The study involved 20 families of *Leucaena trichandra* planted in a trial that had 20 replicates, each divided into small blocks of four plots; each plot consisting of four plots of four trees. The layout is therefore hierarchical, and the results include variance estimates for each layer. There is some variation between replicates, little or none between blocks or plots, and a lot of variation between individual trees. There is also variation between the families. Though small, this can be detected as there are 80 trees of each family.

If we are also interested in the effect of each family, to find the 'best', or to compare those with other particular characteristics, we have some choices. Traditionally we would have simply found the means for each family. However, when families are considered as random there are other estimates that make sense. Think of the problem of estimating just how good the best family is. The one with the highest mean is likely to be a 'good' family, but it may also have a high mean partly because by chance, it was assessed on some good plots. Thus, we might want to reduce (or 'shrink') its means a little, to allow for this. These estimates are sometimes called 'best linear unbiased predictors' or 'blups' and are often more realistic values to use in further work than the simple means. In our example shown in Table 20.12, the blups and the simple means are similar because of the large number of trees involved.

Table 20.12. Comparison of blups and simple means.

Family	Blup	Mean
1	27.65	28.04
2	26.65	27.10
3	24.57	24.59
4	25.36	25.74
5	24.88	24.91
6	29.52	30.30
7	27.83	28.30
8	20.48	19.92
9	22.68	22.78
10	27.64	28.00
11	19.42	18.93
12	23.61	23.45
13	21.97	21.66
14	22.24	22.19
15	24.19	24.21
16	23.38	23.19
17	21.06	20.84
18	19.43	18.74
19	23.02	22.87
20	26.20	26.25

This idea of random effects in the 'pattern' part of the model is important in many contexts. Another example is a study that looked at the effectiveness of community-based organizations (CBOs) in mobilizing household involvement in resource management. A hierarchical survey sampled a number of districts, villages, and households. Households were involved with one or more CBOs. The total number of CBOs was around 25, and the analysis looked at the overall variation due to the CBOs. The study also estimated the effects of each CBO to understand factors that might be contributing to those that were more effective.

The distinction between 'pattern' and 'residual' terms in a model becomes blurred once we start using random terms in the pattern. Sometimes the hierarchy of layers in a study design are described with random 'residual' effects, but then estimates are still produced for particular locations in a layer. For example, a survey measured infiltration in a large number of fields, with three sample locations in each field. A mixed model, with random terms for the layout hierarchy, and 'pattern' terms for factors such as land use and soil type, fitted the data well. The model was then used to produce estimates for each field that would be compared with remotely sensed estimates. In this case individual fields are of interest, and the blups for the fields are the appropriate estimates to use.

20.6.3 Other correlation and variance patterns

If we use a model with random terms to describe a hierarchical design, it is equivalent to assuming certain correlations between observations. For example, if we have plot and farm level residual terms in the model, it implies that all plots in the same farm are (equally) correlated with each other. Thus, a high response on one plot in a farm will tend to mean there will be high responses on other plots in the same farm. Plots on different farms will be independent. Adding a third layer of village will induce a more complex correlation structure, with correlations between plots in the same village as well.

However, there are many other forms of correlation between observations that we may need to allow for in an analysis. The two most common are correlations in space, and correlations in time. In both these situations, observations that are close together may be more highly correlated than those that are further apart. The general ideas of REML modelling for mixed models can also be applied to spatial data and to repeated measurement data.

20.7 Back to Reality

20.7.1 Design still matters

Many experiments are designed using ideas and theory, which were developed when analyses had to be done by hand, and ease of calculation was an important consideration. To some extent, this is true also for survey designs. In the design part of this book, we have emphasized that the restrictions imposed by ease of analysis are no longer relevant. In the analysis part, we have shown that designs that are 'irregular' in various ways, including those with complex hierarchical structures, can be analysed using much the same models as more traditional designs, and with only slightly more computational difficulty. However, design does still matter! If you assume that study design can be largely ignored, and 'everything taken care of at analysis' you will end up with ineffective research. The first two important areas of design are:

- If the analysis is to identify the relationship between a response and an explanatory variable, the explanatory variable must vary! Moreover, the explanatory variable has to vary over a range that at least spans that of interest to the problem.
- If the analysis is to identify the relationship between a response and two or more explanatory variables, it has to be possible to identify their separate effects. This means they must not be confounded or highly correlated with each other.

For example, if an objective is to investigate how ability to participate in community resource management projects (y) depends on wealth (x_1) and education (x_2), we have to collect a sample that contains sufficient spread in those two variables, and in which they are not too highly correlated. No amount of sophistication in statistical analysis will meet the objectives if that has not been achieved. For the case of two variables, it is helpful to visualize the problem as one of estimating a response surface (Fig. 20.20). The sample points lie in the horizontal plane defined by the x_1 and x_2 axes, and the grey response surface is what we are trying to estimate. Casual selection, or simple random sampling of design points, may well end up with the sample represented by the black circles. These do not vary much in either x_1 or x_2. The ranges can be extended by adding the points represented by the open circles. However, the x_1 and x_2 values are still highly correlated and do not allow estimation of the whole response surface. To do that requires the points represented by the crosses. These ideas apply whatever the nature of the response variable and the model being fitted.

As the number of explanatory variables increases, the problem of ensuring a good design that will allow estimation of appropriate models also increases. In Fig. 20.20, just three sample points would allow estimation of a flat, planar response surface. Adding another would allow some simple curvature to appear. A few more (well selected) points allow a more complex curved surface to be estimated. Thus, there is connection between the complexity of the response surface and the design. In large observational studies, many potential x-variables are often identified, but the sample design might not allow estimation of any model other than the flat planar surface. If this does not capture the key features of the problem, then adjustments must be made at the design stage.

In Section 20.6, we described the use of mixed models for analysis of hierarchical data. The design points above apply to every 'layer' in a hierarchy. Furthermore, at each level in a hierarchy we need to estimate not just the equivalent of the response surface in Fig. 20.20, but also an estimate of residual variance, so that the precision of the response can be estimated. The requirements for estimating these variance components are much the same as for simpler models. The quality of the estimated variance depends on its degrees of freedom, and around ten degrees of freedom is a minimum for a reasonable estimate. Suppose the problem of participation in community resource management, described above, has an addition factor to investigate, namely the effect of having an extension agent in the village. This is a 'village level' effect.

Assessing it will require not only sufficient villages with and without extension agents, but at least ten residual village degrees of freedom, irrespective of the number of households or individuals studied within each village. Again, failure to recognize this requirement at the design stage may result in a study from which convincing conclusions cannot be drawn.

Fig. 20.20. Response surface: y (vertical axis) plotted against x_1 and x_2.

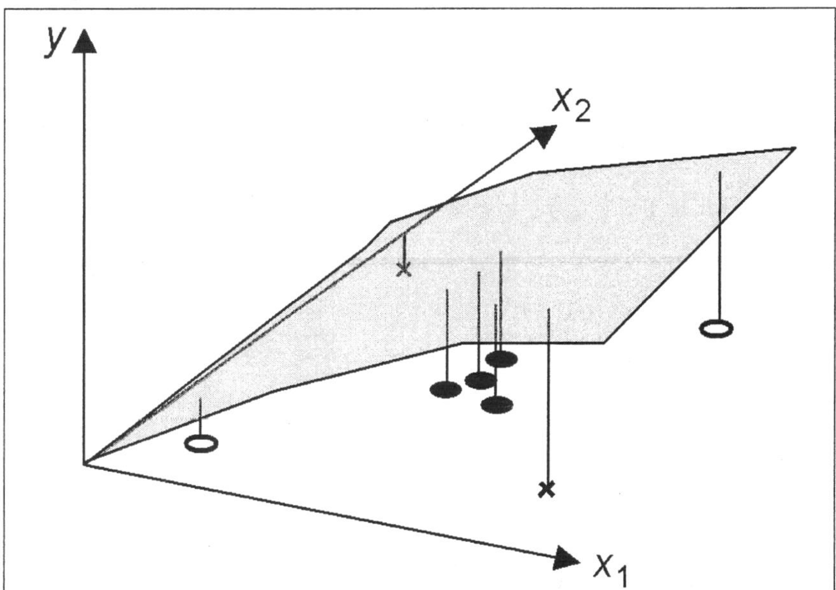

20.7.2 Learning methods, learning software

Before statistics packages became widely available, statistics courses and textbooks were mainly concerned with the teaching of statistical methods. It was vital to understand the method of analysis, if you had to go through all the calculations manually to get the results. One difficulty was that courses often emphasized the details of these methods and omitted much of the context. Users knew how to apply a method, but not when to apply it, nor how to interpret the results or check whether the method had satisfied the objectives of the research.

The availability of statistical software has polarized some training courses and textbooks. There are those that stick grimly to their original topics. If a computer is incorporated, it is merely to assist in what was taught before.

Other courses have changed completely. The course becomes largely one of teaching the software, rather than the methods. They describe how the statistics package chosen for their course, enables the use of all the methods mentioned in this chapter (and many more). This was common when packages were command-driven and learning to use a statistics package meant learning to use the commands (i.e. learning the language).

Now most statistical software is very simple to learn. Training can once again concentrate on teaching statistical methods, with the software being used as an aid to understanding and to applying those methods to datasets of realistic sizes. Understanding the analysis is the key target, and the software is just a tool. We find that even on in-service courses of just a few days we need to devote perhaps only one session to learning the software, and can then return to teaching statistical ideas for the rest of the course.

20.7.3 Keep it simple

The encouragement to 'Keep It Simple', is used in many contexts and is equally relevant to statistical analyses. This chapter has introduced a variety of powerful, modern methods, but this does not mean they should always be used. The rule is to use the simplest appropriate method that meets the objectives. Complexity in statistical analysis is only useful when meeting an objective more successfully, enabling results to be shown more clearly, or by addressing an objective that could not be met by a simple method.

We have called the methods in this chapter the specialist toolbox because some researchers will prefer to use these methods jointly with a specialist. Nevertheless, the ideas and results must still be presented in ways that are clear to non-specialists. Where a specialist states that simple explanations are not possible, we usually find that the analysis does not correspond to an objective of the study. In that case, all that is really needed is a simple method.

We give two examples of common situations in analysis. The first example involves repeated measurements of units; returning to the same unit (person, tree, animal, village) repeatedly, and taking the same measurement. This measurement may be the circumference of each tree, or become an index of poverty. Section 18.5 is titled 'Repeated measures made easy' and outlines elementary ways of processing such data. There are also more complex methods of analysing such data, based on modelling the variation over time; these are relatively easy to implement in some statistics packages. They can be extremely valuable, but they do not always 'pay their way'. Unlike the simple methods, they are sometimes not tailored to the precise objectives of the study and can hide the main messages in the data.

The second example is 'combined analysis' or 'meta-analysis', which is used for analysis of a study that has been repeated in different years or locations. The design may either remain the same or be different in each instance. The first stage in the analysis is usually to process each study individually. The researchers sometimes feel they have to do something more innovative and state a need to do a combined analysis. If a formal combined analysis is undertaken, then the materials in Section 20.6 would often be used. However, it is sometimes just as effective to look simply at the individual analyses and then proceed to make overall conclusions in an informal way. Where a little formality is needed, the key effects of interest for each study can be calculated and the patterns can be observed in the variation of these effects across the different studies.

Chapter 21

Informative Presentation of Tables, Graphs and Statistics

21.1 Introduction

In this chapter, we describe ways in which numerical information can be effectively transmitted in project reports and scientific papers. As an authoritative source of guidance for the presentation of general statistical information, we recommend *Plain Figures* by Chapman and Wykes (2nd edition, 1996). *Plain Figures* is intended particularly for authors addressing non-specialist audiences. This chapter sets out to reinforce and complement it, focusing on points most relevant to those presenting structured research results to an informed audience. We give guidelines for the layout of graphs, charts and tables, as well as for the presentation of results of statistical analyses.

21.2 Nine Basic Points

1. Data can be presented within the text, in a table, or pictorially as a chart, diagram, or graph. Any of these may be appropriate to give information for the reader or viewer, to quickly grasp the content, whilst either reading or listening. This objective, described as 'demonstration', is our main concern in this chapter. For reference purposes, tables are usually the only viable option. Large reference tables are usually put in an appendix, with a summary in the text for demonstration purposes. Chapman and Wykes (1996) provide more detail and numerous examples of reference tables.
2. Text alone should not be used to convey more than three or four numbers. Sets of numerical results can best be presented as tables or figures rather than

included in the text. Well-presented tables and graphs can concisely summarize information that would be difficult to describe in words alone. However, poorly presented tables and graphs can be confusing or irrelevant.

3. When whole numbers (integers) are given in text, numbers less than or equal to nine should be written as words, numbers from 10 upwards should be written in digits. When decimal numbers are quoted, the number of significant digits should be consistent with the accuracy justified by the size of the sample and the variability of the numbers in it.

4. In general, tables are better than graphs for giving structured numeric information, whereas graphs are better for indicating trends and making broad comparisons, or showing relationships.

5. Tables and graphs should be self-explanatory. The reader should be able to understand them without detailed reference to the text, on the grounds that users may wish to gather the information from the tables or graphs without reading the whole text. The title should be informative, and rows and columns of tables, or axes of graphs, should be clearly labelled. The text however, should always refer to the key points in a table or figure: if it does not warrant discussion it should not be there. Write the verbal summary before preparing the final version of the tables and figures; to make sure they illuminate the important points.

6. Descriptions of the information represented in a table or picture should be kept simple, while having sufficient detail to be useful and informative. As with the original data, it is important that summaries make clear *what* was measured, to avoid uncertainty about the definitions and the units; *where* the data were collected, so the extent of the coverage is clear; *when* the data were collected, to define the time frame; and for data quoted from elsewhere, *the source*.

7. Statistical information (e.g. appropriate standard errors) is usually required in formal scientific papers and may not be necessary in articles aimed at a more general audience. As a rule, statistical information should be presented in a way that does not obscure the main message of the table or graph.

8. Conveying information efficiently goes along with careful use of 'non-data ink'. For example, tables do not need to be boxed in with lines surrounding each value. Similarly, pseudo three-dimensional 'perspective' should not be added to two-dimensional charts and graphs, as it impedes quick and correct interpretation.

9. Tabular output from a computer program is not usually ready to be simply cut and pasted into a report. For example, a well-laid-out table need never include vertical lines, and rarely needs as many decimal places.

21.3 Graphs and Charts

The two main types of graphical presentation of research results are line graphs and bar charts. Graphs can be kept small, so multiple plots can be presented on a single page or screen. Line graphs can show more detail than bar charts. They should be used when the horizontal axis represents a continuous quantity, such as time spent weeding or quantity of fertilizer applied. When the horizontal axis is a qualitative factor such as ethnic group, crop variety, or source of protein, bar charts are the more likely choice. In this case, joining up corresponding points in a line graph clarifies which set the points belong to. However, the lines themselves have no interpretive value (see figures below).

21.3.1 Line graphs

Line graphs are useful to display more than one relationship in the same picture, for example the response to fertilizer of three different varieties. While there is no general rule, graphs with more than four or five lines tend to become confusing unless the lines are well separated. In a graph with more than one line, different line styles (e.g. solid line, dashed line), colours, and/or plotting symbols (e.g. asterisks, circles) should be used to distinguish the lines.

In any set of line graphs, plotting symbols and line styles should be used consistently. Also, consider using the same scale on each graph, when comparisons are to be made across graphs.

21.3.2 Bar charts

Bar charts display simple results clearly. They are impractical when dealing with large amounts of structured information. Since the horizontal axis represents a discrete categorization, there is often no inherent order to the bars. In this case, the chart becomes easier to read if the bars are sorted in order of height. For example, the first bar may represent the variety with the highest yield with the next bar displaying the second highest yield and so on. Organizing the bars in the opposite direction, in ascending order, can be equally effective. However, this approach must be modified when presenting a series of charts with the same categories. In such instances, it is preferable to have a consistent bar order throughout the series. Ordering the bars on each chart according to decreasing values of a main response shows more clearly which variables behave like the main response and which are different. In a series of bar charts, the shading of the different bars (e.g. black, grey, diagonal lines) must also be consistent.

It is frequently useful to 'cluster' or group the bars according to the categories they represent to highlight certain comparisons. The method of grouping should be determined by the objective of the chart.

It is easier for readers to make comparisons between adjacent bars than between distant bars, and the chart should be laid out accordingly. Figure 21.1 gives examples of two bar charts displaying the same data. These show grain yields for six groups formed by combinations of three wheat varieties and two cultivation methods (traditional and broadbed). Fig. 21.1a is the better layout for demonstrating differences between cultivation methods, whereas Fig. 21.1c is better for showing variety differences.

Fig. 21.1. Bar charts and line charts showing wheat yield by variety and cultivation method.

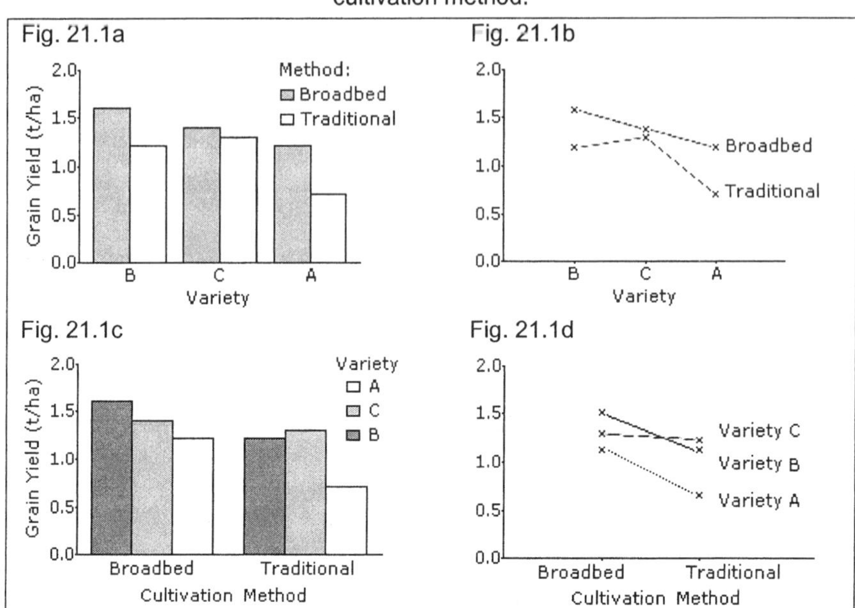

Another way of displaying more complex information on a bar chart is to 'stack' the bars. An example is in Fig. 21.2a, which shows grain yield and straw yield for five wheat varieties. Note that the varieties are sorted according to grain yield. While this graph is good at displaying grain yield and total yield (i.e. grain + straw), it is poor for displaying straw yield alone. For example, it is not obvious that Variety E has the highest straw yield. In this case, if straw yield is important, the stacked bar chart of Fig. 21.2a is unsuitable, and a different presentation, such as the line graph shown in Fig. 21.2b, is needed.

Fig. 21.2. Grain yield and straw yield for five wheat varieties.

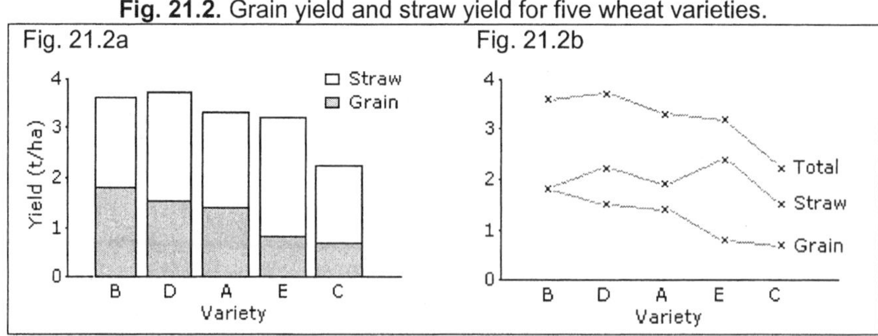

'Perspective' bar charts use shading to give the illusion of a third dimension. These may be superficially attractive, but are generally not as clear as the equivalent simple, two-dimensional chart. True 'three-dimensional bar charts' attempt a two-dimensional representation of a situation where the response is plotted against two classification variables. These are harder to produce and interpret than divided or stacked bar charts and are seldom worth using.

Although bar charts are popular, there are often more appropriate ways to present data in scientific papers or project reports. Consider Fig. 21.1a and Fig. 21.1c, where the bar charts are displaying only six numbers. There are two dimensions here, which the bar charts have forced into a single dimension. Table 21.1, shows the same information, but has the advantage that it is two-dimensional and therefore also allows the marginal means to be presented.

Table 21.1. The information from Fig. 21.1 displayed as a two-dimensional table.

	Cultivation method		
Variety	Broadbed	Traditional	Mean
B	1.59	1.20	1.40
C	1.38	1.30	1.34
A	1.20	0.71	0.96
Mean	1.39	1.07	1.23

If the results are to be presented graphically, then line charts as in Fig. 21.1b and Fig. 21.1d use less ink and are usually clearer. Similarly, the 'stacked bar chart' of Fig. 21.2a is presented as a line chart as in Fig. 21.2b.

21.4 Tables

The type of table used depends on its purpose.

- Reference tables contain information that people will look up; they serve an archival function and often need to be laid out for economy of space, while preserving the data accurately. It is extremely important that they include good metadata. This is the descriptive information that allows the data to be correctly interpreted: usually a comprehensive version of the 'what, where and when' mentioned in Section 21.2 (item 6). Reference tables often appear as appendices.
- Demonstration tables are intended for quick assimilation by the reader or viewer. It is important that they are clear and well presented, using reasonable approximations to reduce numbers to relatively few significant digits. Demonstration tables should be included in the text (because readers following a general argument tend not to bother flipping backwards and forwards).
- Excessively large demonstration tables are intimidating and users tend to give up on them. If it is necessary to display all the information, it should be split into manageable components.
- Omit any column that can be readily calculated from data in other columns. Minor or less important categories can be combined.

21.4.1 Tables of frequencies

The simplest tables arising from surveys, or from coded qualitative information, are those of counts or frequencies. If relatively large counts are to be compared in a table with several rows and columns, it is often helpful to present them as percentages. Percentages add up to 100, either across rows, or down columns, or across the whole table. These facilitate different types of comparison. It is likely that only one type of percentage makes sense for your data. The sizes of sample on which a percentage table is based should be made explicit.

21.4.2 Orientation and order

The orientation of a table can have considerable influence on its legibility. It is much easier for a reader to make comparisons by looking up and down a column of numbers, than when looking across a row. Therefore, if the purpose of a table is to demonstrate differences between treatments or groups for several variables, the groups should define the rows of the table and the variables should define the columns. This is demonstrated in Table 21.2a and Table 21.2b. The data are poorly displayed in Table 21.2a which has the wrong orientation, while Table 21.2b displays the same information with the rows and columns interchanged.

Table 21.2a. A poorly presented table showing mean intake of milk, supplement and water, and mean growth rates for four diets (artificial data).

	Diet[1]			
Variable	I	II	III	IV
Milk intake	9.82	10.48	8.9	9.15
Supplement intake	0	449.5	363.6	475.6
Growth rate	89	145.32	127.8	131.5
Water intake	108.4	143.6	121.29	127.8

[1] Diet I = Control; Diet II = Lucerne supplement; Diet III = Leucaena; Diet IV = Sesbania.

Table 21.2b. Mean growth rate and intakes of supplement, milk and water for four diets (artificial data).

Supplement	Growth rate (g/day)	Supplement intake (g/day)	Milk intake (ml/kg$^{0.75}$)	Water intake (ml/kg$^{0.75}$)
Lucerne	145	450	10.5	144
Sesbania	132	476	9.2	128
Leucaena	128	364	8.9	121
None	89	0	9.8	108

Table 21.2a not only has the wrong orientation, but is also incorrect in a number of other features. There are an unnecessary number of decimal places and this is inconsistent within each variable. The *Diet* labels are also uninformative and the variables do not have units.

In Table 21.2b, the information is the same as Table 21.2a, except that the table has been reoriented, columns and rows have been reordered to highlight differences in growth rates, the number of digits has been reduced and standardized, and more informative labels are used. For simplicity, both tables omit sample size and accuracy information (but this should be stated elsewhere).

The order of the rows in a table is important. In many cases, the order is determined by, for example the nature of the treatments. If there is a series of tables with the same rows or columns, their order should usually be the same for each table. If not, it can be useful to arrange the rows so that the values are in descending order for the most important column, as in Table 21.2b. It is assumed that *growth rate* is the most important variable that is measured and is therefore put in the first column. The rows have been sorted so that the supplement giving the highest growth rate is first. Note that if the four diets had been different levels of the same supplement (e.g. 0, 200, 400 and 600 g of lucerne per day), then the rows would have been left in the natural order.

The labels in Table 21.2b are also more informative. The row labels now give a description of the diets, and the column labels include the units of measurement. The number of digits and decimal places has also been reduced and made consistent.

As a result of these very simple adjustments, the important features of the results stand out more clearly in Table 21.2b.

21.4.3 Quantitative variates defining rows or columns

The rows in Table 21.2b above, represent qualitatively different diets. On occasion quantitative variables may be grouped and used to define table rows or columns.

Table 21.3. Count of farmers grouped by cropping pattern and hours spent weeding per unit area between maize planting and tasselling (180 farmers).

Hours	Sole-crop	Intercropped with beans	Intercropped with pigeon pea	Overall percentage
Less than 10	25	53	18	53
10 to < 15	3	9	-	7
15 to < 20	8	3	5	9
20 to < 30	8	7	2	9
30 or more	20	16	3	22

The row definitions used in Table 21.3 might have seemed reasonable before data collection, but far too many cases (75%) have finished up in the end categories. It is possible that most of the 'Less than 10' cases did 8 or 9 hours weeding, or they may have done none; in the same way, some of the '30 or more' cases may have done many more hours. Possibly sole-crop maize is weeded more, but the information as presented is too vague to be useful. If possible, the table should be reworked with groupings that are more meaningful. Of course, if these weeding categories were used in the data collection, the problem becomes irremediable. Data can be summarized: unavailable detail cannot be created. This is an example where the text describing the conclusions should be drafted before the table is finalized, to check whether a table is necessary, and if so, to ensure the results are clear.

21.4.4 Number of digits and decimal places

The number of digits and decimal places presented in a table should be the minimum that is compatible with the purpose of the table. It is often possible to use as few as two significant digits. For example, in Table 21.2a the *water intake* with values such as 108.4, has three significant digits (the 1 in the 'hundreds' position does not vary and is therefore not significant). This can be reduced to two significant digits (to 108), as in Table 21.2b, without any loss in information, but with an increase in clarity.

Sometimes units of measurement can be changed to make numbers more manageable. For example, numbers such as 12,163 kg/ha could often be better presented as 12.2 t/ha. In other cases, numbers can be multiplied or divided by factors such as one thousand or one million, for clear presentation (e.g. most people would find it much quicker and easier to take in a statistic of '72 HIV deaths per 1000 population per year' than 'a rate of 0.07189').

The number of decimal places should be consistent for each variable presented. In Table 21.2b *milk intake* consistently has one decimal place and *supplement intake* has none. To make comparisons easier, numbers in a column should be aligned according to the decimal point.

21.4.5 More complex tables

Table 21.4. Mean daily gain and feed intake for four breeds and two diets (artificial data).

	Weight gain (g/day)			Feed intake (g/day)		
	Protein level			Protein level		
Breed	Low	High	Mean	Low	High	Mean
Menz	38	56	47	639	952	796
Dubasi	33	57	45	603	1008	806
Wello	28	44	36	591	917	754
Watish	29	40	35	628	889	759
Mean	32	49	41	615	942	779

The arrangement of Table 21.4 facilitates comparisons between four breeds separately for each protein level as well as averaged over both levels. For example, we see that the first two breeds, Menz and Dubasi, have higher weight gains than Wello and Watish. Note that it would be possible to economize on space by combining the two sub-tables into a single table with two observations per cell. This may be justified in reference tables, but is not recommended in demonstration tables, where it can be very confusing. Once again, the table omits sample size or accuracy information.

This kind of table cannot be used effectively when many columns need to be displayed. Blank space is also needed to separate the weight gain results clearly from the feed intake. Without such space, this kind of table becomes hard to read.

Alternative ways of presenting results like this are given in Table 21.5a and Table 21.5b.

21.5 Results of Statistical Analysis

In scientific papers, it is often necessary to present results of statistical analyses. They can indicate the precision of the results, give further description of the data, and demonstrate the statistical significance of comparisons.

Where statistical significance is referred to in text, the reference should be included in such a way as to minimize disruption to the flow of the text. Significance probabilities can either be presented by reference to conventional levels (e.g. ($p < 0.05$)) or, more informatively, by stating the exact probability which is usually derived from a statistical package (e.g. ($p = 0.023$)).

An alternative to including a large number of statements about significance, is to include an overall covering sentence at the beginning of the results section or some other suitable position. An example of such a sentence is: 'All treatment differences referred to in the results are statistically significant at least at the 5% level unless otherwise stated.'

21.5.1 Descriptive statistics

When describing a set of data with summary statistics, useful statistics to present are the mean, the number of observations, a measure of the variation or 'scatter' of the observations, and the units of measurement. The range or the standard deviation (s.d. or SD) are useful measures of the variation in the data. The standard error (s.e. or SE) is not relevant in this context, since it measures the precision with which the mean of the data estimates the mean of a larger population.

If there are a large number of variables to be described, the means etc., should be presented in a table. However, if there are only one or two variables, these results can be included in the text. For example:

'The initial weights of 48 ewes in the study had a mean of 34.7 kg and ranged from 29.2 to 38.6 kg.'

or

'The mean initial weight of ewes in the study was 34.7 kg (n = 48, s.d. = 2.61).'

When quoting a standard deviation (or standard error), a ± sign is irrelevant. As well as being unnecessary, a ± sign is ambiguous if used without explanation in expressions, such as 'Mean = 34.7 ± 3.6 kg'. It is not clear whether the number after the ± sign indicates a standard deviation, a standard error, or a confidence interval.

21.5.2 Results of analyses of variance

The analysis of variance tables are primarily to help the scientist and are not normally included in a report. An exception is a specialized analysis, where the individual mean squares are important in their own right. In such cases, the degrees of freedom and expected mean squares are presented.

In most situations, the only candidates from the analysis of variance table for presentation are the significance probabilities of the various factors and interactions, and sometimes the residual variance or standard deviation. When included, they should be within the corresponding table of means, rather than in a separate table.

In general, authors should present relevant treatment means, a measure of their precision, and possibly significance probabilities. The treatment means are the primary information to be presented, with measures of precision being of secondary importance. The layout of the table should reflect these priorities; the secondary information should not obscure the main message of the data.

The layout of tables of means depends (as is shown below) on the design of the experiment, and in particular on:

- Whether the design is balanced (i.e. equal numbers of observations per treatment).
- Whether the treatments have a factorial structure.
- For factorial designs, whether or not there are interactions.

21.5.3 Measures of precision

The measure of precision should be chosen relative to the objectives of the study and the analysis that has been done. A survey designed to estimate population parameters should be reported with standard errors (or confidence intervals) of those parameter estimates. Experiments are designed to compare treatments, so it is the precision of the treatment *differences* that is important. These are measured by *standard errors of differences* (SEDs) or a *least significant difference* (LSD). Simple standard errors of means from an experiment can be misleading; the study was designed and analysed to focus on differences, and an SE of a mean may give the impression that this is the precision of some population average. Furthermore, the correct SEDs cannot be calculated from the SEs except in the case of the simplest designs.

For data from experiments, we prefer the standard error of differences (SED) over the LSD. The reason is simply that LSD focuses attention on testing rather than estimation – but estimation is usually more useful. However, for most designs the LSDs can be calculated from the SEDs, so the difference is not critical. Note that if an LSD is quoted then it must be accompanied by the significance level, e.g. 5% LSD. Some scientists and journal editors may have a strong preference for either SEDs or LSDs and there is no overriding reason why they need change.

Measures of precision are usually presented with one more decimal place than the means. This is not a strict rule. For example, a mean of 74 with a standard error of 32 is fine, but a mean of 7.4, with a standard error of 0.3, should have the extra decimal place and be given as 0.32.

Some researchers like to include the results of a multiple comparison procedure such as Fisher's LSD. These are added as a column with a series of letters, (a, b, c, etc.) where treatments with the same letter are not significantly different. Often these methods are abused. The common multiple comparison procedures are only valid when there is no 'structure' in the set of treatments (e.g. when a number of different plant accessions or sources of protein are being compared).

The best method of reporting such results should be to sort the treatments into descending order of means, sorting on the most important variable. In addition, a single standard error or LSD should be given in the balanced case, and individual standard errors given in the unbalanced case. This is shown for four treatments in Table 21.5a and Table 21.5b.

A few authors, and the occasional journal editor, will have severe withdrawal symptoms if told that their multiple comparison results are not required. We consider that the results from a multiple comparison procedure should be additional to, and not a substitute for, the reporting of the standard errors. If scientists find that the presence of multiple comparisons is helpful for them to write the text, then they can be included in the table at the draft stage. Once the text is written, the tables should be re-examined. We find that the columns of letters rarely correspond to any of the objectives of the research and hence are not evident in the reporting of the results. If the multiple comparison columns are not referred to within the text, as is often the case, the tables can be simplified by eliminating them.

21.5.4 Single factor experiments

The most straightforward case is a balanced design with simple treatments. Here each treatment has the same precision, so only one SE (SED or LSD) is needed per variable. In the table of results, each row should present means for one treatment; results for different variables are presented in columns as in Table 21.5a. The statistical analysis results are presented as one or two additional rows: one giving SEDs (or LSDs) and the other possibly giving significance probabilities.

If the *F*-probabilities are given, we suggest that the actual probabilities be reported, rather than just the levels of significance (e.g. 5% (0.05), 1% (0.01) or 0.5% (0.005)). In particular, reporting a result was 'not significant' (often written as 'ns') is not helpful. When interpreting the results, it is sometimes useful to know if the level of significance was 6% or 60%. It should also be made clear for what test this is a significance level.

Table 21.5a. Mean growth rate and intakes of supplement, milk and water, for four diets.

Supplement	Growth rate (g/day)	Supplement intake (g/day)	Milk intake (ml/kg$^{0.75}$)	Water intake (ml/kg$^{0.75}$)
Lucerne	145	450	10.5	144
Sesbania	132	476	9.2	128
Leucaena	128	364	8.9	121
None	89	0	9.8	108
SED	15.0	-	1.53	20.6
F-test probability	<0.001	-	0.114	0.023

For unbalanced experiments, as in Table 21.5b, each treatment difference has a different precision. In this case, there are some choices to be made about presentation. The most complete and correct presentation would be to give the full matrix of SEDs. For example, if there are 4 treatments there are 6 differences and hence 6 SEDs to present. This can be unwieldy and may not be necessary. If the variation among the SEDs is not large we can simply quote an average value. A compromise is to give a maximum and minimum SED. If the variation in SED is because of unequal explication, then adding the replicate numbers will help indicate to which contrast different SEDs apply (e.g. the maximum SED will be for comparing the two least replicated treatments).

Table 21.5b. Mean growth rate and intakes of supplement, milk and water, for four diets (an example of presentation of results from an unbalanced experiment).

	Growth rate (g/day)		Supplement intake (g/day)		Milk intake (ml/kg$^{0.75}$)	
Supplement	*n*	Mean	*n*	Mean	*n*	Mean
Lucerne	23	145	22	450	12	10.5
Sesbania	12	132	12	476	7	9.2
Leucaena	32	128	30	346	17	8.9
None	41	89	37	0[1]	20	9.8
Max SED		32.7		113.9		3.16
Min SED		16.4		108.6		1.82

[1] This treatment was excluded from the calculation of the SED.

21.5.5 Factorial experiments

Factorial experiments usually have a lot of information to present. Also the treatment means to be presented will depend on whether there are interactions that are statistically significant and of practical importance.

This section discusses two-factor experiments, but the recommendations can be easily extended to cases that are more complex. A balanced experiment is assumed, with equal numbers of observations for each treatment. However, the recommendations can be combined in a fairly straightforward manner with those in the previous section when confronted with an unbalanced case.

If there is no interaction, then the 'main effect' means should be presented. For example a three × two factorial experiment on sheep nutrition might have three 'levels' of supplementation (None, Medium and High), and two levels of parasite control (None and Drenched), giving six treatments in total. There are five main effect means: three means for the levels of supplementation, averaged over the two levels of parasite control, and two means for the levels of parasite control. In this example, there would also be two SEDs and two significance probabilities for each variable, corresponding to the two factors. Table 21.6a gives a skeleton layout for the presentation of these results.

Table 21.6a. Skeleton results table for a hypothetical lamb nutrition experiment with three levels of supplementation and two levels of parasite control. There is no interaction.

Factor	Final liveweight (kg)	Weight gain 0-120 days (g/day)	Age at puberty (days)
Supplement			
None	-	-	-
Medium	-	-	-
High	-	-	-
SED	-	-	-
Parasite control			
None	-	-	-
Drenched	-	-	-
SED	-	-	-
F-test probabilities			
Supplement (S)	-	-	-
Parasite control (P)	-	-	-

This table could be slightly rearranged. One alternative would be to move the rows giving SEDs to a position immediately above the *F*-test probabilities. Another would be to place the rows of *F*-test probabilities under the SEDs of each of the relevant factors.

If there are interactions that are statistically significant and of practical importance, then main effect means alone, are of limited use. In this case, the individual treatment means should be presented as in Table 21.6b.

For a balanced design, there is now only one SED per variable (except for split plot designs), but three rows giving *F*-test probabilities for the two main effects and the interaction. Additional rows for *F*-test probabilities can be used for results of polynomial contrasts for quantitative factors or other pre-planned contrasts.

When there are just one or two variables, an alternative layout to that in Table 21.6b is as a two-way table as in Table 21.4. The additional information (*F*-test probabilities and SEDs) would be presented underneath the table. The sizes of the main effects, in relation to the interaction, dictate whether it is useful to give the main effects as well, as in Table 21.4, or omit them, as in Table 21.6b.

In situations where there is an interaction that is statistically significant but of relatively minor importance, the results could instead be presented as in Table 21.6a, but with an additional row giving the *F*-test probability for the interaction. A comment in the text adds the necessary explanation.

Table 21.6b. Skeleton layout for a results table for a hypothetical lamb nutrition experiment with three levels of supplementation and two levels of parasite control with significant interaction.

Treatment		Final	Weight gain	Age at
Supplement	Parasite control	liveweight (kg)	0-120 days (g/day)	puberty (days)
None	None	-	-	-
	Drenched	-	-	-
Medium	None	-	-	-
	Drenched	-	-	-
High	None	-	-	-
	Drenched	-	-	-
SED		-	-	-
F-test probabilities				
Supplement (S)		-	-	-
Parasite control (P)		-	-	-
Interaction (S×P)		-	-	-

A graphical display is often useful to show the pattern of the interactions, particularly when some of the factors are quantitative. Examples are in Fig. 21.1b and Fig. 21.1d. The same type of graph is occasionally useful, even when there are no interactions, but when the investigation of a possible interaction was an objective of the study.

21.5.6 Regression analysis

The key results of a linear regression analysis are usually the regression coefficient (*b*), its standard error, the intercept or constant term, the correlation coefficient (*r*), and the residual standard deviation. For multiple regression, there will be a number of coefficients and SEs, and the coefficient of determination (R^2) will replace *r*. If a number of similar regression analyses have been performed, the relevant results can be presented in a table, with one column for each parameter.

If results of just one or two regression analyses are presented, they can be incorporated in the text. This can be done by presenting the regression equation:

'The regression equation relating dry matter yield (DM, kg/ha) to amount of phosphorus applied (P, kg/ha) is DM = 1815 + 32.1P ($r = 0.74$, SE of regression coefficient = 8.9).'

or by presenting individual parameters as in:

'Linear regression analysis showed that increasing the amount of phosphorus applied by 1 kg/ha, increased dry matter yield by 32.1 kg/ha (SE = 8.9). The correlation coefficient was 0.74.'

It is often revealing to present a graph of the regression line. If there is only one line to present on a graph, the individual points should also be included. With more than one line, this is not always necessary and tends to be confusing.

Details of the regression equation(s) and correlation coefficient(s) can be included with the graph if there is sufficient space. If this information would obscure the message of the graph, then it should be presented elsewhere.

21.5.7 Error bars on graphs and charts

Error bars displayed on graphs or charts are sometimes very informative, while in other cases they obscure the trends that the graph is meant to demonstrate. The decision on whether or not to include error bars within the chart, or give the information as part of the caption, should depend on whether they make the graph more clear, or not.

If error bars are displayed, it must be clear whether the bars refer to standard deviations, standard errors, standard errors of differences, confidence intervals or to ranges. Which measure of precision is used should correspond to the objectives, just as when results are presented in tables. Hence, SEDs are usually appropriate for experiments and SEs for surveys. Where error bars represent, say, standard errors, then one of the two methods below should be used.

- The bar is centred on the mean, with one SE above the mean and one SE below the mean (i.e. the bar has a total length of twice the SE).
- The bar appears either completely above or completely below the mean, and represents one SE.

If the error bar has the same length for all points in the graph, then it should be drawn only once, and placed to one side of the graph rather than on the points. This occurs with the results of balanced experiments.

21.6 Data Quality Reporting

Tables such as 22.5a, 22.5b, 22.6a and 22.6b (above), often include additional rows below the standard errors giving coefficients of variation and other measures. Unlike the treatment means, these statistics are normally used to assess data quality. For instance, when assessing yields, scientists might know from past studies, that a coefficient of variation of 20% is acceptable. These statistics are the equivalent of giving non-response rates, or efficiency factors when reporting a survey. Of course they should only be included if they are accompanied by salient interpretation in the text.

Such background information can help readers to interpret the results and the effectiveness of the study design. In a scientific paper, a brief reference to these aspects might be in the materials and methods, or the results section. In a project report, the subject might warrant a complete section.

This type of information ought not to be supplied automatically as a matter of routine. For example, coefficients of variation should not be given for variables (e.g. disease scores), where they are meaningless, because the measurement scale is arbitrary.

Of course it is good practice to include a wider review of factors that may have affected the relevance or quality of parts of the data (e.g. by introducing biases, and of their possible effects).

21.7 In Conclusion

In this chapter, we have devoted more space to the presentation of tabular information than to the production of charts. This is not to negate the importance of charts, or the value of a clear discussion of numerical results in the text.

However, tabular presentation has become the 'Cinderella' method of data presentation, compared to the attention devoted to imaginative methods of graphical presentation. Tables remain important, and it is through the appropriate use of all methods, that research results can be reported in a way that is fair and balanced, at the same time as eye-catching and clear.

Reference

Chapman, M. and Wykes, C. (1996) *Plain Figures*, 2nd Edition. H.M. Stationery Office, London.

Part 5: Where Next?

Chapter 22

Current Trends and their Implications for Good Practice

22.1 Introduction

Good statistical practice can lead to improved project outputs, and hence to more effective NRM research. To conclude our book, we look at some current trends in the areas of planning, data management, and analysis and their implications for good statistical practice in the near future. We also discuss options for training in good statistical practice.

22.2 Trends in Planning

Four trends in NRM project planning have implications for statistical practice.

Firstly, there is an increasing focus on results and impact, with project funders expecting to see a clear logical connection between the planned activities and promised outcomes. If this requirement is followed through systematically, it will result in greater attention being paid to the planning of component studies. 'Good statistical practice' in study design is concerned with meeting the objectives in the most efficient way, so we expect to see more attention paid to the issues discussed in Part 2. However, the relation between certain statistical aspects, such as sample size and design, and the overall effectiveness of a study is often not well understood. Project managers sometimes perceive the statistical aspects of the project as meeting scientific requirements which are somehow different from the real 'information generation' requirements of the project. There should be no such distinction.

Secondly, project designers are increasingly recognizing the importance of the participation of the right set of stakeholders in projects. However, many problems in resource management result from the different perspectives of various interest groups. The objectives of component studies can become vague or contradictory if they try to be 'all things to all people'. Such studies are difficult to plan effectively. There may also be a tendency to focus on the participation rather than on the methods of investigation. Together these can lead to insufficient attention to the statistical aspects of the design, particularly if good design is incorrectly seen as only meeting 'scientific' objectives, rather than being a prerequisite for the efficient generation of valid information.

Thirdly, many projects nowadays are likely to include a dissemination or information exchange component rather than leaving this for a possible later phase. Project teams must therefore be aware of the information needs of all stakeholders, and ensure that studies will actually provide the information in acceptable ways. For example, we know that the information farmers sometimes want to justify changing their management can be very specific to their situation. They want to confirm the effects of proposed interventions in the heterogeneous niches of their land, whereas researchers look for general patterns, not solutions for particular niches on each farm. Also, farmers are often comfortable without control plots, preferring instead to use their own past experience. These have implications for the way the study is designed and the data collected. In addition, researchers may now have to investigate the different ways that information can be disseminated as part of the project. These alternative ways of dissemination can therefore be seen as the experimental 'treatments'.

Finally, a potentially positive trend in projects is the use of spatial information, including new and existing types of remotely sensed data. There has been a tendency for datasets that have been generated by remote sensing to be used separately from other activities. This trend parallels the development of participatory research in the 1980s, when such studies were not integrated with other activities. The value of remotely sensed information will be clearer when proponents view these datasets as contributing to effective NRM research.

22.3 Trends in Data Management

Trends in data management, in the four areas of integration, archiving, access and knowledge management, have implications for the good practice points discussed in Part 3 of the book.

One characteristic of effective NRM research is integration across disciplines and across spatial scales. This requires integration of data from different studies and researchers, so any component can be extracted for analysis. As discussed in Chapter 13, there are data collection requirements (such as geo-referencing all observations), management requirements (such as effective use of a database system), and institutional requirements (such as agreements to share data). These requirements are prerequisites for effective management of the data.

The need for integration does not necessarily imply centralization. There is currently a trend away from having a central data processing unit, towards decentralized data management. In a decentralized system, the scientists managing each study take responsibility for their data, but should do so in a way that facilitates painless integration with others. An advantage over a centralized system is that scientists remain in contact with their own data, and because of this, they retain a strong interest in ensuring the highest quality.

The importance of archiving was discussed in Chapter 10. Archiving the data is often no longer just a matter of good practice, but a contractual requirement imposed by the funding agency. An implication is the need to know both the research and contractual imperatives for archiving and build them into every project from the start. This is related to the next trend, which is in data access.

Some projects are required to make their data available to everyone; for instance by putting their archive in the public domain. Health sciences may be more advanced than NRM researchers in this respect. NIH (2003) states: '*Data sharing is essential for expedited translation of research results into knowledge, products and procedure to improve human health.*' With wider availability of data comes a recognition that the rights and privacy of people who participated in the research must be protected at all times. Hence the data should be free of identifiers that would permit linkage to individuals. A current recommendation (NIH, 2003) is that the data should be released 'in a timely fashion', and no later than the acceptance for publication of the main findings from the final dataset. Where data are collected from large studies, over several discrete time periods or waves, the data should be released as they become available, or as the main findings from waves of data are published. The intention is that the initial investigators should benefit from first and continuing use of the data, but not from prolonged exclusive use.

A similar arrangement would be attractive for NRM research. If project researchers cannot publish their results without making their data available, then they will have to organize the data for outside scrutiny at some stage. The argument to organize the data as early as possible then becomes an easy one to make, since it helps to avoid many of the usual pitfalls of poor data management.

An encouraging trend in scientific publication is that some journals (such as *Ecological Archives*) now accept data as publications. These are datasets together with a full description of how they were collected, but without analysis or interpretation. The idea is that in many research areas, collecting and documenting high quality data is a considerable achievement. Those responsible can get the rightful credit through publication, while at the same time making the data available to those who can add value by analysis.

There are also encouraging trends in access to routinely collected data that many NRM projects should use. This includes data on climate and data collected by national statistical offices. These provide multi-year baseline information that can put research results into a broader context. This is another form of foundation information, or 'table-top', as described in Section 7.9.2. Until recently, many organizations responsible for collecting baseline data have been reluctant to share their information. Sometimes this has been because of issues of confidentiality, but also for less honourable reasons such as the intention to use the data for financial gain.

However there are trends away from this for several reasons. Few users in the developing world were prepared to pay for data, and at the same time many important international datasets on such diverse issues as living standards, forest cover, population projections and water use were becoming freely available over the internet. Recognizing these trends, the Meteorological and Statistical Offices are starting to market their expertise instead. The datasets are available without charge, and it may be cost-effective for research groups to include time for a member of staff from the Meteorological or Statistical Office to tailor the analysis of their data to the needs of the research project.

22.4 Developments in Methods of Analysis

In Part 4 of the book we described some developments in statistical analysis that have made the subject easier and more powerful, as other 'older' statistical methods became 'mainstream'. Becoming mainstream has several facets:

- The terminology becomes general rather than tuned to the particular statistical techniques, and is therefore easier for the non-specialist to learn and understand.
- Methods become incorporated into general statistical software, so they are available to users without their having to obtain and learn special software.
- It becomes possible to use methods correctly and usefully without understanding all the mathematical details.
- Methods become published and accepted in applied science journals, not just in statistical journals.

The term 'mainstream' suggests the analogy of a river, and we find the 'estuary model' a useful description of current and future trends in statistics for NRM research. The stream starts up in the mountains, and includes perhaps analysis of variance and linear regression. Somewhere in another valley is the 'chi-square test' stream. Chi-square analyses of two-way tables of category counts proved useful to researchers for many years. Then in the late 1960s they joined with methods for higher dimensional tables and became log-linear models, a deeper and broader stream that allowed analysis of a much richer class of problems. In the 1970s log-linear models, together with probit analysis and other streams, joined the mainstream of generalized linear models that we describe in Chapter 20. The process continued. The tributaries joining the main flow tended to be shorter; REML only started out in the early 1980s but, according to the criteria above, had joined the mainstream by the mid 1990s.

Now we are reaching the estuary. The mainstream gets wider rapidly, as more and more useful methods become incorporated. And the tributary streams are shorter, with the time between initial development of a method and the confluence with the main stream reduced, due to computing advances and improved communication between researchers. The advantages to the researcher are considerable. No longer is it necessary to flounder around crossing numerous small streams, each with its own terminology, traditions and software. So, down in the estuary you can go where you want to safely as long as you are equipped with a suitable boat, or learn to swim well! A word of caution needs to be added here, though. Now that modern advanced methods are so accessible, analysts (researchers or statisticians) should be sure that they understand the concepts underlying a new method and its implementation, rather than using it blindly just because it is available in their software.

The same point can be made in connection with analytical tools hitherto regarded as distinct from statistics, such as modelling and GIS analysis. The boundaries between these are becoming blurred, to the advantage of the users. GIS analysts are beginning to realize that much of what they do is 'statistics' and there are ideas and methods to learn from statisticians. And statisticians are realizing that the spatial component of many datasets needs more realistic incorporation into analyses. The software is evolving to match, with more GIS functionality in statistical software and vice versa. The same trend is true for process-based systems modelling. This used to be carried out separately from statistical analysis of the relationships that make up such a model. Statistical advances make it possible now to do correct analysis of complex, multi-component models, and the software for this will soon become mainstream.

22.5 Communicating the Results

In Chapter 21 we limited our discussion to statistical practice concerned with the presentation of results in articles designed for fellow-scientists. For NRM projects there is far more to effective communication of the results than writing reports and papers. Writing papers is often the simplest reporting task for the research team, because they are communicating with others who share the same vocabulary and working practices. However, if NRM projects are to have an impact, the team must also make the same information available to farmers, development workers, extension services, policy makers, and others, all of whom have different vocabularies and working practices. This is a more difficult task.

Writing for a non-scientific audience often implies that a simpler presentation, limited to the key points, is needed. Sometimes *simpler* is taken to mean that all ideas of variability are omitted from the presentation of the results. This may not be the appropriate simplification, because farmers and others know that their work involves many risks. So to omit all discussion of variability may lead to reports that are not deemed credible by the recipients.

Even if the outside stakeholders understand all the experimental techniques and analyses from a research programme, they might not find that sufficient data were collected, and they might have different values by which they evaluate findings (Norrish *et al.*, 2001). In some aspects of research, non-scientists have looser criteria than scientists, for example when controlling for errors. But in others they may have more stringent criteria, for example they may want to see several years of data on a proposed new farming system before they are ready to commit resources to it.

Holland (1998) found that some farmers and extension agents rejected analyses and recommendations from researchers because they felt that the outputs had been produced in ways that did not match their own criteria. The farmers saw more dimensions of heterogeneity within and between their fields than they felt were accounted for in research experiments. To these farmers, the research results were anecdotal. What mattered to them was a detailed understanding of the context in which experiments had taken place, measured along their own dimensions. The farmers had been unable to share these requirements with researchers, and remained unconvinced by the results.

Communication strategies must be specified at the start of a project, if the dissemination of the results is to lead to impact. These strategies must include the allocation of responsibilities for dissemination. They must be active, and involve iterative processes that incorporate the communication needs of the wide range of stakeholders who are normally part of the NRM research process. Devising effective communication strategies will itself become a research activity in some projects. These need expertise, as well as planning, management and analysis, i.e. the three components of this book, in the same way as the other research activities.

22.6 Progress in Training Methods

Students in colleges and universities often report the following problems with their statistics training:

- It is dominated by analysis, with little on design or data management.
- A recipe-book approach is used which does not lead to insights and understanding of principles.
- Teachers emphasize hand computation with limited use of computers.
- The presentation is mathematical not conceptual.
- The courses concentrate on methods for formal 'on-station' research.
- No links are made across biophysical, social and economic methods.
- No links are made with participatory and action research or adaptive management.
- The course is taught by someone with no experience of field research, nor interest in the substantive problems students are researching.

The outcome is graduates with a near-universal dislike of statistics, and subsequently a strong demand for in-service training in the subject.

The issue of revising statistics curricula and teaching methods in universities and colleges is well recognized, but is beyond the scope of this book. Some institutions, such as the faculties of agriculture in the universities of Nairobi and Reading, have revised their teaching to incorporate many of the ideas presented here. This book is a useful companion for such courses but is probably not appropriate as a sole course text.

In this section we concentrate on training for researchers and statistical practitioners involved in NRM research. The term 'statistical practitioner' has been used to encompass both the formally trained statistician who is familiar with the theory but not much of the practice, and what we have already referred to as the 'statistical auxiliary' (i.e. a scientist with a reasonable degree of competence in statistics who provides support to colleagues). We contend that training can now be handled more effectively, partly because of the recent ease of use of modern statistical software, combined with the advances in statistical methods. Here we offer some suggestions as to how it might work. The type of training outlined below can be a quick and effective way of improving the quality of natural resources research.

22.6.1 Training the researcher

For in-service training of researchers, we have to balance the need to fill in the gaps in their knowledge of statistics with the time available for training, since statistical knowledge is only one of the skills researchers need. This balance is becoming easier to achieve.

Before computers were widely available, statistical training was limited to problems that could be analysed by hand. More recently, the statistical ideas were often sidetracked by the need to teach computing ideas, not least to teach the commands that were needed to use the statistical software.

Given the wide availability of elementary computing courses, we can nowadays assume that, at the start of a training course in statistics, participants will have some basic computing knowledge. It can even be a prerequisite that they become computer-literate separately from their statistical training.

The ease-of-use of most statistics packages is generally a pleasant surprise to those participants who are familiar only with spreadsheets. Thus we are now able to spend very little time on the use of the package itself, and can concentrate instead on teaching good statistical practice. The software is just a tool to help in the training and then in the researcher's subsequent work.

Our experience is that, for scientists, short training workshops (typically of one or two weeks duration) that link directly with the different stages of the project are most effective. The first may be on the planning of the study (i.e. on some of the topics covered in Part 2 of this book). If the workshop is held before the research begins, participants can bring their current research plans to the workshop. The outputs of the workshop include improved research designs for studies that will then be implemented, not just course exercises.

A second workshop may be on aspects of data management and the third on the analysis (see Parts 3 and 4 respectively). Again, participants could bring their own data. Logically, the analysis is the last of the three phases, but it is sometimes more effective to start a training programme with a workshop on analysis, because participants may have a backlog of data from previous studies. Also, they sometimes believe that all their problems lie in analysis.

Institutes and funding agencies are sometimes cautious about this amount of capacity-building on two counts. The first is the cost; the second is the time away from research. Neither concern is valid if the training is handled well. Workshops are often best done locally, usually 'within country' and possibly 'within station' where it is relatively inexpensive. The direct links to the current research mean that participants often do more of their own planning, data handling and analysis within the workshops than could be undertaken in the same time if there were no workshop. These workshops also assist in related aspects, such as team building and in discussions on sensitive topics, such as data sharing.

22.6.2 Training the statistical practitioner

Our hope is that the materials in this book will tempt formally-trained statisticians to involve themselves more in the full research process for NRM projects. If their own training was mainly on analysis, they should recognize that they start as apprentices on some aspects of planning and data management, and that they may also be out of touch in areas such as on-farm and participatory research.

The demand from statisticians is often for training in the more advanced methods of analysis, such as those mentioned in Chapter 20. While important, perhaps a higher short-term priority is to be able to provide support for the whole research process. Thus, statisticians might attend courses for scientists as trainees, and attach themselves to research groups. Later they can become resource persons as their skills broaden, and they practice communication skills with sceptical scientists.

In the absence of a local statistician, one or two scientists take on the role of unofficial 'statistical auxiliary'. These people are largely self-taught and do not find statistics as daunting as their colleagues. They are also good communicators, and may typically spend 10% to 20% of their time providing support to colleagues. A comparatively small amount of extra training could make them even more valuable in the research process. This training concentrates on their supporting role; it is not to turn them into statisticians. Where training is undertaken, either within individual projects, or for groups of projects, there may be initial 'train-the-trainer' workshops, on a national or regional basis, and these may be for statisticians, together with statistical auxiliaries. These individuals may also have more advanced training that starts where the training for individual scientists stops.

It is important that 'train-the-trainer' courses are recognized as such, and not combined with teaching new statistical techniques to the potential trainers. If we are to teach well in the future, then we must be prepared to spend time learning how to teach, and not just how to analyse.

References

Holland, D.C. (1998) *Empowerment through Agricultural Education: How Science Gets in the Way. The Case of Farmer Field Schools in the Philippines.* Unpublished PhD thesis, University of Reading, UK.

NIH (2003) *Final NIH Statement on Sharing Research Data.* National Institutes of Health, US Department of Health and Human Services, Maryland. grants2.nih.gov/grants/policy/data_sharing/ (accessed 29 August 2003).

Norrish, P., Lloyd Morgan, C. and Myers, M. (2001) Improved communication strategies for renewable natural resource outputs. *Socio-economic Methodologies Best Practice Guideline Series (BPG 8).* Natural Resources Institute (NRI), Chatham, UK.

Chapter 23

Resources and Further Reading

23.1 Introduction

In this chapter we identify some resources related to the topics in this book that complement what we have covered. This is by no means a definitive set of resources; rather it is a list of books and web-based materials that we find useful, that we often recommend to others, and that are sound enough to stand the test of time reasonably well. We have kept the list fairly short, and the reader can view it as a 'bookshelf' on topics relevant to NRM research.

Section 23.2 outlines why we have selected the resources that we have. Section 23.3 contains our suggestions for the bookshelf, with a short comment on each item indicating the intended audience, or a particular aspect that merits its inclusion. Sections 23.4 and 23.5 contain the following web-based materials:

- all the original *Good Practice Guidelines* from which this book evolved;
- other materials available on the SSC and on the ICRAF websites.

Section 23.6 is a list of relevant software packages.

23.2 Adding to this Book

We see this book as a companion to others – one resource among several on the researcher's bookshelf. It is concerned with the statistical concepts and practicalities of NRM research. It does not, for instance,

- explain how to calculate a median or fit a least squares equation;
- give the reader a thorough introduction to an advanced statistical method such as logistic regression or REML modelling;

- show how to carry out standard or more specialized analyses in relevant statistical software.

Resources that complement this book either address some of the above issues, or add further insights into aspects of natural resources research.

The principles underlying experimental design, surveys and sampling have a statistical basis that must be understood when planning NRM research. There are few books on experimental design but many on sampling ideas and surveys. Our choice includes some that describe key principles or shed light on sensible practice, some on the general subject of research methods, and others on qualitative methods.

In Chapter 20, where modern analysis tools are discussed, we observed that each section in the chapter would merit a textbook on its own. We address this now by including books on some of these more advanced methods, selecting those that contain a good mix (or even a minimum) of theory, with practical explanations and examples.

In this book, apart from some aspects of data management, we do not cover the practical, 'how to' issues of planning a study or processing data. Such information is often available in handbooks or as web-based guidelines, rather than in textbooks. We therefore add to our bookshelf a list of resources available from the SSC and ICRAF websites, as well as from other sites. The material includes training notes on data analysis and data management, items that help when using EXCEL, and examples of good statistical practice. Also listed are the original good practice guides that were the springboard for this book. Many of the ideas from the original booklets have been retained, but we have added considerably to the content, so it is often difficult to identify the original booklets in the book.

The user guides and help facilities in statistical software are often a useful resource, so the last section gives web addresses for some relevant packages. This is mainly as a convenient reference, and includes only very brief comments. Some packages are for general statistical analysis, while others are for special methods, for example for ecological applications. Most of the packages listed run under WINDOWS; but some have versions for other platforms (e.g. Apple, Unix).

23.3 The Bookshelf

23.3.1 General: Statistics and research methods

Armitage, P., Berry, G. and Matthews, J.N.S. (2001) *Statistical Methods in Medical Research*, 4th edition. Blackwell Science, Oxford, UK. 832 pp. ISBN 0632052570.

A standard text for medical researchers, with a clarity of explanation that makes it useful more widely.

Ford, D.E. (2000) *Scientific Methods for Ecological Research*. Cambridge University Press, UK. 558 pp. ISBN 0521669731.

This is an unusual book that aims to show how the research process works, and provides great insights into what can be learnt by different types of investigation.

Franzel, S., Cooper, P., Denning, G.L. and Eade, D. (2002) *Development and Agroforestry: Scaling up the Impacts of Research.* Oxfam, Oxford, UK. 196 pp. ISBN 0855984643.

An example of the ideas of what constitutes research, as discussed in Chapter 1.

Greenfield, T. (ed.) (2002) *Research Methods for Postgraduates*, 2nd edition. Hodder Arnold, London. 381 pp. ISBN 0340806567.

Good general advice on research projects.

Manly, B.F.J. (1992) *The Design and Analysis of Research Studies.* Cambridge University Press, UK. 369 pp. ISBN 0521425808.

This book is broader than many on design and analysis, and has a clear practical approach aimed at researchers from a range of backgrounds. Topics include different types of research studies, sampling designs and experimental design, and analysis methods from simple approaches to computer-intensive ones.

Manly, B.F.J. (2001) *Statistics for Environmental Science and Management.* Chapman & Hall/CRC Statistics and Mathematics, UK. 336 pp. ISBN 1584880295.

A useful book for environmental scientists. It covers sampling and data analysis as well as more specialized topics such as environmental monitoring, impact assessment and spatial data analysis.

Mead, R., Curnow, R.N. and Hasted, A.M. (2002) *Statistical Methods in Agriculture and Experimental Biology*, 3rd edition. Chapman & Hall/CRC, UK. 488pp. ISBN 1584881879.

Both traditional analysis methods for experimental data and the role of modern methods are covered here using a minimum of mathematics. Fundamental ideas underlying good experimental design are also addressed.

23.3.2 Design of experiments

Robinson, G.K. (2000) *Practical Strategies for Experimenting.* Wiley (Wiley Series in Probability and Statistics), UK. 282 pp. ISBN 0471490555.

This book links statistical methods with both discussion of research objectives and approaches, and the practical implications of experimenting.

Mead, R. (1990) *The Design of Experiments: Statistical Principles for Practical Applications.* Cambridge University Press. 634 pp. ISBN 0521287626.

A clear explanation, particularly from a statistician's viewpoint, of the principles of good experimental design.

Mooney, H.A., Medina, E., Schindler, D.W., Schulze, E.-D. and Walker, B.W. (1991) *Ecosystem Experiments (Scientific Committee on Problems of the Environment (SCOPE)).* Wiley. 296 pp. ISBN 0471929263.

Various examples of doing experiments at a scale larger than the plot.

Underwood, A.J. (1997) *Experiments in Ecology: their Logical Design and Interpretation using Analysis of Variance.* Cambridge University Press, UK. 504 pp. ISBN 0521556961.
This covers much standard experimental design and analysis but in the context of ecological investigations, rather than more controlled areas of agriculture and forestry, thus making it relevant to many NRM investigators.

23.3.3 Sampling and surveys

Casley, D.J. and Kumar, K. (1987) *Project Monitoring and Evaluation in Agriculture.* Johns Hopkins University Press, Baltimore. 174 pp. ISBN 0801836166.
A rather narrow focus, but useful background reading and not too technical.

Casley, D.J. and Kumar, K. (1988) *The Collection, Analysis and Use of Monitoring and Evaluation Data.* Johns Hopkins University Press, Baltimore. 190 pp. ISBN 0801836697.
Like its earlier companion volume, this offers an accessible account of the handling of monitoring and evaluation issues from a largely quantitative standpoint.

Hayek, L.C. and Buzas, M.A. (1997) *Surveying Natural Populations.* Columbia University Press, New York. 562 pp. ISBN 0231102410.
A guide to basic statistical ideas and data analysis, with good practical advice on ecological surveying.

Kalton, G. (1983) *Introduction to Survey Sampling.* Sage: Quantitative Applications in the Social Sciences, No. 07-035. 96 pp. ISBN 0803921268.
Although rather dated, this book is quite user-friendly compared to other books focusing on formal sampling, being aimed mainly at social scientists.

Lohr, S.L. (1999) *Sampling Design and Analysis.* Duxbury Press, Pacific Grove, California. ISBN 0534353614.
This is a well-written book about sampling methods, for those who want a sound mathematical treatment.

Thompson, S.K. (2002) *Sampling*, 2nd edition. Wiley, New York. 400 pp. ISBN 0471291161.
A modern classic that considers sampling mainly of biological populations, and is best suited to the mathematically-robust reader.

23.3.4 Qualitative research methods / Social research design

Boyatzis, R.E. (1998) *Transforming Qualitative Information: Thematic Analysis and Code Development.* Sage Publications. 266 pp. ISBN 0761909613.
Quite a specialized book from our perspective, about how verbal information can be thought through, and coded for routine analysis.

Chambers, R. (1997) *Whose Reality Counts? Putting the First Last.* ITDG Publishing, London. 297 pp. ISBN 185339386X.
At the core of this book are the methods and approaches of participatory rural appraisal (PRA). PRA has come to affect much development practice,

research, education, training and management, and this has raised wider questions about development and about the human condition. The 'realities' of development professionals, embracing such values as measurement and control, are at odds with those of the poor they seek to assist. Poor people's realities are instead diverse, dynamic and unpredictable, yet the professionals often impose their own realities upon the poor. The book argues that professionals must adapt their approaches to bridge this gap.

Cook, T. and Campbell, D. (1979) *Quasi-Experimentation: Design and Analysis Issues for Field Settings.* Houghton Mifflin (Academic). 405 pp. ISBN 0395307902.

A classic text on the logic of social enquiry, with interesting material on reliability, validity, time series and other themes.

de Vaus, D. (2001) *Research Design in Social Research.* Sage Publications. 296 pp. ISBN 0761953477.

This is a good introduction to thinking about the logic of causation and about choices of approaches to research. Though focused on social applications, it is not difficult for natural scientists to learn important ideas from de Vaus's presentation.

Hakim, C. (2000) *Research Design: Successful Designs for Social and Economic Research*, 2nd edition. Routledge. 272 pp. ISBN 041522313X.

A nicely-written comparison of a range of approaches to social enquiry, which helps to understand how to select the right one in a sensible way.

King, G., Keohane, R. and Verba, S. (1994) *Designing Social Inquiry: Scientific Inference in Qualitative Research.* Princeton University Press. 258 pp. ISBN 0691034710.

Very clear explanation of how research design principles apply in qualitative research studies.

Mason, J. (2002) *Qualitative Researching*, 2nd edition. Sage Publications. 229 pp. ISBN 0761974288.

A text on qualitative approaches which explains issues of concern in a way that the quantitative researcher should be able to relate to and learn from.

Sit, V. and Taylor, B. (editors) (1998) *Statistical Methods for Adaptive Management Studies.* BC Ministry of Forests. Land Management Handbook No. 42. www.for.gov.bc.ca/hfd/pubs/Docs/Lmh/Lmh42.htm (accessed 31 October 2003).

23.3.5 Statistical Analysis

Collett, D. (2002) *Modelling Binary Data*, 2nd edition. Chapman & Hall/CRC Statistics and Mathematics, UK. 408 pp. ISBN 1584883243.

Incorporating methodological and computational developments in the field, this guide to methods for analysing binary data also includes chapters on mixed models for binary data analysis and on exact methods for modelling binary data. A good mix of theory and practical illustrations.

Dobson, A.J. (2002) *An Introduction to Generalized Linear Models*, 2nd edition. Chapman & Hall/CRC Statistics & Mathematics, UK. 231 pp.

ISBN 1584881658.

A good introduction to the subject from both practical and theoretical viewpoints. Practical applications of the models concentrate mainly on Poisson, binary and more general types of categorical data. General linear models are reviewed in detail and illustrated in the generalized linear model framework.

Gelman, A., Carlin, J.B., Stern, H.S. and Rubin, D.B. (2003) *Bayesian Data Analysis*, 2nd edition. Chapman & Hall/CRC Texts in Statistical Science, UK. 640 pp. ISBN 158488388X.

A comprehensive introductory text on data analysis from a Bayesian perspective. Emphasizes practice over theory, with many examples, and includes information on current models for Bayesian data analysis, such as equation models and generalized linear mixed models.

Haining, R. (2003) *Spatial Data Analysis: Theory and Practice.* Cambridge University Press, UK. 452 pp. ISBN 0521774373.

A mathematically-based book, but with plenty of good examples.

Jongman, R.H.G., Ter Braak, C.J.F. and van Tongeren, O.F.R. (1995) *Data Analysis in Community and Landscape Ecology.* Cambridge University Press, UK. 321 pp. ISBN 0521475740.

A clear exposition of methods for handling multivariate species data.

Krzanowski, W. (1998) *An Introduction to Statistical Modelling.* Hodder Arnold. 264 pp. ISBN 0340691859.

A sound practical introduction, covering the general linear model (multiple regression, ANOVA) and extending to the concept of the generalized linear model. Discusses modelling binary data, and counts in the form of contingency tables. The text is elementary but includes essential theoretical aspects.

Krzanowski, W. (2000) *Principles of Multivariate Analysis.* Revised edition. Oxford University Press, UK. 608 pp. ISBN 0198507089.

A detailed account of the subject, both from a practical viewpoint and theoretically. The main emphasis is on a practical approach, and as such, some of the more mathematical parts of the text may be omitted without any detriment to practical understanding.

Manly, B.F.J. (1994) *Multivariate Statistical Methods: a Primer*, 2nd edition. Chapman & Hall/CRC Texts in Statistical Science, UK. 226 pp. ISBN 0412603004.

A good, elementary level introduction to multivariate methods, in which mathematics is kept to a minimum.

McConway, K.J., Jones, M.C. and Taylor, P.C. (1999) *Statistical Modelling Using GENSTAT.* Hodder Arnold. 384 pp. ISBN 0340759852.

A readable text, aimed at undergraduates in applied statistics, but also accessible to a wider audience. It provides a useful background to the ideas of statistical modelling.

Snijders, T.A.B. and Bosker, R.J. (1999) *Multilevel Analysis: an Introduction to Basic and Advanced Multilevel Modelling.* Sage Publications. 272 pp. ISBN 0761958908.

One of the few readable books for non-statisticians on multilevel modelling.

23.3.6 Data management

Berk, K.N. and Carey, P. (2000) *Data Analysis with Microsoft® EXCEL Updated for Office 2000.* Duxbury Press, USA. ISBN 0534362788.

A good book on EXCEL for statistical work, with a useful add-in on a CD.

ICPSR (2002) *Guide to Social Science Data Preparation and Archiving*, 3rd edition. Inter-University Consortium for Political and Social Research. http://www.icpsr.umich.edu/access/dpm.html (accessed 31 October 2003).

Aimed at those engaged in preparing data for deposit in a public archive, but most of the topics are of interest to anyone creating a dataset. An excellent source of information and suggestions; it provides information on issues such as documentation, variable names, codebooks, dealing with missing data, backups and confidentiality.

RSS/UKDA (2003) *Preserving and Sharing Statistical Material: Information for Data Providers.* www.data-archive.ac.uk/home/PreservingSharing.pdf (accessed 8 October 2003).

This document was prepared jointly by the UK Data Archive (www.data-archive.ac.uk) and the Royal Statistical Society (www.rss.org.uk). Both the RSS and UKDA recognise that there are many reasons that seem to prevent organisations from ensuring that their statistical material is preserved in the longer term. These include the perceived costs to the organization, technological constraints and concerns about data confidentiality. Many of these difficulties could be overcome if organizations were to implement a system of best practice to ensure, from the outset of the data collection process, that the datasets, along with accompanying sources and contextual information, are preserved and shared. The document provides advice and information for those who are responsible for data collection, but who may lack experience of organizing data preservation over medium to long periods of time.

Sales Harkins, S., Gerhart, T. and Hansen, K. (1999) *Using Microsoft® ACCESS 2000.* Que Corporation, USA. 734 pp. ISBN 078971604-6.

One of the many books about ACCESS, this is clear and relatively compact.

23.3.7 Presentation and communication of results

Chapman, M. and Wykes, C. (1996) *Plain Figures*, 2nd edition. H.M. Stationery Office, London. 148 pp. ISBN 0117020397.

A useful and colourful reference on how to present results, particularly intended for those addressing non-specialist audiences.

Felsing, M., Haylor, G., Lawrence, A. and Norrish, P. (2000) Developing a communication strategy for the promotion and dissemination of participatory aquaculture research: a case study from Eastern India. *Journal of Extension Systems* 16, 82-106. West Virginia University, USA.

This paper focuses on the rural poor who have limited access to resources and no effective extension support. The relative accessibility of different types of media (video and audiocassettes, posters and leaflets) is presented and costed.

Norrish, P., Lloyd Morgan, C. and Myers, M. (2001) Improved Communication Strategies for Renewable Natural Resource Research Outputs. *Socio-economic Methodologies for NRR: Best Practice Guideline Series (BPG 8).* Natural Resources Institute (NRI), Chatham, UK.
www.nrsp.co.uk
This guide emphasizes the importance of specifying a communications strategy at an early stage of the research process.

Pretty, J.N., Guijt, I., Thompson, J. and Scoones, I. (1995) *Participatory Learning and Action: a Trainers Guide.* International Institute for Environment and Development, London. 270 pp. ISBN 1899825002.
A useful resource for workers in the field.

Stapleton, P., Youdeowei, A., Mukanyange, J. and van Houten, H. (1995) *Scientific Writing for Agricultural Research Scientists.* WARDA/CTA. 128 pp. ISBN 9291130699.
A book that will guide you through all the steps of writing up a paper.

23.3.8 Miscellaneous

American Statistical Association (1999) *Ethical Guidelines for Statistical Practice.*
www.amstat.org/profession/ethicalstatistics.html (accessed 3 October 2003).

Ashby, J.A., Braun, A.R., Gracia, T., del Pilar Guerrero, M., Hernadez, L.A., Quiros, C.A. and Roa, J.I. (2000) *Investing in Farmers as Researchers.* CIAT, Cali. ISBN 9586940306.
www.ciat.cgiar.org/downloads/pdf/Investing_farmers.pdf
(accessed 9 October 2003).
The International Research Institutes have been in the forefront of involving farmers in the research process, with CIAT particularly prominent. This is one such example.

Gardenier, J.S. (2003) Best Statistical Practices to Promote Research Integrity. *Professional Ethics Report*, 16, 1-2 (AAAS publication).
www.aaas.org/spp/sfrl/per/per32.htm (accessed 13 October 2003).
This article provides a taxonomy of statistical practices in science, ranging from *best practices*, through *generally accepted but incompetent statistical practices*, to *research misconduct*.

Salsburg, D. (2001) *The Lady Tasting Tea: How Statistics Revolutionized Science in the Twentieth Century.* Owl Books, New York. 340 pp. ISBN 0805071342.
This highly readable book describes the historical personalities and background to many ideas, such as significance testing, that are now standard practice in statistics.

Upton, G. and Cook, I. (2002) *A Dictionary of Statistics.* Oxford University Press, UK. 420 pp. ISBN 0192801007.
A clear and concise reference book.

23.4 SSC Web-based Resources

(a) Good Practice Guides

www.ssc.rdg.ac.uk/develop/dfid/booklets.html (accessed 13 October 2003).

Abeyasekera, S. and Stern, R.D. (2001) *Modern Methods of Analysis.*
Allan, E.F. (2000) *One Animal per Farm?*
Allan, E.F. and Rowlands, J. (2001) *Mixed Models and Multilevel Data Structures in Agriculture.*
Barahona, C.E. and Leidi, A.A. (2000) *EXCEL for Statistics: Tips and Warnings.*
Garlick, C.A. (2000) *Disciplined Use of Spreadsheets for Data Entry.*
Garlick, C.A. and Stern, R.D. (1998) *Data Management Guidelines for Experimental Projects.*
Garlick, C.A. and Stern, R.D. (2000) *The Role of a Database Package in Research Projects.*
Mead, R. and Stern, R.D. (1998) *Statistical Guidelines for Natural Resources Projects.*
Pearce, S.C. (2000) *The Statistical Background to ANOVA.*
Sherington, J., Stern, R.D., Allan, E.F. and Wilson, I.M. (2000) *Informative Presentation of Tables, Graphs and Statistics.*
Stern, R.D. (2000) *Moving on from MSTAT (to GENSTAT).*
Stern, R.D. and Abeyasekera, S. (2000) *Concepts Underlying the Design of Experiments.*
Stern, R.D. and Allan, E.F. (1998) *On-Farm Trials – Some Biometric Guidelines.*
Stern, R.D. and Allan, E.F. (2001) *Modern Approaches to the Analysis of Experimental Data.*
Wilson, I.M. (1998) *Project Data Archiving – Lessons from a Case Study.*
Wilson, I.M. (2000) *Some Basic Ideas of Sampling.*
Wilson, I.M. and Stern, R.D. (1998) *Guidelines for Planning Effective Surveys.*
Wilson, I.M. and Stern, R.D. (2001) *Approaches to the Analysis of Survey Data.*
Wilson, I.M., Stern, R.D., Abeyasekera, S. and Allan, E.F. (2001) *Confidence & Significance: Key Concepts of Inferential Statistics.*

(b) Case Studies of Good Statistical Practice

www.ssc.rdg.ac.uk/develop/dfid/dfidcase.html (accessed 13 October 2003).

Good practice in well-linked studies using several methodologies, based on a CPP project in the Philippines.
Good practice in survey design and analysis, based on a CPP project in India.
Good practice in on-farm studies, based on a joint NRSP/CPP project in Bolivia.
Good practice in researcher and farmer experimentation, based on an NRSP project in Zimbabwe.

Determining the effectiveness of a proposed sample size, based on work for DFID by a private sector research firm in Bangladesh.

Good practice in data management, based on a bilateral project in Malawi.

Good practice in the preparation of research protocols, based on (i) a Plant Sciences Research programme project in West Africa; (ii) an NRSP project in Nepal; and (iii) an NRSP project in Bangladesh.

Developing a sampling strategy for a nationwide survey in Malawi, based on a bilateral project in Malawi.

(c) Outputs from a project on 'Integrating Qualitative and Quantitative Approaches for Socio-Economic Survey Work'

www.ssc.rdg.ac.uk/develop/dfid/dfidqqa.html (accessed 13 October 2003).

(i) The Methodological Framework

Marsland, N., Wilson, I.M., Abeyasekera, S. and Kleih, U. (2000) A Methodological Framework for Combining Qualitative and Quantitative Survey Methods. *Socio-economic Methodologies Best Practice Guideline Series.* Natural Resources Institute (NRI), Chatham, UK.

(ii) Theme Papers

Abeyasekera, S. (2001) *Analysis approaches in Participatory Work Involving Ranks and Scores.*

Abeyasekera, S., Lawson-McDowell, J. and Wilson, I.M. (2001) *Converting Ranks to Scores for an ad-hoc Assessment of Methods of Communication Available to Farmers.*

Abeyasekera, S. and Marsland, N. (2001) *Comparing Changes in Farmer Perceptions over Time.*

Burn, R.W. (2000) *Quantifying and Combining Causal Diagrams.*

Marsland, N. (2000) *Participation and the Qualitative-Quantitative Spectrum.*

Wilson, I.M. (2000) *Sampling and Qualitative Research.*

(iii) Case Studies

Abeyasekera, S. and Lawson-McDowell, J. (2001) *Computerising and Analysing Qualitative Information from a Study Concerning Activity Diaries.*

Barahona, C.E. (2001) *Generalising Results from Matrices of Scores.*

Conroy, C., Jeffries, D. and Kleih, U. (2001) *The Use of Statistics in Participatory Technology Development - The Case of Seasonal Feed Scarcity for Goats in Semi-arid India.*

Jeffries, D., Warburton, H., Oppong-Nkrumah, K. and Fredua Antoh, E. (2001) *Wealth Ranking Study of Villages in Peri-Urban Areas of Kumasi, Ghana.*

Marsland, N., Golob, P. and Abeyasekera, S. (2001) *Larger Grain Borer Coping Strategies Project.*

Marsland, N., Henderson, S. and Burn, R.W. (2001) *On-going Evaluation of FRP Project 'Sustainable Management of Miombo Woodland Project by Local Communities in Malawi'.*

(d) Other Materials

Barahona, C. and Levy, S. (2003) *How to generate statistics and influence policy using participatory methods in research: reflections on work in Malawi 1999-2002.* IDS Working Papers No. 212, Institute of Development Studies, Brighton, UK.
www.ssc.rdg.ac.uk/partiandstats/intro.html (accessed 31 October 2003).
Using case studies, this paper presents the experience of developing research that integrates qualitative and quantitative approaches. The research was part of an evaluation of a national programme of distribution of inputs to small-holder farmers carried out between 1999 and 2002. The research methodology integrated principles of statistics and participatory approaches in a way that went beyond the more common sequencing of methodologies. The authors claim that 'the statistics generated in these studies are at least as reliable as statistics obtained from traditional methods such as surveys'. The paper attempts to identify the key requirements to achieve an integration of statistics and participatory methodologies.

Leidi, A.A., Stern, R.D. and Grubb, H.J. (2002) *Guidelines for Good Statistical Graphics in EXCEL.* Statistical Services Centre, University of Reading, UK.
www.rdg.ac.uk/~sns97aal/XLgraphics.pdf (accessed 31 October 2003).

Stern, R.D. and Leidi, A.A. (2002) *Good Tables for EXCEL Users.* Statistical Services Centre, University of Reading, UK.
www.rdg.ac.uk/~sns97aal/GoodTablesforExcel.pdf (accessed 31 October 2003).

23.5 ICRAF Web-based Resources

Coe, R., Stern, R.D. and Allan, E.F. (2002) *Data Analysis of Agroforestry Experiments.* World Agroforestry Centre, Nairobi, Kenya.
www.worldagroforestrycentre.org/sites/RSU/dataanalysis/index.html
(accessed 13 October 2003).
These training materials on the analysis of experimental data were designed for users to adapt for their own requirements. One session in this course was the basis for Chapter 19.

Coe, R., Franzel, S., Beniest, J. and Barahona, C.E. (2003) *Designing On-farm Participatory Experiments: Resources for Trainers.* World Agroforestry Centre, Nairobi, Kenya.
www.worldagroforestrycentre.org/sites/RSU/researchdesign/index.html
(accessed 13 October 2003).

Kindt, R. and Burn, R.W. (2003) *A Biodiversity Analysis Package.*
www.worldagroforestrycentre.org/sites/RSU/resources/biodiversity/index.html

(accessed 13 October 2003).
A set of resources and guidelines for the analysis of biodiversity and ecological information. Includes a description of methods of analysis, software and examples of how the software provides results. Downloadable; also available on CD for those without good web access.

Muraya, P., Garlick, C.A. and Coe, R. (2002) *Research Data Management.* World Agroforestry Centre, Nairobi, Kenya.
www.worldagroforestrycentre.org/sites/RSU/datamanagement/index.html
(accessed 13 October 2003).
These course materials cover similar topics to those in Chapters 10 to 13. They may be downloaded and edited as trainers see fit.

23.6 Statistical, Graphical and Data Management Software

Table 23.1. List of some statistical, graphical and data management software.

Software	*Website and comments*
CSPRO	www.census.gov/ipc/www/cspro Package for entering, editing, tabulating, and mapping census and survey data.
DISTANCE	www.ruwpa.st-and.ac.uk/distance Software for the design and analysis of distance sampling surveys of wildlife populations.
EPI INFO	www.cdc.gov/epiinfo Package for entering and processing survey data (especially epidemiological studies), with some statistical facilities.
GENSTAT	www.vsn-intl.com/genstat A general package noted for promoting good practice in modelling and the analysis of designed experiments.
INSTAT	www.ssc.rdg.ac.uk/instat A package for training and promoting good statistical practice; includes special facilities for climatic data.
MARK	www.cnr.colostate.edu/~gnwhite/mark/mark.htm Software for the analysis of capture-recapture data, and for the estimation of population size for closed populations.
MINITAB	www.minitab.com A general package for data analysis; popular in universities for teaching.
MLWIN	www.multilevel.ioe.ac.uk A package for fitting multilevel models for a variety of distributions.
R	www.r-project.org An object-oriented, statistical programming language; open-source software giving access to cutting-edge statistics.
SAS	www.sas.com A general package, with a comprehensive system of components.
S-PLUS	www.insightful.com A general package based on an object-oriented, statistical programming language.
SPSS	www.spss.com A general package, the standard choice for survey data processing.
SSC-STAT	www.ssc.rdg.ac.uk/ssc-stat An EXCEL add-in for data manipulation and basic statistical work.
STATA	www.stata.com A general package, noted for processing survey data.
STATISTICA	www.statsoftinc.com A general package with dynamically-linked graphics.
WINBUGS	www.mrc-bsu.cam.ac.uk/bugs Software for the Bayesian analysis of complex statistical models.

Appendix: Preparing a Protocol

This Appendix uses, as an example, a checklist for the preparation of a protocol established by ICRAF for researchers who are setting up experiments on farms with the close involvement of the farmers. It was originally prepared by members of an AHI workshop on 'Participatory Experimentation' held in Nairobi, 28 June to 3 July 1999, and subsequently compiled by Richard Coe.

Trial Protocols

A protocol for a research activity such as a trial is a detailed and complete description of what should happen. It is first produced during the planning phase and is revised as necessary as the activity progresses.

There are three reasons for producing a detailed written protocol:

1. Before starting, the protocol can be shared with anyone who might be able to comment on it and improve the effectiveness of the research. Consider eliciting comments on the protocol from:

- people who know the farmers and farming systems in the area where you are working;
- people who understand the subject area (soil fertility, animal feed, etc.);
- people with experience of the methods that you plan to use.

Ask anyone who might have something to offer. The worst that can happen is that you get no response! Remember that people working outside your immediate location may have valuable experience that can help your research.

2. Success of the activity cannot depend on the continued presence of any one person. If a detailed protocol is written down, it should be possible for the work to reach a successful conclusion even if the key investigator leaves.

3. After the activity is completed the protocol is a record of what was done, to be referred to in any reporting and to be archived with the data.

The Checklist

Below is a checklist – a list of points to remember when preparing a protocol. To use it, consider each point in the list and make sure that you have addressed it in the protocol. Different institutions require protocols to be prepared in different formats, but any useful protocol will have to cover most of the points in this checklist. The checklist does not give you the answers, or tell you how to carry out the activity. It is simply a reminder of the points to decide on.

The checklist is designed to help develop protocols for single trials or experiments. It does not help in designing projects, which will consist of several linked activities (planning and training meetings, surveys, experiments and so on). It is designed for planning experiments that involve farmers. Other types of research activities require protocols with different information.

Abstract

A summary of the trial is useful. Note the format of the whole protocol may have to conform to institutional requirements.

1. Reference or Number

Use a unique number or reference code for the study. This is needed to keep track of the experiment and to avoid confusion with other similar studies.

2. Title

Choose a short, memorable title that people can quickly learn and relate to your study.

3. Location

Part of the identification of the study is where it takes place. Details (e.g. exactly which villages) come later.

4. Investigators

4.1 Team Leader (Principal Investigator) and his/her institution
Remember that the PI is the person responsible for the design and implementation of the work. However the team operates, there should be one person who takes overall responsibility.

4.2 Team members and their institution(s)
It is often useful to list the professional area of each individual.

5. Background and Justification

You have to justify spending money on this work! In each of the following sections you must make clear:

- what the (farming) problem is;
- how your work will help solve the problem;
- what the next step (when the experiment is concluded) is expected to be;
- the target group and why it was chosen;
- on whom the work will make an impact.

The information must be specific to the location in which you are working and the problem on which you are working.

5.1 Summary of literature
Review what is known about the problem and possible solutions. This will often come from other locations within the country and elsewhere.

5.2 Past research by you and others in this area
Results from PRAs, surveys, other experiments, etc.

5.3 Links to other parts of the same project
Describe how your trial links to other activities of the same or related projects, such as other trials, farmer training activities, or planning meetings. Describe existing institutional linkages and relationships with farmers and farmer groups.

5.4 Hypotheses
Even initial and exploratory studies have hypotheses! Hypotheses are statements that you believe to be true and when they are confirmed by the study allow the work to progress. They may concern biophysical, social or economic processes, or the links between these processes.

5.5 Potential impact
If the work goes as planned, what will the effects be once the hypotheses are (or are not) confirmed. Who will benefit, both on and off the farm? How will they benefit, and by how much? Will anyone suffer a negative impact? How sustainable will any impacts be?

6. Objectives

The whole of the rest of the design, and hence the protocol, depends on the objectives. List them clearly, completely and in sufficient detail to leave no room for subsequent doubts about any aspect of the study. Include a description of who the resulting information is aimed at.

The objectives must be consistent with each other and capable of being met with a single trial.

Farmers should have been involved in deciding on the objectives, and the process described in the justification (Section 5 above). If that has not yet been done, it is probably too early to write a detailed protocol.

7. Methods

Give enough detail under this heading for the protocol to be useful for the following:

- Anyone to see what you plan to do, so that suggestions for improvements can be made.
- As the permanent record of what should be done, to be referred to during implementation. It should be good enough for this purpose even if the principal investigator leaves.

Farmers will be involved in deciding many of the details of the protocol. Under each item below, describe how it was done (if completed), or how it will be done.

7.1 Trial type

There are several ways of summarizing the type of trial. ICRAF uses Types 1, 2 and 3. The 'Contractual–Consultative–Collaborative–Collegial' distinction is useful. Try to find a simple description that summarizes the approach used.

7.2 Duration

Be realistic! The start date must be far enough in advance to make proper preparations. The trial must be long enough to get outputs (e.g. for farmers to realize benefits) but short enough to keep everyone interested. The appropriate length will depend on the objectives.

7.3 Location

Describe how and why the locations are selected. Methods may be random, stratified, selected to follow a known gradient, and so on. Descriptions of locations may well be hierarchical e.g. districts chosen because of the mandate area, locations chosen randomly, sub-locations chosen to cover a range of distances from the main road. Describe the sampling scheme(s) and sample sizes.

7.4 Farmers

Describe how farmers (or households, fields, etc.) are selected to be in the study. Examples are volunteers at a village meeting, introductions made by extension staff, contact with farmers from a previous study, farms chosen during a mapping exercise. Section 7.10 (below) points out that it is important to find out who the farmers are (gender, household type, wealth category, etc.) and how they relate to the general population. State how many farmers will be involved.

7.5 Treatments

Describe both the treatments to be compared and the method of arriving at these. Make it clear what is determined by farmers and what by researchers. If the farmer makes the decision, find out what they are comparing with (it may not be physically adjacent and it could be from another season). If 'farmer's practice' is included as a treatment make clear exactly how this is defined and if it varies between farmers.

7.6 Layout

Describe where the treatments will be applied and how these are chosen. Describe plot locations within a farm, sizes, method of allocating treatments to plots. Make clear what is determined by farmers and what by researchers.

7.7 Inputs

Describe what inputs (e.g. seeds) are needed and how (and on what terms) these will be supplied.

7.8 Management

Identify who is responsible for deciding on management activities (e.g. planting, weeding, spraying, harvesting). Who is responsible for carrying them out? Describe each management decision and who is making it. Do not generalize. Distinguish between decisions about the management (e.g. how many times to weed) and those about carrying out the work (e.g. doing the weeding).

7.9 Non-experimental variables

Describe key variables and state if they are fixed. If so, at what level and by whom? How are farmers involved in deciding the level at which to fix any non-experimental variables?

7.10 Data collection

Data may be collected on 'response' variables such as:

- agronomic performance;
- economic performance;
- farmer's assessment;
- ecological impact.

It is also necessary to record 'design' variables, which include things like:

- location of participants (possibly using GPS);
- household and farmer characteristics;
- layout details (what treatments are selected by farmers and why, what plots or niches are used and why);
- levels of non-experimental variables.

For each variable describe:

- exactly what variables or indicators will be used;

- who decides on them and how farmers are involved;
- the weights they should be given in data analysis (perhaps also elicited from farmers);
- the measurement tool (e.g. survival by researcher count, farmer's assessment by questionnaire);
- the measurement unit (e.g. plant, plot, farm, village) and sampling scheme.

Describe the monitoring process (e.g. visits to the farmers that do not involve any planned data collection).

7.11 Data management

Describe who will be collecting data. If farmers are doing data collection, explain how this will be organized and the training necessary. Describe how and where data will be looked after. How will it be computerized? Who will have access? How and where will the datasets be archived?

7.12 Data analysis, reporting and feedback

Describe methods to be used for analysing, interpreting and reporting the data. Describe farmer involvement, including how results will be reported to farmer-collaborators.

8. Implementation Plan

8.1 Outputs

List 'hard' outputs.

8.2 List of tasks

Include such activities as monitoring and evaluation, reporting and reviewing – things which take time but that are often forgotten at the planning stage.

8.3 Timing

8.4 List of partners/team members

8.5 Roles of all partners

Make it clear exactly who is responsible for what.

8.6 Budget

9. References

10. Version

Keep the protocol up-to-date, both while it is being revised during planning and when details change during implementation.

Index